Extreme Hydroclimatic Events and Multivariate Hazards in a Changing Environment

EXTREME HYDROCLIMATIC EVENTS AND MULTIVARIATE HAZARDS IN A CHANGING ENVIRONMENT

A Remote Sensing Approach

Edited by

VIVIANA MAGGIONI
Sid and Reva Dewberry Department of Civil, Environmental & Infrastructure Engineering, George Mason University, Fairfax, VA, United States

CHRISTIAN MASSARI
Research Institute for Geo-Hydrological Protection (IRPI), National Research Council (CNR), Perugia, Italy

ELSEVIER

Elsevier
Radarweg 29, PO Box 211, 1000 AE Amsterdam, Netherlands
The Boulevard, Langford Lane, Kidlington, Oxford OX5 1GB, United Kingdom
50 Hampshire Street, 5th Floor, Cambridge, MA 02139, United States

British Library Cataloguing-in-Publication Data
A catalogue record for this book is available from the British Library

Library of Congress Cataloging-in-Publication Data
A catalog record for this book is available from the Library of Congress

ISBN: 978-0-12-814899-0

For Information on all Elsevier publications
visit our website at https://www.elsevier.com/books-and-journals

Publisher: Candice Janco
Acquisition Editor: Laura Kelleher
Editorial Project Manager: Karen Miller
Production Project Manager: Omer Mukthar
Cover Designer: Matthew Limbert

Typeset by MPS Limited, Chennai, India

This book is dedicated to the memory of all the lives lost to natural disasters, and to the people who work to understand, predict, and prevent these catastrophes.

Contents

List of Contributors

Ron Abileah
Jomegak, San Carlos, CA, United States

Emmanouil N. Anagnostou
Department of Civil and Environmental Engineering, University of Connecticut, Mansfield, CT, United States

Oscar Manuel Baez-Villaneuva
Institute for Technology and Resources Management in the Tropics and Subtropics (ITT), TH Köln, Cologne, Germany; Faculty of Spatial Planning, TU Dortmund University, Dortmund, Germany

Jérôme Benveniste
European Space Agency (ESA-ESRIN), Directorate of Earth Observation Programmes, EO Science, Applications and Climate Department, Frascati, Italy

Marco Borga
Department of Land, Environment, Agriculture and Forestry, University of Padova Agripolis, Legnaro, Italy

Luca Brocca
Research Institute for Geo-Hydrological Protection of the National Research Council (CNR-IRPI), Perugia, Italy

Jean-Christophe Calvet
CNRM (University of Toulouse, Meteo-France, CNRS), Toulouse, France

Stefania Camici
Research Institute for Geo-Hydrological Protection of the National Research Council (CNR-IRPI), Perugia, Italy

Paolo Cipollini
Telespazio VEGA UK Ltd. for ESA-ECSAT, Harwell, United Kingdom

Wade T. Crow
USDA ARS Hydrology and Remote Sensing Laboratory, Beltsville, MD, United States

Yagmur Derin
Department of Civil and Environmental Engineering, University of Connecticut, Mansfield, CT, United States

Ashraf Dewan
School of Earth and Planetary Sciences, Curtin University, Perth, WA, Australia

Celso Moller Ferreira
George Mason University, Fairfax, VA, United States

Sven Fuchs
Institute of Mountain Risk Engineering, University of Natural Resources and Life Sciences, Vienna, Austria

Manuela Girotto
Department of Environmental Science, Policy, and Management, University of California, Berkeley, CA 94720, United States; Global Modeling and Assimilation Office, NASA Goddard Space Flight Center, Greenbelt, MD, United States; GESTAR, Universities Space Research Association, Columbia, MD, United States

Jesús Gómez-Enri
University of Cadiz, Cadiz, Spain

Thomas R.H. Holmes
Hydrological Science Lab, NASA Goddard Space Flight Center, Greenbelt, MD, United States

Paul R. Houser
Department of Geography and GeoInformation Science, George Mason University, Fairfax, VA, United States

Kexiang Hu
School of Earth and Planetary Sciences, Curtin University, Perth, WA, Australia

Mohammad Kamruzzaman
Institute of Water Modelling, Dhaka, Bangladesh

Margreth Keiler
Institute of Geography, University of Bern, Bern, Switzerland

Christopher Kidd
Earth System Science Interdisciplinary Center, University of Maryland, MD, United States; NASA Goddard Space Flight Center, Greenbelt, MD, United States

Vincenzo Levizzani
Institute of Atmospheric Sciences and Climate of the National Research Council (CNR-ISAC), Bologna, Italy

Efthymios Nikolopoulos
Department of Civil and Environmental Engineering, University of Connecticut, Mansfield, CT, United States

Md. Rafi Uddin
Bangladesh University of Engineering and Technology, Dhaka, Bangladesh

Mohammad Rezaur Rahman
Bangladesh University of Engineering and Technology, Dhaka, Bangladesh

Ali Mohammad Rezaie
George Mason University, Fairfax, VA, United States

Matthew Rodell
Hydrological Science Lab, NASA Goddard Space Flight Center, Greenbelt, MD, United States

Andrea Scozzari
Consiglio Nazionale delle Ricerche (CNR-ISTI), Pisa, Italy

Xinyi Shen
Department of Civil and Environmental Engineering, University of Connecticut, Mansfield, CT, United States

Sergey Sokratov
Faculty of Geography, M.V. Lomonosov Moscow State University, Moscow, Russian Federation

Angelica Tarpanelli
Research Institute for Geo-Hydrological Protection, National Research Council, Perugia, Italy

Stefano Vignudelli
Biophysics Institute of the National Research Council (CNR-IBF), Pisa, Italy

Mauricio Zambrano-Bigiarini
Department of Civil Engineering, Universidad de la Frontera, Temuco, Chile; Center for Climate and Resilience Research, Universidad de Chile, Santiago, Chile

Foreword

As a hydrologist living in Ellicott City, Maryland, the topic of this book hits close to home. May 27, 2018 brought devastating flooding again to Ellicott City—just 2 years after the July 30, 2016 flood. Both floods are 1000-year events that resulted in widespread destruction and loss of life in the city's beloved historic district. Historically, the city has had 15 floods since 1768, but these recent floods were larger, more frequent, and more destructive. Their increased frequency and destruction are arguably the result of multiple forcings. Persistent storm patterns that saturated the area followed by very intense thunderstorms potentially linked to climate change, combined with intensive urbanizing land use change and Ellicott City's flood-prone location all combined to create these events.

After the 2016 storm, huge efforts and resources were spent on rebuilding and reopening businesses, but virtually nothing was done to improve stormwater infrastructure. Now that a second event within 2 years is on the books, city leaders are beginning to address infrastructure. However, these events illustrate a more fundamental disconnect between science and practice that also needs to be addressed in this era of rapid change. At issue is the practice or engineering hydrology that uses past rainfall-runoff statistics to design infrastructure for the future. Current engineering standards use historical rainfall-runoff information and assume hydrologic stationarity to design and build infrastructure. To address multihazards in a changing world, we will need to advance innovative nonstationary scientific methods into engineering design and mitigation efforts.

Even more complexity arises as we look toward coastal regions, for example, the tidal Potomac and Chesapeake Bay. Tidal flooding occurs most commonly in these areas from coastal storms (nor'easters and tropical storms), which push a surge of water from the Atlantic up the Chesapeake Bay and Potomac River, which—combined with riverine, stormwater, and sea level rise—produces complex multihazard flooding. As demonstrated by the impacts of Hurricane Sandy in 2012 and Florence in 2018, these coastal regions, and most other coastal regions of the world are highly susceptible to complex multihazard flooding.

Changing climate, sea level rise, subsidence, and land use changes have profound implications for the combined impact of storm surge and riverine flooding in sensitive tidal and coastal zones by changing precipitation, infiltration, abstraction, and drainage characteristics. With the predicted sea level rise of two to four feet over the next century, increases in runoff from urbanizing watersheds, land subsidence, and changes in precipitation intensity and duration, these multihazard floods are expected to significantly exceed levels observed in the past. Complex multihazard floods will affect tidal and coastal zones around the world, impacting as much as one fourth of the world's population.

It remains a challenge to observe, understand, predict, mitigate, and recover from the impacts from a single hazard. So multihazard events pose tremendous challenges for scientists, engineers, and first responders to anticipate, mitigate, and recover from the combined impacts of multiple changing hazard drivers. Addressing these multihazard challenges of monumental scale and complexity requires a comprehensive, long-term strategy that involves partnership between scientific breakthroughs, governments, the private sector, and citizens. These efforts must include advanced understanding, assessment, and prediction of natural and human-induced variations in our environment, enabling retooled policies and planning, allocation of resources, and partnership strategies. Whether the cause is a terrorist, an accident, a natural disaster, or a combination of hazards, we must be able to identify and assess the magnitude of current threats, to evaluate various preventative and corrective actions, and to predict future threats.

Breakthrough advances in earth observation techniques to observe global and regional precipitation, surface soil moisture, snow, surface soil freezing and thawing, surface inundation, river flow, and total terrestrial water storage changes, combined with better estimates of evaporation, now provide the basis for a concerted water monitoring effort in support of hazard security. The Global Precipitation Measurement mission is at the heart of this effort. A system that combines observations from a constellation of microwave imaging spacecraft provides frequent sampling of instantaneous precipitation rates. As the primary input of water to the land surface, precipitation defines the terrestrial water cycle. After sea surface temperature, soil moisture is the most important surface boundary condition for weather and climate prediction. Knowledge of soil moisture is also vital to understanding the Earth system cycling of water, energy, and carbon through its control of transport of these quantities over land. Dedicated spaceborne soil moisture observations are available from the 2009 Soil Moisture and Ocean Salinity (SMOS) and 2015 Soil Moisture Active Passive (SMAP) satellites. Satellites that measure changes in gravity, such as the Gravity Recovery and Climate Experiment (GRACE), have provided much-needed data to understand and predict total water storage changes. However, there are still significant gaps in our satellite water observations, namely snow, evapotranspiration, and streamflow.

These global measurements of water storage and fluxes are powerful tools for advancing multihazard science and mitigation, especially when integrated by assimilating land surface modeling systems such as NASA Global Land Data Assimilation System (GLDAS). The combination of adequate soil moisture and precipitation observations, coupled with improved understanding of climatic phenomena such as ocean circulation anomalies, will greatly enhance weather and climate forecast skill, thus enabling us to better cope with future multihazard issues. The increasing availability of earth observations also enables us to refine and constrain environmental prediction models that forecast weather and climate and their associated water availability and atmospheric transport. Weather prediction centers are continuously improving hurricane and tornado forecasts; Web-based resources include short- and long-term forecast information, graphical watches, location-specific warnings, and various types of observations, such as radar, satellite, and lightning data. These observation and prediction resources are vital for hazard security.

The studies in this volume show significant advancements towards describing multivariate hazards in a nonstationary environment and their climatological, seasonal, and real-time prediction from earth observations, climate dynamics, and models. These studies provide a detailed overview of earth observations and current and future satellite missions useful for hydrologic studies and water resources engineering, an overview of hydroclimatic hazards and the ability of monitoring, modeling, and predicting these extremes, and multivariate hazards case studies that analyze the combination of natural hazards and their impact on the natural and built environment.

Over the past few decades, we have made substantial progress in our ability to monitor, assess, and predict complex multivariate hazards. However, the scientific community's efforts have had marginal returns for hazard management and operations. The often significant delay in implementing new scientific understanding into management and operations, often called the "Valley of Death," is caused largely by processional and legal precedence, and a lack of partnership between the scientific and practice communities. Now, more than ever, we must take action to make the necessary links for improved multihazard mitigation through knowledge-added disaster preparation, assessment, and mitigation. We must adopt strategies at the global and multinational levels, encompassing natural disasters and potential adversities from human induced change.

A traditional disconnect between the environmental management and the scientific research communities has prevented the definition of a mutually beneficial research agenda and the free flow of information to address new threats. As a result, a significant time lag occurs before scientific advancements are implemented to the benefit of society. Environmental management policy is often based on outdated knowledge and technology. Further, scientific research is often performed without understanding stakeholder needs. This paradigm lock has come about because the two main groups have become isolated: scientists by the lack of proven utility of their findings and stakeholders by legal and professional precedence and by disaggregated institutions. For example, global change research is largely focused on mean climate impacts (such as global temperature) of century-scale greenhouse gas changes, while environmental managers need a reliable prediction of extreme event variations (such as floods and droughts) in the seasonal to decadal time frame to make informed decisions.

We must take decisive action to eliminate this paradigm lock. It is essential that we continuously modernize and integrate our earth observation, assessment, and prediction tools to provide reliable and timely information for improved hazard security. But this information is meaningless unless accompanied by timely and adequate mitigation action. Communication must be established to transmit information to users quickly, to evaluate various response options in a prediction system, to enable planning, and to take decisive mitigation action. We must encourage and demonstrate bridge-building dialogue between scientists and policy makers to establish real pathways to define a society-relevant research agenda and to transfer state-of-the-art data and tools to the users who need them.

Paul R. Houser
Department of Geography and GeoInformation Science,
George Mason University, Fairfax, VA, United States

Preface

Floods, droughts, landslides, and other hydroclimatic hazards cause thousands of deaths every year. Only in the United States, droughts, floods, and severe storms have led to more than 5000 fatalities since 1980 [NOAA National Centers for Environmental Information (NCEI), 2018]. From 1904 to 2004, landslides and debris flows in Asia and Europe have killed more than 30,000 people (Nadim et al., 2006), whereas storm surges due to hurricanes and tropical cyclones have caused an average of about 13,000 annual deaths per year around the world in the last 60 years (Dilley et al., 2005). With the predicted sea level rise of half to one meter over the next century, increases in runoff from urbanizing watersheds, land subsidence, and changes in precipitation intensity and duration, future hydroclimatic hazards are expected to significantly exacerbate these numbers. Changing climate, sea level rise, and land use changes have profound implications on the combined impact of natural hazards in sensitive zones (e.g., coastal areas, mountainous basins) by modifying precipitation, infiltration, abstraction, and drainage characteristics.

Monitoring and predicting extreme events around the world is a difficult task particularly within poorly gauged areas of the planet, where the scarcity of in situ measurements is the main obstacle for implementing models and early warning systems. Where present, in situ gauging networks represent a viable tool for quantifying precipitation intensities, soil moisture, instantaneous water volume in river channels, and other relevant hydrological variables. However, because of the inherent point scale type of measurement of in situ gauges, our knowledge of the spatial and temporal dynamics of these variables is still extremely limited. This poses a serious challenge, as many hydroclimatic hazards are strongly connected with atmospheric circulation, ocean surface, and land conditions, and have therefore to be addressed at scales that transcend national borders.

In the last decades, a number of spaceborne sensors have changed the way we assess and predict hydroclimatic hazards. These new instruments are able to monitor hydroclimatic variables at spatiotemporal scales relevant for floods, droughts, typhoons, inundations, and many other climate-induced disasters. We are now at a critical juncture in the study of hydrological extremes and past and new satellite missions can provide unprecedented tools and data to monitor and predict hydroclimatic hazards in a changing environment.

This book describes multivariate hazards in a nonstationary environment and their long-term/seasonal/real-time prediction from Earth observations, long-term climate dynamics, and models. Specifically, this book is designed to provide readers with (1) a detailed overview of the key hydrological variables that play a role in the triggering and evolution of hydroclimatic hazards and methods to estimate them (including remotely sensed observations, models, and current and future satellite missions); and (2) a discussion of the main challenges along with practical examples of monitoring and predicting hydroclimatic extremes with remote sensing and modeling techniques.

The book is intended to be a unique reference for hydrologists, meteorologists, and water resources scientists, faculty, and managers on the availability and use of remotely sensed datasets and models to monitor and predict hydroclimatic extremes. It is also of interest to scientists whose research focus is on sustainability, land use changes, disaster management, risk-assessment, and risk reduction. Professionals working with water resources operational plans, water related policies, water resources management (e.g., agriculture managers, dam regulators, land planners, etc.) are encouraged to use this book to learn about new instruments, datasets, and models that could be adopted in their current work and projects. The book is suitable for specific training courses for industry, Civil Protection Departments, insurance companies, and international organizations and advanced undergraduate and graduate students in water resources engineering and science, environmental, atmospheric, climate science, physical geography, and/or remote sensing.

The book consists of two parts. Part I stresses the importance of global and quasi-global observations of geophysical variables for hydroclimatic hazard monitoring and management. Chapters in Part I aim to guide the reader to understand the link between the specific geophysical variable and its role in hydroclimatic risk. Chapter 1, *Quantitative precipitation estimation from satellite observations*, introduces the main satellite sensors and instruments for estimating precipitation (rainfall and snowfall) and available datasets and discusses their advantages and limitations. Chapter 2, *Terrestrial Water Storage*, provides an overview of the GRACE satellite mission dedicated to measure water storage variations and describes some regional to global applications for monitoring extremes, including droughts and flooding events. Chapter 3, *Utility of soil moisture data products for natural disaster applications*, describes the different methods to measure soil moisture and reviews ongoing efforts to integrate soil moisture in natural disaster monitoring and prediction. Chapter 4, *Water surface elevation in coastal and inland waters using satellite radar altimetry*, provides an overview of how water surface elevation products are (and will be) obtained from current (and future) satellite altimetry measurement systems and their importance in forecasting hydrometeorological extreme events. Chapter 5, *Remote sensing techniques for estimating evaporation*, introduces modeling and remote sensing methods to measure evaporation. Chapter 6, *Vegetation*, discusses how satellite observations and model estimates can be merged together to improve the estimation of vegetation variables (e.g., land cover, leaf area index, vegetation biomass) globally.

Part II provides examples and applications of currently (and soon to be) available satellite products and modeling techniques to study multihydroclimatic hazards, including floods, droughts, and snow avalanches. The first two chapters focus on precipitation extremes. Chapter 7, *Estimating extreme precipitation using multiple satellite-based precipitation products*, evaluates the ability of six global-scale high-resolution satellite-based precipitation products of estimating extreme events in nine mountainous regions around the world. Chapter 8, *Evaluating the spatiotemporal pattern of concentration, aggressiveness and seasonality of precipitation over Bangladesh with time—series Tropical Rainfall Measuring Mission Data*, studies precipitation aggressiveness, concentration, and seasonality over various hydrological regions in Bangladesh. Chapter 9, *Characterizing meteorological droughts in data scare regions using remote sensing estimates of precipitation*, introduces meteorological droughts and presents a case study on the use of remote sensing estimates of precipitation for monitoring droughts in Chile. The following three chapters address the use of satellite observations and hydrologic/hydraulic models to monitor and predict floods. Specifically, Chapter 10, *Recent advances in remote sensing of precipitation and soil moisture products for riverine flood prediction*, focuses on the most recent advances and challenges in precipitation and soil moisture remote sensing for riverine flood forecasting. Chapter 11, *On the potential of altimetry and optical sensors for monitoring and forecasting river discharge and extreme flood events*, discusses the potential of altimetry and optical sensors for measuring river discharge during extreme events. Both chapters provide examples of the application of these datasets and models in flood monitoring. Chapter 12, *Inundation mapping by remote sensing techniques*, discusses satellite-based inundation mapping for near real-time simulations and forecasts of flood extent. This chapter reviews theories and algorithms of flood-inundation mapping using space born radar data and discusses their strengths and limitations. Chapter 13, *Storm surge and sea level rise: threat to the coastal areas of Bangladesh*, provides an overview of the most important physical processes related to storm surge and sea level rise. The chapter also discusses coastal flooding risks in the coastal areas of Bangladesh, existing coastal protection measures, and potential strategies to improve coastal resilience in the region. Chapter 14, *Hazard assessment and forecasting of landslides and debris flows: a case study in Northern Italy*, introduces physical processes triggering shallow landslides and debris flows and discusses methods for their assessment and forecasting. Finally, Chapter 15, *Snow avalanches*, focuses on snow avalanche characteristics and on how remote sensing can be used to gain insight in hazard magnitude and frequency.

This book shows how remote sensing observations and models can provide continuous and consistent estimates of several hydrological variables and therefore play a role in monitoring and predicting hydroclimatic hazards. Nevertheless, the spatial and temporal resolution of such estimates is often inadequate to fully capture the fine-scale variability of hydrological processes, particularly in heterogeneous areas, like mountains and coasts, which are also amongst the most vulnerable regions to natural hazards. In the recent past, the research community has been embracing the challenge of hyper-resolution ($\sim$1-km or finer) hydrologic and land surface modeling, but more work is needed to provide detailed information about the storage, movement, and quality of terrestrial carbon, energy, and water (Wood et al., 2011).

Moreover, several satellite-based products and models have been developed for research purposes and their transition to operations may not be straightforward. If on one hand there is large potential for using satellite observations and model simulations in hydroclimatic hazard forecasting and monitoring, on the other hand there are many roadblocks to developing an operational system, which include interaction with stakeholders (that sometimes are transboundary agencies), staff training, and allocation of resources.

Viviana Maggioni[1] and Christian Massari[2]
[1]Sid and Reva Dewberry Dept. of Civil, Environmental & Infrastructure Engineering, George Mason University, Fairfax, VA, United States
[2]Research Institute for Geo-Hydrological Protection (IRPI), National Research Council (CNR), Perugia, Italy

References

Dilley, M., Chen, R.S., Deichmann, U., Lerner-Lam, A.L., Arnold, M., 2005. Natural disaster hotspots: a global risk analysis. The World Bank.
Nadim, F., Kjekstad, O., Peduzzi, P., Herold, C., Jaedicke, C., 2006. Global landslide and avalanche hotspots. Landslides 3 (2), 159–173.
NOAA National Centers for Environmental Information (NCEI), 2018. U.S. Billion-Dollar Weather and Climate Disasters. <https://www.ncdc.noaa.gov/billions/>.
Wood, E., Roundy, J.K., Troy, T.J., van Beek, R., Bierkens, M., Blyth, E., et al., 2011. Hyper-resolution global land surface modeling: meeting a grand challenge for monitoring Earth's terrestrial water. Water Resour. Res. 47, W05301. Available from: https://doi.org/10.1029/2010WR010090.

Acknowledgments

The editors are grateful to all the contributors who made this project possible by giving their time and effort. The editors would also like to thank the Editorial Project Manager Karen Miller for her endless support and help through this journey.

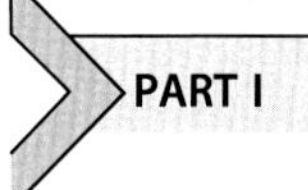

Water cycle variables for monitoring hydroclimatic hazards: State-of-the-art and future directions

Quantitative precipitation estimation from satellite observations

Christopher Kidd[1,2] and Vincenzo Levizzani[3]
[1]Earth System Science Interdisciplinary Center, University of Maryland, MD, United States
[2]NASA Goddard Space Flight Center, Greenbelt, MD, United States
[3]Institute of Atmospheric Sciences and Climate of the National Research Council (ISAC-CNR), Bologna, Italy

1.1 Introduction

The International Meteorological Vocabulary of the World Meteorological Organization defines "precipitation" as: *Hydrometeor consisting of a fall of an ensemble of particles. The forms of precipitation are: rain, drizzle, snow, snow grains, snow pellets, diamond dust, hail and ice pellets.* Precipitation is a key variable in the Earth's water and energy cycles. As a fundamental element water may take the form of liquid, solid, or vapor, and may exist in all three forms at the same time. It is the transition of water between these different phases that make it so important and difficult to measure. Water vapor is fundamental in the formation of clouds, which are composed of liquid and solid (frozen) water, and from which rainfall and snowfall may precipitate.

The generation of global precipitation products is important for a range of scientific and societal applications (Kirschbaum et al., 2017; Kucera et al., 2013). These include observing and monitoring flash floods, assessing groundwater storage, as well as forecasting crop yields and combating water-borne diseases (Kirschbaum and Patel, 2016). Such applications require spatial scales ranging from local to global and temporal scales from instantaneous to climate (Michaelides et al., 2009). While conventional observations of precipitation from surface-based measurements such as rain gauges and weather radar often form the foundation of precipitation observing systems, their distribution over the Earth's surface is

Extreme Hydroclimatic Events and Multivariate Hazards in a Changing Environment.
DOI: https://doi.org/10.1016/B978-0-12-814899-0.00001-8

somewhat limited (Kidd et al., 2017) and availability of data from them is often problematic. Satellite observations of clouds and precipitation have been exploited to provide a range of products that may be used to monitor precipitation occurrence and amounts at a range of spatial and temporal scales.

Perhaps the most important single parameter for precipitation with regard to hydroclimatic hazards is the volume of water over time. The importance of this becomes apparent when dealing with satellite observations, which have specific spatial and temporal sampling characteristics. While Earth observation satellites are capable of providing high-resolution imagery (< 1 m) and imagery as often as every few seconds, the combination of high spatial resolution and high temporal resolution is not possible for any single sensor. For operational quantitative precipitation estimation from satellite sensors, the best spatial resolution is typically in the order of several kilometers with a temporal resolution of 15 minutes or more. However, the temporal and spatial resolution of any observation and subsequent product also has to take into account the directness of the precipitation measurement; the high spatial/temporal resolution products tend to be less directly related to surface precipitation than the poorer spatial/temporal resolution products.

Thus, the relevance of satellite systems for observing precipitation and applications for hazard monitoring has to consider the:

1. resolution (temporal and spatial) of the satellite observations;
2. latency, or availability of observational data and/or products within a certain amount of time;
3. accuracy of the results as determined through validation of data products; and
4. usefulness of the resulting products to user community.

1.2 Satellites and instruments

Since the launch of the first meteorological satellite in 1960, a progression of satellites have been launched that carry sensors capable of providing observations from which surface precipitation may be derived. The majority of these satellite sensors are not necessarily specifically designed for the measurement of precipitation per se. Not until the launch of the Tropical Rainfall Measuring Mission (TRMM; Kummerow et al., 1998; Simpson et al., 1988) in December 1997, later followed by the Global Precipitation Measurement (GPM) mission (Hou et al., 2014) in February 2014, did dedicated satellite precipitation missions exist.

1.2.1 Satellite platforms

Satellites provide an unparalleled view of the Earth and its atmosphere allowing them to observe processes within the Earth system, including those relating to precipitation. The precipitation-capable missions typically comprise of two orbital types: the Low Earth Orbiting (LEO) satellites that circle the Earth at about 850 km altitude or lower, and the Geostationary (GEO) satellites that view the Earth from an altitude of about 36,000 km.

The LEO satellites are often sun-synchronous allowing observations to be made at the same local time and are typical of operational satellites (see Table 1.1), particularly where the overpass time is deemed to be critical. Observations from these satellites are typically available a maximum of twice per day at the Equator, thus limiting the usefulness of any single satellite system for monitoring rapidly changing phenomena such as precipitation, although with multiple satellites more frequent observations may be possible. Satellites such as the European Organization for the Exploitation of Meteorological Satellites (EUMETSAT) Meteorological Operational satellite (MetOp) series provide stringently maintained orbits, ensuring consistent overpass times. The similar National Oceanic and

Table 1.1 Current operational LEO meteorological satellites.

Country	Satellite	Sensor	Local time
China	FY-3E	MERSI-2	06:00 desc
US	NOAA-15	AVHRR/3	06:28 desc
US	NOAA-18	AVHRR/3	07:32 desc
Russia	Meteor-M N2-2	MSU-MR	09:00 desc
Russia	Meteor-M N2	MSU-MR	09:10 desc
Europe	Metop-A	AVHRR/3	09:30 desc
Europe	Metop-B	AVHRR/3	09:30 desc
Europe	Metop-C	AVHRR/3	09:30 desc
US	NOAA-20	VIIRS	13:25 asc
US	SNPP	VIIRS	13:25 asc
China	FY-3D	MERSI-2	14:00 asc
China	FY-3B	VIRR (FY-3)	14:45 asc
US	NOAA-19	AVHRR/3	15:44 asc

Atmospheric Administration (NOAA) series of satellites in contrast are less tightly managed and are allowed to drift over time, thus the overpass times will vary slightly; this is in common with other satellite missions such as the Defense Meteorological Satellite Program (DMSP) series (Curtis and Adams, 1987). A number of LEO satellites are placed into nonsun-synchronous orbits that result in overpasses across all the hours of the day, albeit, over an extended period. Such satellites include the TRMM, GPM, and Megha-Tropiques missions; measuring the diurnal variations in precipitation being a key element in their implementation. A consequence of a nonsun-synchronous orbit is that their orbital path will intersect the orbital paths of the sun-synchronous satellites, thus allowing the intercomparison and/or cross-calibration of instantaneous satellite observations from different satellite sensors.

GEO satellites remain (nearly) stationary with respect to their subsatellite position on the Earth's surface. From their orbital location frequent and regular images can be acquired over a full disc of the Earth, although to obtain (quasi-) global coverage a constellation of satellites located around the Equator is required. The main geostationary satellites currently providing this global coverage include the US NOAA Geostationary Operational Environmental Satellite (GOES)-15/16 (Goodman et al., 2012; Schmit et al., 2017), the European Meteosat Second Generation (MSG) Meteosat-8/9/10 (Schmetz et al., 2002), and the Japanese Himawari (Bessho et al., 2016), with additional geostationary satellites provided by the China Meteorological Agency (CMA) Feng-Yun (FY)-series and the Indian National Satellite System (INSAT)-3 series (see Table 1.2).

1.2.2 Sensors

Satellite platforms host one or more instruments that provide the sensing capabilities. The development of sensors for observing precipitation has largely concentrated upon visible (VIS), infrared (IR), and microwave (MW) systems, the latter covering both passive microwave (PMW) radiometers and active microwave (AMW, or radar) instruments. At present the geostationary platforms only carry VIS/IR instruments, while the LEO platforms may carry one or more VIS/IR, PMW or AMW sensors.

Development of satellite-based precipitation measurements were first based upon the VIS/IR observations; the identification of clouds being fundamental in observing and monitoring precipitation, or at least precipitation-related systems. However, VIS/IR observations are indirectly

Table 1.2 Current geostationary meteorological satellites (35,786 km altitude).

Country	Name	Acronym	launch	until	position
Korea	Communication, Oceanography and Meteorology Satellite	COMS-1	26-Jun-10	≥ 2019	128.2 °E
Russia	Electro	Electro-L N2	11-Dec-15	≥ 2022	76.1 °E
China	Feng-Yun—2	FY-2E	23-Dec-08	≥ 2018	86.5 °E
China	Feng-Yun—2	FY-2G	31-Dec-14	≥ 2018	105.0 °E
China	Feng-Yun—4	FY-4A	10-Dec-16	≥ 2021	104.7 °E
China	Gao Fen	GF-4	28-Dec-15	≥ 2023	106.0 °E
US	Geostationary Operational Environmental Satellite (2nd generation)	GOES-15	04-Mar-10	≥ 2020	134.9 °W
US	Geostationary Operational Environmental Satellite (3rd generation)	GOES-16	19-Nov-16	≥ 2027	75.2 °W
US	Geostationary Operational Environmental Satellite (3rd generation)[a]	GOES-17	01-Mar-18	≥ 2029	89.5 °W
Japan	Himawari (3rd generation)	Himawari-8	07-Oct-14	≥ 2029	140.68 °E
India	Indian National Satellite—3	INSAT-3D	25-Jul-13	≥ 2021	82.0 °E
India	Indian National Satellite—3	INSAT-3DR	08-Sep-16	≥ 2024	74.0 °E
India	Kalpana	Kalpana-1	12-Sep-02	≥ 2018	74.0 °E
Europe	Meteosat Second Generation (MSG)	Meteosat-8	28-Aug-02	≥ 2019	41.5 °E
Europe	Meteosat Second Generation (MSG)	Meteosat-9	21-Dec-05	≥ 2019	9.5 °E
Europe	Meteosat Second Generation (MSG)	Meteosat-10	05-Jul-12	≥ 2019	0.0°

[a]Currently undergoing commissioning.
Source: WMO OSCAR database.

related to surface precipitation since they only relate to information obtained from the cloud top properties. Nevertheless, such techniques are commonly used due to the relative ease of accessing the data and subsequent analysis together with the frequency and resolution of the imagery acquired, particular when obtained from geostationary satellite sensors.

More direct observations of precipitation are possible through PMW sensors that sense the upwelling radiation from the Earth's surface. Over radiometrically cold surfaces (such as water), precipitation enhances the signal, while over radiometrically warm surfaces (e.g., land) precipitation can scatter the upwelling radiation, resulting in a decrease in the received signal. Although more direct than the VIS/IR observations, PMW observations are only available from LEO satellites, and thus provide only intermittent samples. In addition, since the level of radiation emanating from the Earth and the atmosphere in the MW region is small, the spatial resolution of the observations is much poorer than that of VIS/IR sensors.

PMW sensors include both "imagers" (those operating at frequencies within atmospheric window channels) and "sounders" (those that operate at, or close to, atmospheric absorption bands), with many newer instruments encompassing both imaging and sounding channels. Conically-scanning instruments (typically imagers) are usually preferred for precipitation retrievals due to consistent Earth incidence angle (EIA), polarization, and resolution. Cross-track instruments (typically sounders) scan perpendicular to the satellite's track and consequently have a variable EIA that affects the spatial resolution and polarization of the observations; however, since their observations are typically close to atmospheric absorption bands they are less sensitive to the surface background. Tables 1.3 and 1.4 summarize the PMW satellites and sensors since 1978.

The most direct satellite-based observations of precipitation come from AMW measurements. The basis of such measurements is similar to surface-based radar systems, converting backscattered radiation to a precipitation measurement. However, the small number of available sensors and limited swath width results in poor temporal sampling, although the longevity of the TRMM mission (Kummerow et al., 1998, 2000; Simpson et al., 1988) with its Precipitation Radar (PR; Kozu et al., 2001), and the current GPM mission (Hou et al., 2014) with the Dual-frequency Precipitation Radar (DPR; Kojima et al., 2012) now provide a long-term record of precipitation across the Tropics. In addition, CloudSat (Stephens et al., 2002) provides valuable information, particularly at the higher latitudes, on light rainfall and snowfall due to the sensitivity of the Cloud Profiling Radar (CPR).

Table 1.3 Passive microwave imaging radiometers commonly used for precipitation estimation; dates represent full extent of collected data.

Sensor	SMMR	SSM/I	SSMIS	AMSR	AMSR2	TMI	GMI
Satellite	Seasat, Nimbus-7	DMSP-F08, F10-F15	DMSP-F16, F17-F19	AQUA, ADEOS-II	GCOMW1	TRMM	GPM
Dates	1978–88	1987-present	2003-present	2002–2011	2012-present	1997–2015	2014-present
Orbit	Sun-sync	Sun-sync	Sun-sync	Sun-sync	Sun-sync	Nonsun sync	Nonsun sync
Scan	Conical	Conical	Conical	Conical	Conical	Conical	Conical
Frequencies (GHz)	6.6VH	–	–	6.925VH	6.925/7.3VH	–	–
	10.7VH	–	–	10.65VH	10.65VH	10.65VH	10.65VH
	18.0VH	19.35VH	19.35VH	18.70VH	18.70VH	18.70VH	18.70VH
	21.0VH	22.235 V	22.235 V	23.80 VH	23.80VH	23.80VH	23.80V
	37.0VH	37.0VH	37.0VH	36.5VH	36.5VH	36.5VH	36.5VH
	–	–	50.3–63.3VH	–	–	–	–
	–	85.5VH	91.65VH	89.0VH	89.0VH	89.0VH	89.0VH
	–	–	150 H	–	–	–	165.6VH
	–	–	183.31(2)H	–	–	–	183.31 V(2)
	–	–	–	–	–	–	–

Table 1.4 Passive microwave sounding radiometers commonly used for precipitation estimation; dates represent full extent of collected data.

Sensor	**AMSU-B**	**MHS**	**SAPHIR**	**ATMS**
Satellite	**NOAA 15,16,17**	**NOAA18,19, MetOp-A, B**	**Megha-Tropiques**	**NPP**
Dates	**1998–2010**	**2005–present**	**2011–present**	**2011–present**
Orbit	**Sun-sync**	**Sun-sync**	**Non sun-sync**	**Sun-sync**
Scan	**Cross-track**	**Cross-track**	**Cross-track**	**Cross-track**
Frequencies (GHz)	–	–	–	–
	–	–	–	–
	–	–	–	–
	–	–	–	23.8 V
	–	–	–	31.4 V
	–	–	–	50.3–57.3 H
	89 V	89 V	–	87–91 V
	150 V	157 V	–	164–167 H
	183.31(3)V	183.31 H(2)	183.31 H(6)	183.31(5)H
	–	190.31 V	–	–

1.2.3 Current precipitation observing systems

To maximize the availability of precipitation data, a range of different satellite systems are used to provide observations. These can be split broadly into the GEO IR-based observing network and PMW/AMW LEO satellites that form the GPM constellation. The GEO suite of sensors is maintained to provide a consistent set of observations around the Equator and extending from about 60°N to 60°S. Although individual satellites are capable of providing higher temporal and spatial resolution, a combined IR brightness temperature (Tb) product is available every 30 minutes at a nominal resolution of 4 km (Janowiak et al., 2001).

The GPM constellation consists of a suite of international partner satellites with PMW radiometers and AMW instruments, with the GPM Core Observatory (CO; Hou et al., 2014) acting as a calibration reference (Draper et al., 2015; Wentz and Draper 2016) to ensure consistent observations, and subsequent precipitation retrievals thereafter (Wilheit et al., 2015; Berg et al., 2016). The constellation currently consists of 10 PMW sensors, namely the Advanced Microwave Scanning Radiometer (AMSR-2), Special Sensor Microwave Imager/Sounder (SSMIS, 3 currently operational), Microwave Humidity Sounder (MHS, 4 currently operational), Sondeur Atmosphérique du Profil d'Humidité Intertropicale par Radiométrie (SAPHIR), and the Advanced Technology Microwave Sounder (ATMS).

Many techniques have been developed to exploit these observations, both through the utilization of individual channels and through channel combinations: these are described in detail in the following section.

1.3 Observations to estimates

A good number of algorithms, techniques, and schemes exist to extract precipitation from satellite observations. Many of these rely upon simple relationships between the satellites' observations and surface precipitation to generate the final precipitation product (Kidd et al., 1998). Although generally simple (i.e., not necessarily reliant upon radiative transfer modeling), they do have a sound physical basis, and are nevertheless still capable of generating useful products (e.g., Kidd and Levizzani, 2011; Tapiador et al., 2012). Techniques have also been developed or adapted to consider the product latency time in order to meet user requirements: the typical data delivery time for global IR and MW data is around 2–3 hours, though some regional applications exploit direct broadcast capabilities and significantly reduce the latency to the order of 10–15 minutes. It should be noted however that no satellite product is capable of producing a perfect estimate of surface precipitation. Indeed, the accuracy of many products vary and depend upon the perceived application, such as global products resulting from a different approach through optimal use of the IR and/or MW data available, versus instantaneous precipitation estimates in real time. Ultimately many precipitation retrieval techniques seek a compromise between providing a reasonable estimate and surface precipitation within a set of developer and user requirements.

1.3.1 VIS/IR

The development of techniques utilizing VIS/IR observations is now well established. Such observations benefit from the frequent and consistent temporal sampling available from GEO-based sensors, together with spatial resolutions that are commensurate with many user applications. Although VIS imagery provides the best resolution, it is limited to daytime only and consequently the majority of the development work has concentrated upon producing precipitation products from IR imagery. The IR-based techniques rely upon the general notion that taller clouds have colder cloud tops and precipitate more. However, since only the temperature of the cloud tops is measured, it is indirectly related to the precipitation near to the Earth's surface. The use of additional infrared channels can help to better-define the cloud top characteristics, but the indirectness of cloud-top properties to surface precipitation still remains.

Several satellite precipitation retrieval techniques are based on IR-only measurements, some using real-time or archival MW data for calibration (see Section 3.4 below). These techniques seek to infer precipitation from observed geostationary IR brightness temperatures (since these provide the best temporal and spatial resolution of precipitation-capable satellite sensors). However, their main disadvantage is the very indirect nature of the precipitation estimates. A benchmark technique is the GOES Precipitation Index (GPI; Arkin and Meisner, 1987), which simply assigns 3 mm/h to any cloud top temperature below 235K. Although simple, this works surprisingly well particularly at monthly/2.5-degree resolution scales. Among the range of other IR-based precipitation products currently available, the Hydro-estimator (Scofield and Kuligowski, 2003) distinguishes between intense convective cores and less precipitation-producing stratiform areas and thus accounts for some of the cloud top to surface precipitation variations. Meanwhile the PERSIANN-CCS technique (Hong et al., 2004) uses cloud texture information such as variability, minimum, etc. to better define the surface precipitation. However, all IR products have problems in terms of detection of precipitation since they may confuse high (but thin) clouds with high deep clouds resulting in precipitation where there is none, or conversely miss precipitation where clouds are shallow but nevertheless precipitating. Overall, IR-based products perform better in tropical regimes where cold cloud-top temperature to surface precipitation relationships are more evident.

1.3.2 PMW

In contrast to IR techniques, observations from PMW sensors are more sensitive to the ice and liquid water within the atmosphere, with higher-frequency channels being more sensitive to solid (ice) particles and lower-frequency channels being more directly sensitive to the liquid hydrometeors. Due to the effect of surface emissivity the lower-frequency channels are usually confined to precipitation retrievals over the ocean, while high frequencies can be used over both land and ocean. Over the ocean, increases in the Tb of the low frequency channels are strongly linked to increased rainfall, while over land decreases in Tbs in high frequency channels are linked to the precipitation-size ice in clouds, and thus surface precipitation. These basic criteria have been exploited by many techniques to provide a measure of precipitation at the surface. One drawback is that the resolutions from PMW sensors are much poorer than IR sensors: high-frequency channels have resolutions about 5 km at best, while low frequency channels are typically about 30 km. In addition, at present all MW instruments are on low-Earth-orbiting satellites that provide a maximum of only two observations per day per satellite, and thus require a constellation of sensors to adequately resolve the variability of precipitating systems.

Techniques for the retrieval of precipitation have been built upon basic radiometric properties of the interaction of the precipitation-sized hydrometeors with the radiation sensed by the radiometer. A number of basic techniques have achieved a degree of success through linking this received radiation with the precipitation at the surface. However, providing an unambiguous retrieval of precipitation is often difficult due to the variability of the surface background, or the nonunique spectral signature to hydrometeor profile/surface rainfall relationships. This, together with a desire to provide vertical profiles of precipitation has led to the development of physically-based precipitation retrieval schemes, as typified by the Goddard Profiling (GPROF) scheme (Kummerow et al., 2015). The GPROF technique is designed to provide estimates of surface precipitation and vertical profiles utilizing model information (2-m temperature and total precipitable water) and ancillary datasets (surface types) to better constrain the set of possible precipitation retrievals. Importantly the GPROF scheme is designed to work across the range of PMW observations to provide consistent precipitation estimates. Similarly, You et al. (2015) utilized surface conditions and cloud vertical structure to stratify PMW observations over land in precipitation retrieval algorithms. Kidd (2017) developed the Precipitation Retrieval and Profiling Scheme (PRPS) for the PMW SAPHIR sensor, using the GPM DPR measurements as a calibrator (Table 1.5).

A 1-D variational approach was adopted by Boukabara et al. (2011) while building their Microwave Integrated Retrieval System (MiRS). MiRS is an iterative physical inversion system starting from the radiative transfer equation to find radiometrically appropriate profiles of temperature, moisture, liquid cloud, and hydrometeors, as well as the surface emissivity spectrum and skin temperature. It is applicable globally with differences between land, ocean, sea ice, and snow background. The system is thus more than a simple rainfall rate retrieval and is applicable as an assimilation system when combined with a forecast field.

Over ocean the retrieval methods show better performances given the relative homogeneous emissivity background. Several methods have been

Table 1.5 The sensors contributing to the current set of FCDRs.

	SMMR	F08	F16	F19	TMI	N15	N18	SAP
		F10/11	F17		AMSR2	N16	MOA	ATMS
		F13/14/15	F18		GMI	N17	MOB	
CSU1		Y	Y	Y				
CSU2			Y	Y	Y		Y	Y
CSU3		Y	Y	Y	Y	Y	Y	Y
NCEI						Y	Y	
CM-SAF V001		Y						
CM-SAF V002		Y	Y					
CM-SAF V003	Y	Y	Y	Y				

developed using conically scanning MW sensors. The Unified Microwave Ocean Retrieval Algorithm (UMORA; Hilburn and Wentz, 2008) simultaneously retrieves sea surface temperature, surface wind speed, columnar water vapor, columnar cloud water, and surface rain rate. It parameterizes the beamfilling effect, cloud and rain water partitioning, and effective rain layer thickness. Meanwhile the Hamburg Ocean Atmosphere Parameters and Fluxes from Satellite Data (HOAPS-3; Andersson et al., 2010) provides fields of turbulent heat fluxes, evaporation, precipitation, freshwater flux, and related atmospheric variables over the global ice-free ocean. The strength of the method is a careful intersensor calibration, which ensures a homogeneous time series with dense data sampling. This characteristic makes HOAPS-3 suitable for water cycle studies.

MW-derived rainfall products are also made available by EUMETSAT's Satellite Application Facility on Support to Operational Hydrology and Water Management (H-SAF; Mugnai et al., 2013). A Cloud Dynamics and Radiation Database (Casella et al., 2013) precipitation retrieval algorithm was conceived to improve upon ambiguities of conventional cloud radiation database (CRD) approaches by using a regional/mesoscale model, applied in cloud resolving model (CRM) mode, to produce a large set of numerical simulations of precipitating storms and extended precipitating systems. The simulations are used for the selection of millions of meteorological/microphysical vertical profiles within which surface rainfall is identified. The improvements are evident over land surface. Casella et al. (2015) then introduced a novel algorithm based on canonical correlation analysis (CCA) applicable to all PMW radiometers of the GPM constellation. The method is based on the definition of a threshold to be applied to the resulting linear combination of the brightness temperatures in all available channels. The Passive microwave Neural network Precipitation Retrieval v2 (PNPR; Sanò et al., 2016) is designed to retrieve instantaneous surface precipitation rate from the advanced spectral capabilities of the ATMS and new generation PMW sensors. It is based on a single neural network and works over all surface backgrounds, being trained by a large database of cloud-resolving model simulations over Europe and Africa. Simpler methods were also designed based on thresholding approaches of the Tb depression at the various PMW sensor channels (e.g., Laviola and Levizzani, 2011): these methods are very simple to implement and particularly suitable for an operational environment after an appropriate validation with ground-based data (Laviola et al., 2013).

1.3.3 AMW

The development and implementation of precipitation radars began with the TRMM Precipitation Radar (Nakamura et al., 1990) in 1997 and continued with the GPM Dual-frequency Precipitation Radar (DPR) from 2014. The current DPR sensor consists of Ku- and Ka-band radars that provide cross-track swaths of approximately 245 and 125 km respectively with a resolution of about 5.2 km. The minimum detectable radar reflectivity is equal to about 0.5 mm/h rainfall intensity, although studies have shown minimum detectable rain rates of about 0.2 mm/h are possible. The Ku-band and Ka-band channels are processed first, followed by the processing of Ku + Ka combined channels. The basic processing is essentially the same: conversion of the return power into radar reflectivity, or normalized surface radar cross-section (depending on atmosphere or surface returns); rain/no-rain discrimination; determination of stratiform/convective, and determination of additional information, such as presence of bright-band, snowfall at surface, hail detection, etc. An algorithm using the surface return estimates the path integrated attenuation, which is used in the retrieval algorithm to estimate the water content, rainfall rate, and parameters of the particle size distribution (see Iguchi et al., 2015). However, the retrieval capabilities from such radar systems depend on the Ku-Ka-band sensitivities to precipitation (e.g., Toyoshima et al., 2015).

Although the AMW/radar provides the most direct measure of precipitation, these products are limited by long revisit times (on the order of days), and latitudinal extent (37°N–37°S for the TRMM PR and 68°N–68°S for the GPM DPR). However, these sensors are capable of providing detailed three-dimensional information of precipitation systems, and given the longevity of the TRMM PR, together with the GPM DPR, for establishing climatological distributions of precipitation rates (see Liu and Zipser, 2015). An example of the utility of such data is shown in Fig. 1.1, which shows the distribution of the occurrence of precipitation features with DPR reflectivities greater than 40 dBZ above 10 km, together with the contribution of the accumulation of this precipitation: the mid-west of the United States, Argentina, central Africa, and Pakistan having the greatest occurrence and contribution of intense precipitation events.

1.3.4 Combined and merged products

There is an overall desire to improve precipitation retrievals and products through the utilization of multiple datasets, both through a greater

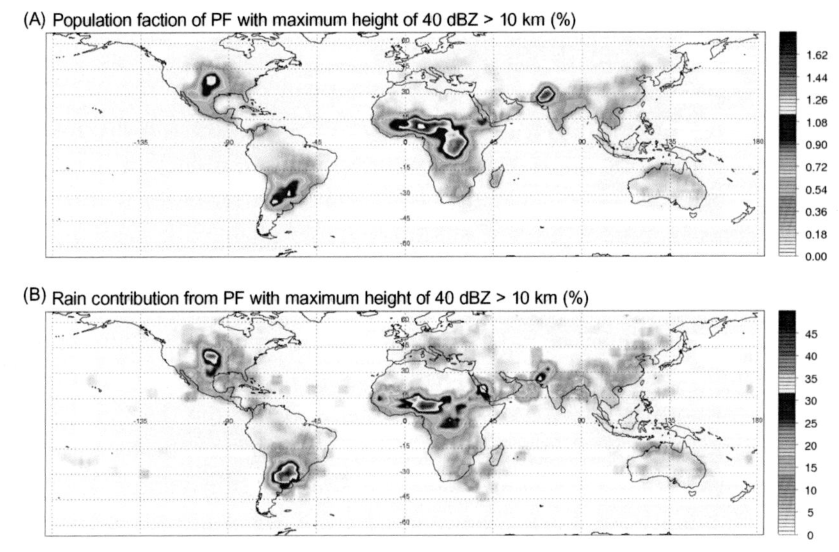

Figure 1.1 Distribution of the occurrence of precipitation features with DPR reflectivities higher than 40 dBZ above 10 km, together with the contribution of the accumulation of this precipitation. *Courtesy Dr. Chuntao Liu, Texas A&M University, Corpus Christi.*

understanding of the interaction of the observed radiation and the precipitation itself, and through providing better temporal and spatial sampling for the final precipitation product.

Improving precipitation products through a better understanding of the observation-to-retrieval process is exemplified in the GPM COmbined Radar-Radiometer algorithm (CORRA), whereby PMW observations from the GMI and AMW from the DPR are used in a combined approach to try and reduce ambiguities in the physical modeling of precipitation (Grecu et al., 2016), building upon a similar scheme that was implemented for TRMM (Haddad et al., 1997).

The combination of multisensor satellite observations is perhaps more usually associated with IR and PMW observations. As noted above, the GPI product is surprisingly good at the relatively coarse monthly/2.5-degree scales, although the technique is known to have regional and seasonal biases. Various schemes have been developed to modify and improve the GPI, such as that of Xu et al. (1999) who developed a MW-IR threshold technique. Kidd et al. (2003) outlined the combined PMW-IR algorithm, which matched the rainfall intensity distributions from PMW rainfall retrievals to the IR cloud top temperature thresholds.

Similarly, the Self-Calibrating Multivariate Precipitation Retrieval (SCaMPR) techniques (Kuligowski, 2002, 2010) routinely updates IR-to-rainfall relationships based upon PMW calibration.

However, the main thrust of current combined retrievals is the generation of consistent, timely, and frequent precipitation products at the (near) instantaneous scale. The generation of precipitation retrievals at intermittent instants in time from PMW observations, often with very uneven sampling around the diurnal cycle, can poorly represent precipitation, which is highly variable in space and time. Furthermore, most user applications require regular precipitation products. It is therefore logical that more complete precipitation products can be generated through the combination of both IR and PMW measurements, exploiting the strengths of each. At present three main products combine PMW and IR data to provide estimates of surface precipitation at scales in the region of about 10 km/30 minutes or finer. These are CMORPH (Joyce et al., 2011), GSMaP (Aonashi et al., 2009; Kubota et al., 2007; Ushio et al., 2009), and IMERG (Huffman et al., 2014).

These three schemes are based upon similar three-step approaches. First, the individual PMW measurements are generated (or obtained) to detect and estimate any surface precipitation; second, wind vectors or changes in cloud top temperatures derived from IR data (or more recently models) move (or morph) the precipitation between the individual PMW overpasses, and; finally (if required) IR-derived estimates are combined with the morphed PMW estimates. Thus, the final product has the potential to produce a continuous retrieval of precipitation over time. However, errors in the products arise from the accuracy of the PMW retrieval itself and the ability to faithfully move/morph the precipitation between the individual PMW retrievals. Indeed, development of clouds in the IR might not necessarily be associated with a similar development in (the intensity) of precipitation at the surface, which is particularly true with orographic cloud development/precipitation. The main differences between these schemes result from options taken at each of the different stages. GSMaP uses the changes in the IR temperatures rather than the IR-generated wind vectors (CMORPH) to guide the evolution between PMW observations. The IMERG technique uses forward and backward morphing as opposed to CMORPH. Long gaps between PMW overpasses are also treated differently. All these three products use, in some way, surface precipitation gauge data to produce a final, research-quality version that ensures that the satellite estimates match the monthly surface precipitation gauge analysis. More recently, the Multisource Weighted-Ensemble Precipitation (MSWEP; Beck et al., 2017a) scheme has generated a combined 3-hourly product at 0.1 degree resolution based upon the combined products themselves.

Until recently the vast majority of precipitation retrieval schemes have been designed to estimate "rainfall," with little research into identifying and retrieving snowfall per se. It should be noted however that estimating liquid precipitation directly is essentially limited to over-ocean using PMW; over land the scattering signal in PMW observations is from frozen hydrometeors which is then interpreted as liquid/rain. More recently, measuring snowfall has become an important research topic not least due to the availability of data from the GPM mission.

1.3.5 Measuring solid precipitation

Measuring solid precipitation is crucial for several reasons including weather monitoring and forecasting, hydrology of high-latitude basins, but especially for closing the terrestrial water budget. Snowfall is hard to measure through ground-based gauges due to difficulty in discriminating between falling and resuspended snow, snow accumulation, icing of the sensors and error definition problems (e.g., Levizzani et al., 2011).

Snowfall observations from satellite have suffered from all the problems affecting rainfall estimates, but additional problems are related to:

- distinguishing ice hydrometeors from water drops making use also of the high frequency channels above 100 GHz;
- the complexity of the radiative properties of snowflakes and ice crystals;
- the poorly known cloud vertical structure and ice water content;
- the incomplete understanding of snow microphysics in mixed-phase clouds.

With the advent of high-frequency PMW radiometers such as the Advanced Microwave Sounding Unit-B (AMSU-B) snowfall has started to be detectable, although limitations persist since the ice content of a cloud is not known a priori. Physical models were developed to discriminate snowfall over land (e.g., Skofronick-Jackson et al., 2004). A most important problem to be solved is the detection of the snowfall signature over snow-covered ground. The advent of the GPM has recently shown that the combination of low- (10–19 GHz) and high-frequency (89–166 GHz) spectral bands of the GMI provides the maximum amount of information for snowfall detection (Ebtehaj and Kummerow, 2017). Operational algorithms have started to become available such as the one of NOAA (Meng et al., 2017), which has four components: cloud properties retrieval, computation of ice particle terminal velocity, ice water content adjustment, and the determination of snowfall rate. The algorithm is based on a 1D-Var approach and has been validated against radar and other ground-based observations. Several studies are being conducted using such radars and the GMI to quantify the detection skills of the instruments. You et al. (2017) have shown that scattering signatures are essential for snowfall detection. A statistical approach was adopted by Liu and Seo (2013) to correct precipitation retrievals where concentrations of high water vapor above the precipitation layer negated the scattering signatures in certain snowfall events. Intercomparison exercises are being held to determine the potential of snowfall detection by the various algorithms (e.g., Laviola et al., 2015).

However, it is only with the combined availability of observations from the CPR on CloudSat and the GPM DPR that the research in this field has received a substantial boost. Skofronick-Jackson et al. (2013) determined the thresholds of detection for various active and passive sensor channel configurations and falling snow events over land surfaces and lakes, through model simulations of the minimum amount of snow using nonspherical snowflake shapes. Studies conducted on the snowfall detection capabilities of the DPR and more will become available soon (e.g., Casella et al., 2017; Panegrossi et al., 2017). The availability of CloudSat's CPR allowed Liu (2008) to conceive a derivation of snow-cloud characteristics in two steps: (1) a snow-rain threshold based on multiyear land station and shipboard present weather reports; and (2) a second part based on backscatter computations of nonspherical ice particles and in situ measured particle size distributions. The author found that the characteristics of the vertical distribution of snowfall rate are quite similar for over-ocean and over-land snow clouds, except that over-land snow clouds seem to be somewhat shallower than those over ocean. CloudSat has contributed to a global census of such shallow cumuliform snow clouds (Kulie et al., 2016; Kulie and Milani, 2018).

One more aspect concerns the ability to detect hail-fall. Recent work by Cecil (2009, 2011) and Cecil and Blankenship (2012) have demonstrated a strong relationship between the occurrence of hail and the MW Tb, primarily at 37 GHz, but also at 85 GHz. These studies were performed with the TRMM Microwave Imager (TMI) and the Advanced

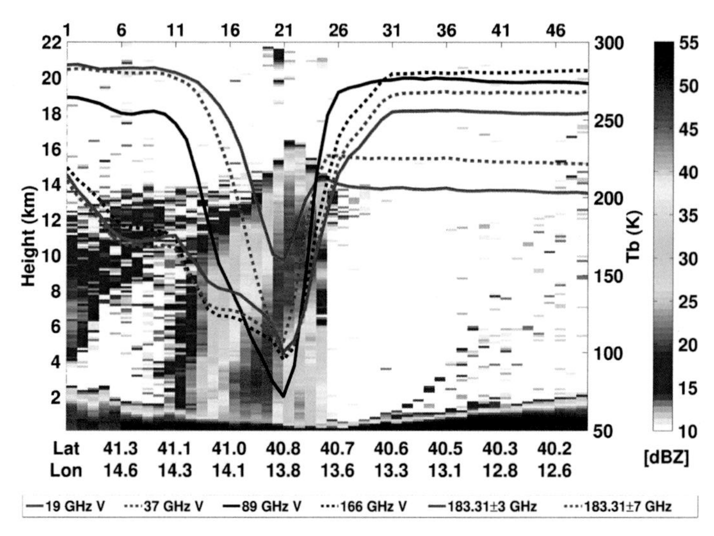

Figure 1.2 3 June 2015 2050 UTC. DPR cross-track section of estimated rainfall at Ku-band of a violent hailstorm which generated 7–10 cm diameter hailstones over Nebraska, (US). Corresponding GMI Tbs at six selected window and water vapor channels are also shown as lines of different colors.

Microwave Scanning Radiometer for EOS (AMSR-E); the climatologies derived from them were consistent with those observed from surface observations. Recently, Ferraro et al. (2015) have developed a prototype AMSU-based hail detection algorithm and a 12-year climatology (2000–11) of hail occurrence over the Continental United States. The new observing capabilities of the GPM CO are also instrumental for intense hailstorm detection (e.g., Marra et al., 2017). An example is shown in Fig. 1.2 where the cross-section of the DPR Ku-band rain rates and the Tbs at six selected window and water vapor channels of the GMI are shown for a large hailstorm over Nebraska, US on 3 June 2014.

1.4 Errors and uncertainties

It is inevitable that any observational dataset will have errors and uncertainties associated with it. This is particularly problematic for any satellite dataset that provide (quasi-) global products from a number of observational samples. For precipitation the problem is more acute: for instantaneous precipitation the distribution is very much skewed towards zero, with the modal value of instantaneous precipitation being zero. Accumulation of these instantaneous samples through time and space provides a more normal distribution, although other errors and uncertainties contribute, such as the representativeness of the samples. At the same time an error decomposition is fundamental to trace the degree of improvement of algorithms that are being modified and renewed relatively frequently (e.g., Gebregiorgis et al., 2018); this allows further advances in algorithm development on one side and societal applications on the other.

The quantification of errors and uncertainties is needed to enable the correct use of satellite precipitation products in hydrological applications, climate studies, and water resource management (e.g., Maggioni et al., 2016; Maggioni and Massari, 2018). This is attributable to the directness of the observation to the retrieved parameters. In particular this relates to the frequencies/channels utilized, the type of precipitation being observed, and the heterogeneity of the precipitation within the footprint of the sensor. The frequency/channel used essentially determines whether information about the cloud/precipitation is derived from cloud tops, ice above (any) freezing level, and liquid water content; IR channels only provide observations on cloud-top characteristics, while high-frequency PMW channels provide information on precipitation-related ice content, and low-frequency PMW channels provide information on water content. Many of the more complex techniques use combinations of channels/frequencies with different weightings, thus changing retrieval errors associated with each individual observation.

Precipitation at the instantaneous scale has a highly skewed distribution towards zero, which, together with the small occurrence of precipitation, impacts the sampling of precipitation events by satellite observing systems. The sampling errors are typically greatest at the 1 hour to monthly scale; below that the spatial/temporal autocorrelation relationships are generally sufficient, while at or above the monthly scale sufficient samples are generally available to provide a reasonably robust estimate. The impact of sampling errors is usually determined by assuming a standard set of criteria for how often the precipitation is sampled within a certain time frame. A number of studies have attempted to quantify the uncertainties of satellite precipitation estimates through comparison surface rain gauges (e.g., Katsanos et al., 2004; Villarini et al., 2008), surface radar networks (e.g., Hossain and Anagnostou, 2006; Krajewski et al., 2010), or multiple satellite estimates (Tian and Peters-Lidard, 2010). However, no overall consensus is apparent since precipitation uncertainty tends to be dependent on region, season, and rain rate (Tian and Peters-Lidard, 2010). Nevertheless, precipitation accumulated over larger spatial and temporal domains has lower errors and uncertainties due to a greater number of samples. However, it should be noted that instantaneous scale precipitation is highly variable by precipitation type/system.

Little work has been done on the contribution of errors associated with the diurnal cycle of precipitation, and in particular the impact of the time-sampling of the diurnal cycle using satellite observations (Tan et al., 2017; Oliveira et al., 2016). Many satellite orbits drift through time, thus affecting the overpass times, thus affecting the time at which each sample is collected. In addition, different sensors collect data at different times of day. For regions that experience a strong diurnal cycle this has significant implications since over time satellite observations will sample different parts of the diurnal cycle. Even with a simple diurnal cycle of rainfall significant biases together with step-changes may occur as sensors become available or retire from long-term datasets. Furthermore, small, but not insignificant variations occur after the launch of TRMM and change after the launches of Megha-Tropiques and GPM due to the interaction of the nonsun-synchronous sampling. To correct these diurnal biases, care is needed when combining information from multiple satellites to ensure that at the very least, time-weighted averages are used to ensure correct daily-monthly estimates.

Another important aspect of error quantification concerns the identification of the influence of surface characteristics. Carr et al. (2015) found that a decreased algorithm performance is apparent over dry and sparsely vegetated regions, a probable result of the surface radiation signal mimicking the scattering signature associated with frozen hydrometeors. Complex terrain represents a source of error limiting the quantitative use of satellites estimates in many areas of the world (Bartsotas et al., 2018; Maggioni et al., 2017). High-resolution radar data are often required for a thorough error analysis over small mountain catchments (e.g., Derin et al., 2018). At the same time, the dependence of the algorithm performance on the physical character of precipitation is important. Carr et al. (2015) results show that algorithms generally perform better in pure stratiform and convective regimes, compared to events containing a mixture of the two.

The generation of errors and uncertainties associated with precipitation products has been very limited, not least since there is no clear understanding or standardization of what such numbers mean in reality, how such values are transferred from one scale to another (both temporally and spatially) and how the user community can easily implement and include such errors and uncertainties in their applications. Work needs to be done to more clearly identify and categorize the error sources thus bridging between gridded rainfall and instantaneous satellite estimates (Tan et al., 2016).

1.5 Data products

Satellite precipitation products are numerous and are available at no cost over the internet from many of the developers, thus allowing the widespread use for a range of different applications. Searching of starting points for appropriate products appropriate to their needs can be undertaken by visiting the International Precipitation Working Group (IPWG) web page on Data & Products (http://www.isac.cnr.it/~ipwg/data.html). The classification adopted by IPWG divides the products in to four different categories:

- Combined satellite-gauge datasets;
- Combined satellite datasets;
- Satellite-based single-sensor datasets, and;
- Precipitation gauge analyses.

Although this is not the only viable classification, it is the most widespread in terms of dataset development. Classifications based on the sensor or the application are also possible, for example.

It should be noted that products created for a specific region might not perform well beyond the region or time frame of their development. For example, one of the earliest IR methods, the GOES Precipitation Index (GPI; Arkin and Meisner, 1987), trained for tropical regions over large time and space domains, performs poorly outside the tropics. The development of global products, however, is addressing this problem and products from the constellation of satellites nowadays perform reasonably well on a global perspective. Very recent products such as MSWEP (Beck et al., 2017a) include not only gauge data, but also model reanalyses. While this approach contributes to better global performances, validation becomes difficult and the contribution of the satellite products is generally lost.

The availability of long-term satellite-based precipitation data is of great importance to the climate community, not least since it has the potential to provide an independent source of information with which to validate climate-scale models (e.g., Tapiador et al., 2017). One of the most used and longer datasets is that produced by the Global Precipitation Climatology Project, which has recently issued v.2.3 (Adler et al., 2018). This product combines multisensor information from satellites together with surface gauge data to provide a long-term precipitation climate data record.

Further research into long-term climate-scale precipitation data products necessitate the generation of Fundamental Climate Data Records (FCDRs) of PMW Tbs. These FCDRs provide a time series of observations of sufficient length, consistency, and continuity necessary for climate analysis studies and have been carefully developed to improve homogenization and intercalibration of the Tbs to ensure long-term stability of the FCDRs.

Three sources of intercalibrated FCDR radiometer data for the passive microwave instruments are available. The first, containing three versions, is available from NOAA and/or Colorado State University, and consists of an initial version covering the period from July 1987 to the present for the SSM/I and SSMIS sensors, calibrated to the F13 sensor (Berg et al., 2013; Sapiano et al., 2013). The second version includes all TRMM/GPM constellation sensors and is produced by the NASA/GSFC Precipitation Processing System ("V05") starting 1 October 2017, and is intercalibrated to the GMI. Note that this dataset also contains all the sounder (MHS, ATMS, SAPHIR) instrument data (Berg et al., 2016; Semunegus et al., 2010; Wilheit et al., 2015). This will be extended back in version 3 to July 1987. Alongside these FCDRs NOAA's National Centers for Environmental Information (NCEI) has been producing a FCDR for the sounding instruments.

The EUMETSAT Satellite Application Facility on Climate Monitoring (CM-SAF) currently develops, produces, archives, and disseminates satellite-based products in support of climate monitoring. The three satellite observational datasets, namely the FCDR of Special Sensor Microwave/Imager (SSM/I) Tbs (V001) is available from July 1987 through to December 2008 for the F08, F10, F11, F13, F14, and F15 SSM/I sensors (Fennig et al., 2013); the FCDR of SSMI/Special Sensor Microwave Imager Sounder (SSMIS) Tbs (V002) which extends the V001 dataset to include the F16, F17, and F18 SSMIS sensors satellites (Fennig et al., 2015); and the FCDR of Microwave Imager Radiances (V003) which provides intercalibrated Tbs from the SMMR, SSM/I, and

SSMIS from October 1978 to December 2015, extending the dataset back to the SMMR radiometer aboard Nimbus-7 satellite.

Other long-term datasets exist for applications ranging from continental scale to global and dwelling on the coverage of GEO-IR sensors. These datasets tend to exploit the coverage of the IR bands at the expense of an overall lower instantaneous precipitation quality. However, they are quite useful for studies on temporal trends and spatial analyses. Among these global datasets we will mention the Climate Hazards InfraRed Precipitation with Stations (CHIRPS; Funk et al., 2015), which is widely used for climatic studies and for drought- and famine-related monitoring. An example of such studies is shown in Fig. 1.3, which shows the anomalies of rainfall during the short-rains season over East Africa for the period 1983–2010. On the African continent two more datasets are being used and maintained: the African Rainfall Climatology v2 (Novella and Thiaw, 2013) and the TAMSAT African Rainfall Climatology and Time series (TARCAT; Maidment et al., 2017). Several single-product or comparative evaluations have been conducted over the whole continent (e.g., Thiemig et al., 2012) or over specific areas such as East Africa (e.g., Cattani et al., 2016; Dinku et al., 2018).

1.6 Validation

The validation of the satellite-based precipitation estimates is an essential part of the development and assessment of retrieval techniques. Datasets derived from rain (or snow) gauges and weather radar form the basis of many surface reference datasets, covering a range of temporal and spatial scales, although both primarily limited to land surfaces. The physical nature of these surface reference data dictates their usefulness in validating satellite precipitation products. Gauges provide a time-integrated point measurement and remain the de facto source of validation data. However, due to their uneven distribution, gauge data are usually aggregated into coarser spatial/temporal resolution products and thus tend to be used at the daily 1×1 degree scales or coarser. Radar data, being instantaneous spatial measurements, may be used to provide validation/comparisons at instantaneous scales with resolutions commensurate with those of the satellite spatial resolutions. At the instantaneous scale surface

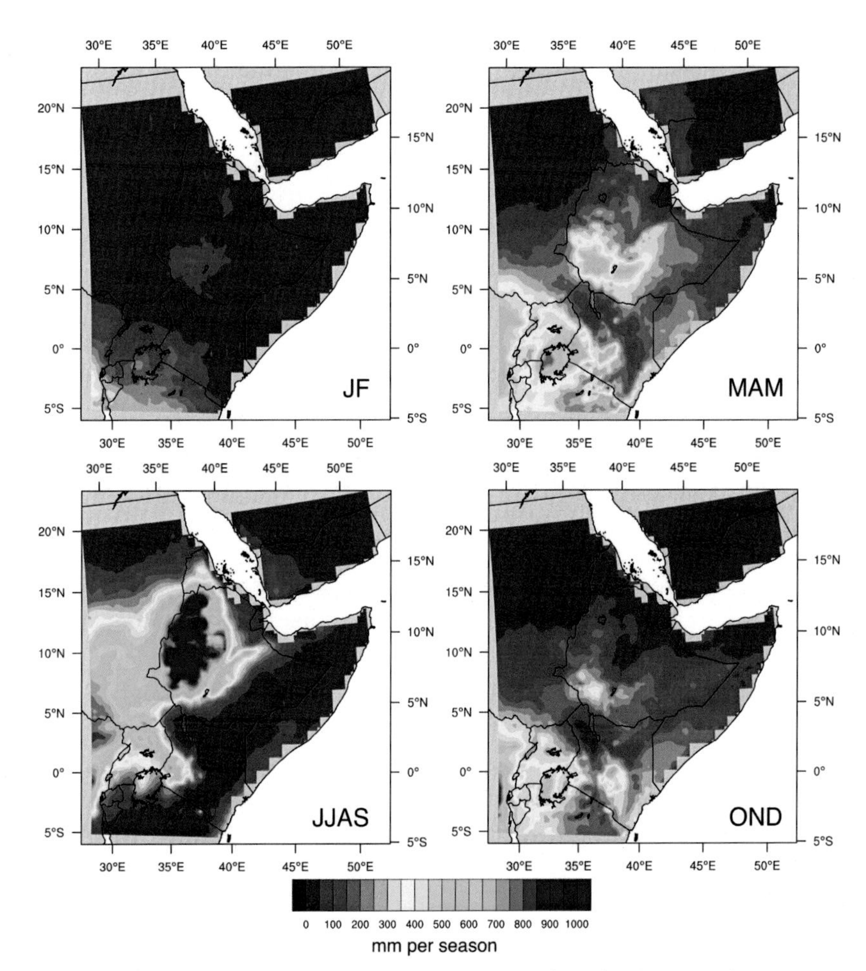

Figure 1.3 Total rainfall (mm per season) over East Africa for the period 1983–2017 in the four seasons using CHIRPS data. The data refer to January-February (JF; upper left), March-April-May (MAM; upper right), June-July-August-September (JJAS; lower left), and October-November-December (OND; lower right). *Courtesy of Dr. Elsa Cattani, CNR-ISAC.*

radar and dense gauge networks provide an essential means for verifying satellite retrievals due to the available high spatial and temporal resolutions (Kirstetter et al., 2015).

Past intercomparison exercises (e.g., Adler et al., 2001) have shown that despite the multitude of precipitation-related observations and accompanying retrievals and products, it is often difficult to ascribe any single precipitation estimate to be better than any other estimate. In general, they have pointed to merged data products (including rain gauge information) as providing the overall best results. It should be noted that this study has concentrated upon the European and US regions, and used different reference datasets from those regions; the performance of these products over other regions and seasons with unique precipitation regimes is likely to be different.

However, to match the availability of the surface reference data, together with the need for validation at different scales and requirements, a multitier approach is now usually adopted: validation of precipitation products across all scales is a key objective of the IPWG and is summarized in Levizzani et al. (2018). Concerted fine scale multisensor validation is carried out at the local scale to improve our understanding of the microphysics of precipitation. Next, local to regional scale validation is done using gauge and/or radar data to investigate local or regional performance of satellite products, and finally large scale or global validation is carried out to assess these satellite products for global/climate scale studies.

Fine temporal/spatial short-period local comparisons are conducted to provide physical validation of the satellite observations and retrievals, such as those carried out by the GPM ground validation (GV) program (e.g., Petersen et al., 2016). These studies, usually carried out over a limited region for a limited time, often incorporate airborne data to provide information on the microphysical nature of the precipitation system to better understand the precipitation-observation relationships. The GPM has conducted a series of dedicated GV campaigns primarily aimed at collection of detailed information from multiple sensors on multiple platforms, including satellites, aircraft, and surface (Hou et al., 2014). This approach enables a comprehensive picture of the precipitation to be captured enabling a better understanding of the relationship between the in situ precipitation and the satellite observations (e.g., Jensen et al., 2016; Skofronick-Jackson et al., 2015; Petersen et al., 2016; Houze et al., 2017). The core GV data products used to assess GPM satellite products over the continental US typically include 2 and 30-minute radar rain rate products (rain gauge bias-adjusted) and precipitation types (rain/snow) adapted/modified from the NOAA/OU multiradar multisensor (MRMS) product (Zhang et al., 2016). Over the ocean polarimetric radar estimates of rain rate are collected from the K-Pol radar at the Kwajalein Atoll in the Marshall Islands and the Middleton Island WSR-88D radar located in the Gulf of Alaska. In addition, field campaign data includes site-specific disdrometer-measured rain/snow size distribution (DSD), phase and fall speed information used to derive polarimetric radar-based DSD retrievals, and snow water equivalent rates (SWER) for comparison with coincident GPM-estimated DSD and precipitation rates/types.

Validation is conducted at the local to regional scales to assess the satellite precipitation products to address the needs of the user community, particularly hydrology and water resources. Rain gauges provide the basis of many validation studies, although these tend to be limited to more developed regions such as the United States, Europe, and SE Asia, where sufficient gauges are available to reduce spatial sampling errors. The availability of precipitation derived from weather radar, national and international radar networks (e.g., Zhang et al., 2011) now provides quantitative precipitation estimates at temporal and spatial scales necessary for the validation of satellite estimates (Chen et al., 2013; Kirstetter et al., 2012, 2014). Some of these networks have been usefully employed in the validation or intercomparison of precipitation products, although generally at daily, 0.25×0.25 degree scales (see Ebert et al., 2007; Kidd et al., 2012).

Improvements in the quality of radar data now means that it can be effectively used to assess instantaneous precipitation estimates allowing the performance of retrieval techniques to be made. Kidd et al. (2018) used high quality surface reference datasets over Europe and the United States based upon surface radar and gauge networks to validate the latest GPM V05 precipitation products from the GMI, AMSR-2, SSMIS, MHS, and ATMS radiometers together with the DPR-Ku precipitation product. The results showed generally good agreement between the satellite products and the surface datasets although systematic differences between the seasonal performance of the products between the United States and the European regions were noted. However, the results were consistent between satellite-radar and satellite-gauge comparisons suggesting more fundamental underlying issues relating to the identification and retrieval of precipitation under differing meteorological regimes.

The latest satellite sensors provide unprecedented data to tackle studies on the global distribution of precipitation that are fundamental for future trend studies. Examples of such studies are available on oceanic precipitation distribution (Behrangi et al., 2014), tropical and subtropical convection (Houze et al., 2015), deep convection distribution (Liu and Liu, 2016), and intense precipitating systems (Liu and Zipser, 2015).

Large-scale (global/monthly) validation is necessary to assess their performance on a regional or global basis, over extended periods to capture the full range of precipitation systems and seasonality. However, global studies based upon gauge analyses, such as that of the GPCC, suffer from the paucity of data over the oceans and the coarseness of the spatial and temporal resolutions. Beck et al. (2017b) provide a useful introduction to the statistical performance of several major datasets across land areas, including model reanalysis products, using surface precipitation gauges as the reference. Global products have varying degrees of fidelity compared to the surface gauges in different areas, and consequently could be used as an initial assessment of a product's relative utility in a particular location. Beck et al. (2017b) also demonstrate that large areas of the globe have such sparse (or nonexistent) gauge networks that satellite-based datasets are the only practical observation source of precipitation data.

1.7 Conclusions

The estimation of precipitation from satellite observations is now well established, utilizing a range of sensors to provide data products at a range of scales from instantaneous to climate and from local to global. These data products may be used in their own right, or as an additional source of information that may be used to augment existing precipitation datasets or provide estimates where surface data are deficient or missing. While no current or planned satellite systems are capable of providing continuous global precipitation estimates, through a combination of multiple satellite sensors, the physical characteristics of the clouds and precipitation structures together with their inherent spatial and temporal resolutions, quasi-continuous, quasi-global precipitation estimates can be generated such that the spatial resolutions of the data products are now generally commensurate with hydrometeorological extremes. Such data currently have a variety of user applications, ranging from the monitoring of droughts and flash floods, through to disease and health applications.

Users should however be aware of the suitability of precipitation products to their particular application. For example, the capturing of extreme events from satellite observations is limited primarily by the temporal resolution of the product and of the observations used in the generation of that product: this impacts short-term studies the greatest, while longer-term, climatological studies will be more representative. Likewise, products that incorporate gauge information are likely to be the most accurate for regions where gauges are plentiful. However, these same techniques might be less accurate where gauge information is sparse or nonexistent. Ultimately, the suitability of satellite-derived precipitation data products for applications such as extremes and hazards is dependent upon a number of factors, including the availability of a product for the particular region of interest and the temporal and spatial resolution of the data product. Precipitation products with a known heritage, performance, and consistency and which have undergone rigorous evaluation and validation should be considered to be the most reliable, although other techniques may be used that fulfill the necessary individual user requirements.

References

Adler, R.F., Kidd, C., Petty, G., Morissey, M., Goodman, H.M., 2001. Inter-comparison of global precipitation products: the third Precipitation Intercomparison Project (PIP-3). Bull. Amer. Meteor. Soc. 82, 1377–1396. Available from: http://dx.doi.org/10.1175/1520-0477(2001)082<1377:IOGPPT>2.3.CO;2.

Adler, R.F., Sapiano, M.R.P., Huffman, G.J., Wang, J.-J., Gu, G., Bolvin, D., et al., 2018. The Global Precipitation Climatology Project (GPCP) monthly analysis (new version 2.3) and a review of 2017 global precipitation. Atmosphere 9, 138. Available from: https://doi.org/10.3390/atmos9040138.

Andersson, A., Fennig, K., Klepp, C., Bakan, S., Grassl, H., Schulz, J., 2010. The hamburg ocean atmosphere parameters and fluxes from satellite data—HOAPS-3. Earth Syst. Sci. Data 2, 215–234. Available from: https://doi.org/10.5194/essd-2-215-2010.

Aonashi, K., Awaka, J., Hirose, M., Kozu, T., Kubota, T., Liu, G., et al., 2009. GSMaP passive microwave precipitation retrieval algorithm: algorithm description and validation. J. Meteor. Soc. Japan 87A, 119–136. Available from: https://doi.org/10.2151/jmsj.87A.119.

Arkin, P.A., Meisner, B.N., 1987. The relationship between large scale convective rainfall and cold cloud over the western hemisphere during 1982-84. Mon. Wea. Rev. 115, 51–74. Available from: http://dx.doi.org/10.1175/1520-0493(1987)115<0051:TRBLSC>2.0.CO;2.

Bartsotas, N.S., Anagnostou, E.N., Nikolopoulos, E.I., Kallos, G., 2018. Investigating satellite precipitation uncertainty over complex terrain. J. Geophys. Res. Available from: https://doi.org/10.1029/2017JD027559.

Beck, H.E., van Dijk, A.I.J.M., Levizzani, V., Schellekens, J., Miralles, D.G., Martens, B., et al., 2017a. MSWEP: 3-Hourly 0.25° global gridded precipitation (1979–2015) by merging gauge, satellite, and reanalysis data. Hydrol. Earth Syst. Sci. 21, 589–615. Available from: https://doi.org/10.5194/hess-21-589-2017.

Beck, H.E., Vergopolan, N., Pan, M., Levizzani, V., van Dijk, A.I.J.M., Weedon, G., et al., 2017b. Global-scale evaluation of 22 precipitation datasets using gauge observations and hydrological modeling. Hydrol. Earth Syst. Sci. 21, 6201–6217. Available from: https://doi.org/10.5194/hess-21-6201-2017.

Behrangi, A., Stephens, G., Adler, R.F., Huffman, G.J., Lambrigtsen, B., Lebsock, M., 2014. An update on the oceanic precipitation rate and its zonal distribution in light of advanced observations from space. J. Climate 27, 3957–3965. Available from: https://doi.org/10.1175/JCLI-D-13-00679.1.

Berg, W., Sapiano, M.R.P., Horsman, J., Kummerow, C., 2013. Improved geolocation and Earth incidence angle information for a Fundamental Climate Data Record of the SSM/Isensors. IEEE Trans. Geosci. Remote Sens 51, 1504–1513. Available from: https://doi.org/10.1109/TGRS.2012.2199761.

Berg, W., Bilanow, S., Chen, R.Y., Datta, S., Draper, D., Ebrahimi, H., et al., 2016. Intercalibration of the GPM microwave radiometer constellation. J. Atmos. Oceanic Technol. 33, 2639–2654. Available from: https://doi.org/10.1175/JTECH-D-16-0100.1.

Bessho, K., Date, K., Hayashi, M., Ikeda, A., Imai, T., Inoue, H., et al., 2016. An introduction to Himawari-8/9—Japan's new-generation geostationary meteorological satellites. J. Meteor. Soc. Japan 94, 151–183. Available from: https://doi.org/10.2151/jmsj.2016-009.

Boukabara, S.-A., Garrett, K., Chen, W., Iturbide-Sanchez, F., Grassotti, C., Kongoli, C., et al., 2011. MiRS: an all-weather 1DVAR satellite data assimilation and retrieval system. IEEE Trans. Geosci. Remote Sens. 49, 3249–3272. Available from: https://doi.org/10.1109/TGRS.2011.2158438.

Carr, N., Kirstetter, P.E., Hong, Y., Gourley, J.J., Schwaller, M., Petersen, W., et al., 2015. The influence of surface and precipitation characteristics on TRMM Microwave Imager rainfall retrieval uncertainty. J. Hydrometeor. 16, 1596–1614. Available from: https://doi.org/10.1175/JHM-D-14-0194.1.

Casella, D., Panegrossi, G., Sanò, P., Dietrich, S., Mugnai, A., Smith, E.A., et al., 2013. Transitioning from CRD to CDRD in Bayesian retrieval of rainfall from satellite passive microwave measurements: Part 2. Overcoming database profile selection ambiguity by consideration of meteorological control on microphysics. IEEE Trans. Geosci. Remote Sens. 51, 4650–4671. Available from: https://doi.org/10.1109/TGRS.2013.2258161.

Casella, D., Panegrossi, G., Sano, P., Milani, L., Petracca, M., Dietrich, S., 2015. A novel algorithm for detection of precipitation in tropical regions using PMW radiometers. Atmos. Meas. Tech. 8, 1217–1232. Available from: https://doi.org/10.5194/amt-8-1217-2015.

Casella, D., Panegrossi, G., Sanò, P., Marra, A.C., Dietrich, S., Johnson, B.T., et al., 2017. Evaluation of the GPM-DPR snowfall detection capability: comparison with CloudSat-CPR. Atmos. Res. 197, 64–75. Available from: https://doi.org/10.1016/j.atmosres.2017.06.018.

Cattani, E., Merino, A., Levizzani, V., 2016. Evaluation of monthly satellite-derived precipitation products over East Africa. J. Hydrometeor 17, 2555–2573. Available from: https://doi.org/10.1175/JHM-D-15-0042.1.

Cecil, D., 2009. Passive microwave brightness temperatures as proxies for hailstorms. J. Appl. Meteor. Climatol. 48, 1281–1286. Available from: https://doi.org/10.1175/2009JAMC2125.1.

Cecil, D., 2011. Relating passive 37-GHz scattering to radar profiles in strong convection. J. Appl. Meteor. Climatol. 50, 233–240. Available from: https://doi.org/10.1175/2010JAMC2506.1.

Cecil, D., Blankenship, C.B., 2012. Toward a global climatology of severe hailstorms as estimated by satellite passive microwave imagers. J. Climate 25, 687–703. Available from: https://doi.org/10.1175/JCLI-D-11-00130.1.

Chen, S., Kirstetter, P.E., Hong, Y., Gourley, J.J., Tian, Y.D., Qi, Y.C., et al., 2013. Evaluation of spatial errors of precipitation rates and types from TRMM space-borne radar over the southern CONUS. J. Hydrometeor. 14, 1884–1896. Available from: https://doi.org/10.1175/JHM-D-13-027.1.

Curtis, J.A., Adams, L.J., 1987. Defense meteorological satellite program. IEEE Aerospace Electronic Syst. Mag. 2, 13–17. Available from: https://doi.org/10.1109/MAES.1987.5005348.

Derin, Y., Anagnostou, E.N., Anagnostou, M.N., Kalogiros, J., Casella, D., Marra, A.C., et al., 2018. Passive microwave rainfall error analysis using high-resolution X-band dual-polarization radar observations in complex terrain. IEEE Trans. Geosci. Remote Sens 56, 2565–2586. Available from: https://doi.org/10.1109/TGRS.2017.2763622.

Dinku, T., Funk, C., Peterson, P., Maidment, R., Tadesse, T., Gadain, H., et al., 2018. Validation of the CHIRPS satellite rainfall estimates over eastern ofAfrica. Quart. J. Roy. Meteor. Soc. Available from: https://doi.org/10.1002/qj.3244.

Draper, D.W., Newell, D.A., Wentz, F.J., Krimchansky, S., Skofronick-Jackson, G., 2015. The Global Precipitation Measurement (GPM) Microwave Imager (GMI): instrument overview and early on-orbit performance. IEEE J. Sel. Topics Geosci. Remote Sens. 8, 3452–3462. Available from: https://doi.org/10.1109/JSTARS.2015.2403303.

Ebert, E.E., Janowiak, J.E., Kidd, C., 2007. Comparison of near real time precipitation estimates from satellite observations and numerical models. Bull. Amer. Meteor. Soc. 88, 47–64. Available from: https://doi.org/10.1175/BAMS-88-1-47.

Ebtehaj, A.M., Kummerow, C.D., 2017. Microwave retrievals of terrestrial precipitation over snow-covered surfaces: a lesson from the GPM satellite. Geophys. Res. Lett. 44, 6154–6162. Available from: https://doi.org/10.1002/2017GL073451.

Fennig, K., Andersson, A., Schröder, M., 2013. Fundamental climate data record of SSM/I brightness temperatures. Satellite Application Facility on Climate Monitoring. Available from: https://doi.org/10.5676/EUM_SAF_CM/FCDR_SSMI/V001.

Fennig, K., Andersson, A., Schröder, M., 2015. Fundamental climate data record of SSM/I/SSMIS brightness temperatures. Satellite Application Facility on Climate Monitoring. Available from: https://doi.org/10.5676/EUM_SAF_CM/FCDR_MWI/V002.

Ferraro, R., Beauchamp, J., Cecil, D., Heymsfield, G., 2015. A prototype hail detection algorithm and hail climatology developed with the advanced microwave sounding unit (AMSU). Atmos. Res. 163, 24–35. Available from: https://doi.org/10.1016/j.atmosres.2014.08.010.

Funk, C., Peterson, P., Landsfeld, M., Pedreros, D., Verdin, J., Shukla, S., et al., 2015. The climate hazards infrared precipitation with stations – a new environmental record for monitoring extremes. Sci. Data 2, 150066. Available from: https://doi.org/10.1038/sdata.2015.66.

Gebregiorgis, A.S., Kirstetter, P.-E., Hong, Y.E., Gourley, J.J., Huffman, G.J., Petersen, W.A., et al., 2018. To what extent is the day 1 GPM IMERG satellite precipitation estimate improved as compared to TRMM TMPA-RT? J. Geophys. Res. 123, 1694–1707. Available from: https://doi.org/10.1002/2017JD027606.

Goodman, S.J., Gurka, J., DeMaria, M., Schmit, T.J., Mostek, A., Jedlovec, G., et al., 2012. The GOES-R proving ground: accelerating user readiness for the next-generation Geostationary Environmental Satellite System. Bull. Amer. Meteor. Soc. 93, 1029–1040. Available from: https://doi.org/10.1175/BAMS-D-11-00175.1.

Grecu, M., Olson, W.S., Munchak, S.J., Ringerud, S., Liao, L., Haddad, Z.S., et al., 2016. The GPM combined algorithm. J. Atmos. Oceanic Technol. 33, 2225–2245. Available from: https://doi.org/10.1175/JTECH-D-16-0019.1.

Haddad, Z.S., Smith, E.A., Kummerow, C.D., Iguchi, T., Farrar, M.R., Durden, S.L., et al., 1997. The TRMM "Day-1" radar/radiometer combined rain-profiling algorithm. J. Meteor. Soc. Japan 75, 799–809. Available from: https://doi.org/10.2151/jmsj1965.75.4_799.

Hilburn, K.A., Wentz, F.J., 2008. Intercalibrated passive microwave rain products from the Unified Microwave Ocean Retrieval Algorithm (UMORA). J. Appl. Meteor. Climatol. 47, 778–794. Available from: https://doi.org/10.1175/2007JAMC1635.1.

Hong, Y., Hsu, K.L., Sorooshian, S., Gao, X., 2004. Precipitation estimation from remotely sensed imagery using an artificial neural network cloud classification system. J. Appl. Meteor. 43, 1834–1853. Available from: https://doi.org/10.1175/JAM2173.1.

Hossain, F., Anagnostou, E.N., 2006. A two-dimensional satellite rainfall error model. IEEE Trans. Geosci. Remote Sens. 44, 1511–1522. Available from: https://doi.org/10.1109/TGRS.2005.863866.

Hou, A.Y., Kakar, R.K., Neeck, S.A., Azarbarzin, A., Kummerow, C.D., Kojima, M., et al., 2014. The global precipitation measurement mission. Bull. Amer. Meteor. Soc. 95, 701–722. Available from: https://doi.org/10.1175/BAMS-D-13-00164.1.

Houze, R.A., Rasmussen, K.L., Zuluaga, M.D., Brodzik, S.R., 2015. The variable nature of convection in the tropics and subtropics: a legacy of 16 years of the tropical rainfall measuring mission satellite. Rev. Geophys. 53, 994–1021. Available from: https://doi.org/10.1002/2015RG000488.

Houze, R.A., McMurdie, L.A., Petersen, W.A., Schwaller, M.R., Baccus, W., Lundquist, J., et al., 2017. The olympic mountains experiment (OLYMPEX). Bull. Amer. Meteor. Soc. 98, 2167–2188. Available from: https://doi.org/10.1175/BAMS-D-16-0182.1.

Huffman, G.J., Bolvin, D.T., Braithwaite, D., Hsu, K., Joyce, R., Xie, P., 2014. GPM integrated multi-satellite retrievals for GPM (IMERG) algorithm theoretical basis document (ATBD) version 4.4. PPS, NASA/GSFC, 30 pp. Available at http://pmm.nasa.gov/sites/default/files/document_files/IMERG_ATBD_V4.4.pdf.

Iguchi, T., Seto, S., Meneghini, R., Yoshida, N., Awaka, J., Le, M., et al., 2015. GPM/DPR level-2. Algorithm Theoretical Basis Doc., 68 pp. Available online at http://pps.gsfc.nasa.gov/Documents/ATBD_DPR_2015_whole_a.pdf.

Janowiak, J.E., Joyce, R.J., Yarosh, Y., 2001. A real-time global half-hourly pixel-resolution infrared dataset and its applications. Bull. Amer. Meteor. Soc. 82, 205–217. Available from: http://dx.doi.org/10.1175/1520-0477(2001)082 < 0205:ARTGHH > 2.3.CO;2.

Jensen, M.P., Petersen, W.A., Bansemer, A., Bharadwaj, N., Carey, L.D., Cecil, D.J., et al., 2016. The midlatitude continental convective clouds experiment (MC3E). Bull. Amer. Meteor. Soc. 97, 1667–1686. Available from: https://doi.org/10.1175/BAMS-D-14-00228.1.

Joyce, R.J., Xie, P., Janowiak, J.E., 2011. Kalman filter based CMORPH. J. Hydrometeor. 12, 1547–1563. Available from: https://doi.org/10.1175/JHM-D-11-022.1.

Katsanos, D., Lagouvardos, K., Kotroni, V., Huffmann, G.J., 2004. Statistical evaluation of MPA-RT high-resolution precipitation estimates from satellite platforms over the central and eastern Mediterranean. Geophys. Res. Lett. 31, L06116. Available from: https://doi.org/10.1029/2003GL019142.

Kidd, C., 2017 (last updated 2018). GPROF CNES/ISRO Megha-Tropiques SAPHIR radiometer Precipitation Retrieval and Profiling Scheme, Level 2A precipitation product. NASA/GSFC, Greenbelt, MD, USA. doi:10.5067/GPM/SAPHIR/MT1/PRPS/2A/05.

Kidd, C., Levizzani, V., 2011. Status of satellite precipitation retrievals. Hydrol. Earth Syst. Sci. 15, 1109–1116. Available from: https://doi.org/10.5194/hess-15-1109-2011.

Kidd, C., Kniveton, D., Barrett, E.C., 1998. Advantage and disadvantages of statistical/empirical satellite estimation of rainfall. J. Atmos. Sci. 55, 1576–1582. Available from: http://dx.doi.org/10.1175/1520-0469(1998)055 < 1576:TAADOS > 2.0.CO;2.

Kidd, C., Kniveton, D.R., Todd, M.C., Bellerby, T.J., 2003. Satellite rainfall estimation using a combined passive microwave and infrared algorithm. J. Hydrometeor. 4, 1088–1104. Available from: http://dx.doi.org/10.1175/1525-7541(2003)004 < 1088:SREUCP > 2.0.CO;2.

Kidd, C., Bauer, P., Turk, F.J., Huffman, G.J., Joyce, R., Hsu, K.-L., et al., 2012. Intercomparison of high-resolution precipitation products over Northwest Europe. J. Hydrometeor 13, 67–83. Available from: https://doi.org/10.1175/JHM-D-11-042.

Kidd, C., Becker, A., Huffman, G.J., Muller, C.L., Joe, P., Skofronick-Jackson, G., et al., 2017. So, how much of the Earth's surface is covered by rain gauges? Bull. Amer. Meteor. Soc. 98, 69–78. Available from: https://doi.org/10.1175/BAMS-D-14-00283.1.

Kidd, C., Tan, J., Kirstetter, P.-E., Petersen, W.A., 2018. Validation of the version 05 level 2 precipitation products from the GPM core observatory and constellation satellite sensors. Quart. J. Roy. Meteor. Soc. Available from: https://doi.org/10.1002/qj.3175.

Kirschbaum, D.B., Patel, K., 2016. Precipitation data key to food security and public health. *EOS*, https://eos.org/meeting-reports/precipitation-data-key-to-food-security-and-public-health.

Kirschbaum, D.B., Huffman, G.J., Adler, R.F., Braun, S., Garrett, K., Jones, E., et al., 2017. NASA's remotely-sensed precipitation: a reservoir for applications users. Bull. Amer. Meteor. Soc. 98, 1169–1194. Available from: https://doi.org/10.1175/BAMS-D-15-00296.1.

Kirstetter, P.-E., Hong, Y., Gourley, J.J., Cao, Q., Schwaller, M., Petersen, W.A., 2014. A research framework to bridge from the Global Precipitation Measurement mission core satellite to the constellation sensors using ground radar-based national mosaic QPE. In: Lakshmi, V. (Ed.), Remote Sensing of the Terrestrial Water Cycle. John Wiley & Sons, Inc, Hoboken, NJ, AGU Books Geophysical Monograph Series Chapman Monograph on Remote Sensing.

Kirstetter, P.E., Hong, Y., Gourley, J.J., Chen, S., Flamig, Z., Zhang, J., et al., 2012. Toward a framework for systematic error modeling of spaceborne Precipitation Radar with NOAA/NSSL ground radar-based national mosaic QPE. J. Hydrometeor. 13, 1285–1300. Available from: https://doi.org/10.1175/JHM-D-11-0139.1.

Kirstetter, P.E., Gourley, J.J., Hong, Y., Zhang, J., Moazamigoodarzi, S., Langston, C., et al., 2015. Probabilistic precipitation rate estimates with ground-based radar networks. Water Resour. Res. 51, 1422–1442. Available from: https://doi.org/10.1002/2014WR015672.

Kojima, M., Miura, T., Furukawa, K., Hyakusoku, Y., Ishikiri, T., Kai, H., et al., 2012. Dual-frequency precipitation radar (DPR) development on the Global Precipitation Measurement (GPM) core observatory. Proc. SPIE 8528. Available from: https://doi.org/10.1117/12.976823. Earth Observing Missions and Sensors: Development, Implementation, and Characterization II, 85281A.

Kozu, T., Kawanishi, T., Kuroiwa, H., Kojima, M., Oikawa, K., Kumagai, H., et al., 2001. Development of precipitation radar onboard the tropical rainfall measuring mission satellite. IEEE Trans. Geosci. Remote Sens. 39, 102–116. Available from: https://doi.org/10.1109/36.898669.

Krajewski, W.F., Villarini, G., Smith, J.A., 2010. RADAR-rainfall uncertainties: where are we after thirty years of effort? Bull. Amer. Meteor. Soc. 91, 87–94. Available from: https://doi.org/10.1175/2009BAMS2747.1.

Kubota, T., Shige, S., Hashizume, H., Aonashi, K., Takahashi, N., Seto, S., et al., 2007. Global precipitation map using satellite-borne microwave radiometers by the GSMaP Project: production and validation. IEEE Trans. Geosci. Remote Sens. 45, 2259–2275. Available from: https://doi.org/10.1109/TGRS.2007.895337.

Kucera, P.A., Ebert, E.E., Turk, F.J., Levizzani, V., Kirschbaum, D., Tapiador, F.J., et al., 2013. Precipitation from space: advancing Earth system science. Bull. Amer. Meteor. Soc. 94, 365–375. Available from: https://doi.org/10.1175/BAMS-D-11-00171.1.

Kulie, M.S., Milani, L., 2018. Seasonal variability of shallow cumuliform snowfall: a CloudSat perspective. Quart. J. Roy. Meteor. Soc. Available from: https://doi.org/10.1002/qj.3222.

Kulie, M.S., Milani, L., Wood, N.B., Tushaus, S.A., Bennartz, R., L'Ecuyer, T.S., 2016. A shallow cumuliform snowfall census using spaceborne radar. J. Hydrometeor. 17, 1261–1279. Available from: https://doi.org/10.1175/JHM-D-15-0123.1.

Kuligowski, R.J., 2002. A self-calibrating real-time GOES rainfall algorithm for short-term rainfall estimates. J. Hydrometeor. 3, 112–130. Available from: http://dx.doi.org/10.1175/1525-7541(2002)003 < 0112:ASCRTG > 2.0.CO;2.

Kuligowski, R.J., 2010. The self-calibrating multivariate precipitation retrieval (SCaMPR) for high-resolution, low-latency satellite-based rainfall estimates. In: Gebremichael, M., Hossain, F. (Eds.), Satellite Rainfall Applications for Surface Hydrology. Springer, pp. 39–48. Available from: http://dx.doi.org/10.1007/978-90-481-2915-7-3.

Kummerow, C., Simpson, J., Thiele, O., Barnes, W., Chang, A.T.C., Stocker, E., et al., 2000. The status of the Tropical Rainfall Measuring Mission (TRMM) after two years in orbit. J. Appl. Meteor. 39, 1965–1982. Available from: http://dx.doi.org/10.1175/1520-0450(2001)040 < 1965:TSOTTR > 2.0.CO;2.

Kummerow, C.D., Barnes, W., Kozu, T., Shiue, J., Simpson, J., 1998. The Tropical Rainfall Measuring Mission (TRMM) sensor package. J. Atmos. Oceanic Technol. 15, 809–817. Available from: http://dx.doi.org/10.1175/1520-0426(1998)015 < 0809: TTRMMT > 2.0.CO;2.

Kummerow, C.D., Randel, D.L., Kulie, M., Wang, N.-Y., Ferraro, R., Munchak, S.J., et al., 2015. The evolution of the Goddard PROFiling algorithm to a fully parametric scheme. J. Atmos. Oceanic Technol. 32, 2265–2280. Available from: https://doi.org/10.1175/JTECH-D-15-0039.1.

Laviola, S., Levizzani, V., 2011. The 183-WSL fast rainrate retrieval algorithm. Part I: retrieval design. Atmos. Res. 99, 443–461. Available from: https://doi.org/10.1016/j.atmosres.2010.11.013.

Laviola, S., Levizzani, V., Cattani, E., Kidd, C., 2013. The 183-WSL fast rainrate retrieval algorithm. Part II: validation using ground radar measurements. Atmos. Res. 134, 77–86. Available from: https://doi.org/10.1016/j.atmosres.2013.07.013.

Laviola, S., Dong, J., Kongoli, C., Meng, H., Ferraro, R., Levizzani, V., 2015. An intercomparison of two passive microwave algorithms for snowfall detection over Europe. IEEE Geosci. Remote Sensing Symp 886–889. Available from: https://doi.org/10.1109/IGARSS.2015.7325907.

Levizzani, V., Laviola, S., Cattani, E., 2011. Detection and measurement of snowfall from space. Remote Sens. 3 (1), 145–166. Available from: https://doi.org/10.3390/rs3010145.

Levizzani, V., Kidd, C., Aonashi, K., Bennartz, R., Ferraro, R.R., Huffman, G.J., et al., 2018. The activities of the International Precipitation Working Group. Quart. J. Roy. Meteor. Soc. Available from: https://doi.org/10.1002/qj.3214.

Liu, C., Zipser, E.J., 2015. The global distribution of largest, deepest, and most intense precipitation systems. Geophy. Res. Lett. 42, 3591–3595. Available from: https://doi.org/10.1002/2015GL063776.

Liu, G., 2008. Deriving snow cloud characteristics from CloudSat observations. J. Geophys. Res. 113, D8. Available from: https://doi.org/10.1029/2007JD009766.

Liu, G., Seo, E.-K., 2013. Detecting snowfall over land by satellite high-frequency microwave observations: the lack of scattering signature and a statistical approach. J. Geophys. Res. 118, 1376–1387. Available from: https://doi.org/10.1002/jgrd.50172.

Liu, N., Liu, C., 2016. Global distribution of deep convection reaching tropopause in 1 year GPM observations. J. Geophys. Res. 121, 3824–3842. Available from: https://doi.org/10.1002/2015JD024430.

Maggioni, V., Massari, C., 2018. On the performance of satellite precipitation products in riverine flood modeling: a review. J. Hydrol. 558, 214–224. Available from: https://doi.org/10.1016/j.jhydrol.2018.01.039.

Maggioni, V., Meyers, P.C., Robinson, M.D., 2016. A review of merged high-resolution satellite precipitation product accuracy during the Tropical Rainfall Measuring Mission (TRMM) era. J. Hydrometeor. 17, 1101–1117. Available from: https://doi.org/10.1175/JHM-D-15-0190.1.

Maggioni, V., Nikolopoulos, E.I., Anagnostou, E.N., Borga, M., 2017. Modeling satellite precipitation errors over mountainous terrain: the influence of gauge density, seasonality, and temporal resolution. IEEE Trans. Geosci. Remote Sens. 55, 4130–4140. Available from: https://doi.org/10.1109/TGRS.2017.2688998.

Maidment, R., Grimes, D., Black, E., Tarnavsky, E., Young, M., Greatrex, H., et al., 2017. A new, long-term daily satellite-based rainfall dataset for operational monitoring in Africa. Sci. Data 4, 170063. Available from: https://doi.org/10.1038/sdata.2017.63.

Marra, A.C., Porcù, F., Baldini, L., Petracca, M., Casella, D., Dietrich, S., et al., 2017. Observational analysis of an exceptionally intense hailstorm over the Mediterranean area: role of the GPM core observatory. Atmos. Res. 192, 72–90. Available from: https://doi.org/10.1016/j.atmosres.2017.03.019.

Meng, H., Dong, J., Ferraro, R., Yan, B.H., Zhao, L.M., Kongoli, C., et al., 2017. A 1DVAR-based snowfall rate retrieval algorithm for passive microwave radiometers. J. Geophys. Res. 122, 6520–6540. Available from: https://doi.org/10.1002/2016JD026325.

Michaelides, S., Levizzani, V., Anagnostou, E.N., Bauer, P., Kasparis, T., Lane, J.E., 2009. Precipitation: measurement, remote sensing, climatology and modeling. Atmos. Res. 94, 512–533. Available from: https://doi.org/10.1016/j.atmosres.2009.08.017.

Mugnai, A., Casella, D., Cattani, E., Dietrich, S., Laviola, S., Levizzani, V., et al., 2013. Precipitation products from the hydrology SAF. Nat. Hazards Earth Syst. Sci. 13, 1959–1981. Available from: https://doi.org/10.5194/nhess-13-1959-2013.

Nakamura, K., Okamoto, K., Ihara, T., Awaka, J., Kozu, T., Manabe, T., 1990. Conceptual design of rain radar for the Tropical Rainfall Measuring Mission. Int. J. Satellite Comm. 8, 257–268. Available from: https://doi.org/10.1002/sat.4600080318.

Novella, N.S., Thiaw, W.M., 2013. African rainfall climatology version 2 for Famine Early Warning Systems. J. Appl. Meteor. Climatol. 52, 588–606. Available from: https://doi.org/10.1175/JAMC-D-11-0238.1.

Oliveira, R., Maggioni, V., Vila, D., Morales, C., 2016. Characteristics and diurnal cycle of GPM rainfall estimates over the Central Amazon Region. Remote Sensing – Special Issue on "Uncertainties in Remote Sensing" 8 (7), 544. Available from: https://doi.org/10.3390/rs8070544.

Panegrossi, G., Rysman, J.-F., Casella, D., Marra, A.C., Sanò, P., Kulie, M.S., 2017. CloudSat-based assessment of GPM Microwave Imager snowfall observation capabilities. Remote Sens. 12, 1263. Available from: https://doi.org/10.3390/rs9121263.

Petersen, W.A., Houze, R.A., McMurdie, L., Zagrodnik, J., Tanelli, S., Lundquist, J., et al., 2016. The Olympic Mountains Experiment (OLMPEX): from ocean to summit. Meteor. Technol. Int. 22–26.

Sanò, P., Panegrossi, G., Casella, D., Marra, A.C., Di Paola, F., Dietrich, S., 2016. The new Passive microwave Neural network Precipitation Retrieval (PNPR) algorithm for the cross-track scanning ATMS radiometer: description and verification study over Europe and Africa using GPM and TRMM spaceborne radars. Atmos. Meas. Tech. 9, 5441–5460. Available from: https://doi.org/10.5194/amt-9-5441-2016.

Sapiano, M.R.P., Berg, W.K., McKague, D.S., Kummerow, C.D., 2013. Toward an intercalibrated Fundamental Climate Data Record of the SSM/Isensors. IEEE Trans. Geosci. Remote Sens 51, 1492–1503. Available from: https://doi.org/10.1109/TGRS.2012.2206601.

Schmetz, J., Pili, P., Tjemkes, S., Just, D., Kerkmann, J., Rota, S., et al., 2002. An introduction to Meteosat Second Generation (MSG). Bull. Amer. Meteor. Soc. 83, 977–992. Available from: http://dx.doi.org/10.1175/1520-0477(2002)083 < 0977: AITMSG > 2.3.CO;2.

Schmit, T.J., Griffith, P., Gunshor, M.M., Daniels, J.M., Goodman, S.J., Lebair, W.J., 2017. A closer look at the ABI on the GOES-R series. Bull. Amer. Meteor. Soc. 98, 681–698. Available from: https://doi.org/10.1175/BAMS-D-15-00230.1.

Scofield, R.A., Kuligowski, R.J., 2003. Status and outlook of operational satellite precipitation algorithms for extreme-precipitation events. Wea. Forecasting 18, 1037–1051. Available from: http://dx.doi.org/10.1175/1520-0434(2003)018 < 1037:SAOOOS > 2.0.CO;2.

Semunegus, H., Berg, W., Bates, J.J., Knapp, K.R., Kummerow, C., 2010. An extended and improved Special Sensor Microwave Imager (SSM/I) period of record. J. Appl. Meteor. Climatol. 49, 424–436. Available from: https://doi.org/10.1175/2009JAMC2314.1.

Simpson, J.R., Adler, R.F., North, G.R., 1988. A proposed Tropical Rainfall Measuring Mission (TRMM) satellite. Bull. Amer. Meteor. Soc. 69, 278–295. Available from: http://dx.doi.org/10.1175/1520-0477(1988)069 < 0278:APTRMM > 2.0.CO;2.

Skofronick-Jackson, G., Kim, M.-J., Weinman, J.A., Chang, D.-E., 2004. A physical model to determine snowfall over land by microwave radiometry. IEEE Trans. Geosci. Remote Sens. 42, 1047–1058. Available from: https://doi.org/10.1109/TGRS.2004.825585.

Skofronick-Jackson, G., Johnson, B.T., Munchak, S.J., 2013. Detection thresholds of falling snow from satellite-borne active and passive sensors. IEEE Trans. Geosci. Remote Sens. 15, 4177–4189. Available from: https://doi.org/10.1109/TGRS.2012.2227763.

Skofronick-Jackson, G., Hudak, D., Petersen, W., Nesbitt, S.W., Chandrasekar, V., Durden, S., et al., 2015. Global Precipitation Measurement Cold Season Precipitation Experiment (GCPEx): for measurement sake let it snow. Bull. Amer. Meteor. Soc. 96, 1719–1741. Available from: https://doi.org/10.1175/BAMS-D-13-00262.1.

Stephens, G.L., Vane, D.G., Boain, R.J., Mace, G.G., Sassen, K., Wang, Z., et al., 2002. The CloudSat mission and the A-Train: a new dimension of space-based observations of clouds and precipitationand the CloudSat Science Team Bull. Amer. Meteor. Soc. 83, 1771–1790. Available from: https://doi.org/10.1175/BAMS-83-12-1771.

Tan, B.-Z., Petersen, W.A., Tokay, A., 2016. A novel approach to identify sources of errors in IMERG for GPM ground validation. J. Hydrometeor. 17, 2477–2491. Available from: https://doi.org/10.1175/JHM-D-16-0079.1.

Tan, B.-Z., Petersen, W.A., Kirstetter, P.E., Tian, Y., 2017. Performance of IMERG as a function of spatiotemporal scale. J. Hydrometeor. 18, 307–319. Available from: https://doi.org/10.1175/JHM-D-16-0174.1.

Tapiador, F.J., Turk, F.J., Petersen, W., Hou, A.Y., García-Ortega, E., Machado, L.A.T., et al., 2012. Global precipitation measurement: methods, datasets and applications. Atmos. Res. 104-105, 70–97. Available from: https://doi.org/10.1016/j.atmosres.2011.10.021.

Tapiador, F.J., Navarro, A., Levizzani, V., García-Ortega, E., Huffman, G.J., Kidd, C., et al., 2017. Global precipitation measurements for validating climate models. Atmos. Res. 197, 1–20. Available from: https://doi.org/10.1016/j.atmosres.2017.06.021.

Thiemig, V., Rojas, R., Zambrano-Bigiarini, M., Levizzani, V., De Roo, A., 2012. Validation of satellite-based precipitation products over sparsely-gauged African river basins. J. Hydrometeor. 13, 1760–1783. Available from: https://doi.org/10.1175/JHM-D-12-032.1.

Tian, Y., Peters-Lidard, C.D., 2010. A global map of uncertainties in satellite-based precipitation measurements. Geophys. Res. Lett. 37, L24407. Available from: https://doi.org/10.1029/2010GL046008.

Toyoshima, K., Masunaga, H., Furuzawa, F.A., 2015. Early evaluation of Ku- and Ka-band sensitivities for the Global Precipitation Measurement (GPM) Dual-Frequency Precipitation Radar (DPR). SOLA 11, 14–17. Available from: https://doi.org/10.2151/sola.2015-004.

Ushio, T., Sasashige, K., Kubota, T., Shige, S., Okamoto, K., Aonashi, K., et al., 2009. A Kalman filter approach to the Global Satellite Mapping of Precipitation (GSMaP) from combined passive microwave and infrared radiometric data. J. Meteor. Soc. Japan 87A, 137–151. Available from: https://doi.org/10.1175/JHM-D-16-0174.1.

Villarini, G., Mandapaka, P.V., Krajewski, W.F., Moore, R.J., 2008. Rainfall and sampling uncertainties: a rain gauge perspective. J. Geophys. Res. 113, D11. Available from: https://doi.org/10.1029/2007JD009214.

Wentz, F.J., Draper, D., 2016. On-orbit absolute calibration of the Global Precipitation Measurement Microwave Imager. J. Atmos. Oceanic Technol. 33, 1393–1412. Available from: https://doi.org/10.1175/JTECH-D-15-0212.1.

Wilheit, T., Berg, W., Ebrahami, H., Kroodsma, R., McKague, D., Payne, V., et al., 2015. Intercalibrating the GPM constellation using the GPM Microwave Imager (GMI). IEEE Int. Geosci. Remote Sens. Symp. (IGARSS). Available from: https://doi.org/10.1109/IGARSS.2015.7326996.

Xu, L., Gao, X., Sorooshian, S., Arkin, P.A., Imam, B., 1999. A microwave infrared threshold technique to improve the GOES Precipitation Index. J. Appl. Meteor. 38, 569–579. Available from: http://dx.doi.org/10.1175/1520-0450(1999)038 < 0569: AMITTT > 2.0.CO;2.

You, Y., Wang, N.-Y., Ferraro, R., 2015. A prototype precipitation retrieval algorithm over land using passive microwave observations stratified by surface condition and precipitation vertical structure. J. Geophys. Res. 120, 5295–5315. Available from: https://doi.org/10.1002/2014JD022534.

You, Y., Wang, N.-Y., Ferraro, R., Rudlosky, S., 2017. Quantifying the snowfall detection performance of the GPM Microwave Imager channels over land. J. Hydrometeor. 18, 729–751. Available from: https://doi.org/10.1175/JHM-D-16-0190.1.

Zhang, J., Howard, K., Langston, C., Vasiloff, S., Kaney, B., Arthur, A., et al., 2011. National Mosaic and multi-sensor QPE (NMQ) system: description, results, and future plans. Bull. Amer. Meteor. Soc. 92, 1321–1338. Available from: https://doi.org/10.1175/2011BAMS-D-11-00047.1.

Zhang, J., Howard, K., Langston, C., Kaney, B., Qi, Y., Tang, L., et al., 2016. Multi-Radar Multi-Sensor (MRMS) quantitative precipitation estimation: initial operating capabilities. Bull. Amer. Meteor. Soc. 97, 621–638. Available from: https://doi.org/10.1175/BAMS-D-14-00174.1.

Further reading

Wenhaji Ndomeni, C., Cattani, E., Merino, A., Levizzani, V., 2018. An observational study of the variability of East African rainfall with respect to sea surface temperature and soil moisture. Quart. J. Roy. Meteor. Soc. Available from: https://doi.org/10.1002/qj.3255.

CHAPTER TWO

Terrestrial water storage

Manuela Girotto[1,2,3] and Matthew Rodell[4]
[1]Global Modeling and Assimilation Office, NASA Goddard Space Flight Center, Greenbelt, MD, United States
[2]GESTAR, Universities Space Research Association, Columbia, MD, United States
[3]Department of Environmental Science, Policy, and Management, University of California, Berkeley, CA 94720, United States
[4]Hydrological Science Lab, NASA Goddard Space Flight Center, Greenbelt, MD, United States

2.1 Terrestrial water storage components

Terrestrial water storage (TWS) is a dynamic component of the hydrological cycle that exerts important controls over the water, energy, and biogeochemical fluxes, thereby playing a major role in Earth's climate system (Syed et al., 2008; Famiglietti, 2004). TWS is defined as the summation of all water stored above and below the land surface. This includes surface waters, soil moisture, groundwater, snow, ice, and water stored in vegetation. The variability of TWS tends to be dominated by surface waters in wet, tropical regions such as the Amazon, by soil moisture and groundwater in mid-latitudes, and by snow and ice in polar and alpine regions (Rodell and Famiglietti, 2001). The variations in water stored in vegetation are small compared to those of the other TWS components, therefore they are typically assumed to be negligible in TWS budget analyses (Rodell et al., 2005).

Surface waters include rivers, inland water bodies, wetlands, and inundated floodplains. Surface waters supply agricultural and energy production for most regions in the tropics, and because they exist at the land–atmosphere interface they affect hydrometeorological and biogeochemical processes. Getirana et al. (2017) found that changes in surface water storage are a substantial component of TWS variability in the wet tropics (e.g., the Amazon), where major rivers flow over arid regions (e.g., the Nile River in Egypt) and in subpolar regions. Conversely, changes in surface water storage are negligible in the Western United States, Northern Africa, Middle East, and Central Asia.

Soil moisture is defined as the water stored in the unsaturated zone of the soil. It controls the partitioning of rainfall into runoff and infiltration,

Extreme Hydroclimatic Events and Multivariate Hazards in a Changing Environment.
DOI: https://doi.org/10.1016/B978-0-12-814899-0.00002-X

constrains evapotranspiration, and influences the occurrence of flood water extremes, because when the soil is saturated it cannot be infiltrated by any more water (Milly and Dunne, 1994; Crow et al., 2005; Zehe et al., 2005). Soil moisture governs the partitioning of the turbulent fluxes (latent and sensible heat); hence it affects global and regional hydrometeorological processes (Entekhabi et al., 1996; Seneviratne et al., 2010). In parts of the tropics and in the mid-latitudes, soil moisture variations are generally the largest component of seasonal terrestrial water storage changes.

Groundwater is the other major component of TWS in the mid-latitudes, where it provides domestic water to a billion people and plays a central role in agriculture and energy production (Gleeson et al., 2012; Famiglietti, 2014). This resource has received increasing attention in recent years because of the accelerated rate of depletion of major global aquifers (e.g., Rodell et al., 2009; Wada et al., 2010; Famiglietti et al., 2011; Richey et al., 2015; Girotto et al., 2017). Groundwater storage changes are rarely the dominant component of TWS variations on a seasonal basis, but they are often the dominant component on interannual to decadal timescales (Rodell and Famiglietti, 2001; Li et al., 2015).

Finally, the water stored in the form of snow or ice is the primary component of TWS in high latitude regions and mountainous regions of the mid-latitudes (Rodell and Famiglietti, 2001; Seneviratne, 2003; Getirana et al., 2017). This is because snow water storage induces significant variations in TWS due to the mass that accumulates during winter and the mass that melts during springtime. These fluctuations also affect the surface energy balance because of the changes in surface albedo (Hall et al., 1988). Accurate global estimates of snow across space and time are currently lacking due to the complexity of the terrain or inaccessibility of the regions where snow/ice is stored. For this reason, large-scale efforts for estimating snow properties must rely heavily on remote sensing (Schmugge et al., 2002; Girotto et al., 2014). Existing efforts to estimate snow water equivalent (SWE) from remote sensing include using microwave, visible and near infrared, radar, and gravimetric measurements. A dedicated snow-measuring satellite mission has not yet been launched, but efforts have begun (e.g., NASA SnowEx, https://snow.nasa.gov/campaigns/snowex) in which international teams of experts are working to devise satellite instruments capable of measuring SWE and/or snow depth from space.

2.2 Overview of the GRACE mission

The Gravity Recovery and Climate Experiment (GRACE) maps the gravity potential of the Earth. From these gravity maps scientists can infer global and regional changes of the terrestrial water storage, ice loss, and sea level change caused by the addition of water to the ocean providing a unique view of Earth's climate and have far-reaching benefits to society and the world's population. GRACE is jointly operated by NASA and the German Aerospace Center. It was launched on March 17, 2002, and it continued to perform past its nominal mission lifetime of 5 years until battery failure caused the end of the science mission in 2017. It consisted of two satellites in a tandem orbit about 200 km apart at 450–500 km altitude. Fig. 2.1 shows in a simplistic way the principle behind the gravity measurements. The two GRACE satellites flew in a tandem orbit over the Earth. In the case shown in Fig. 2.1, a mountain represents a positive mass anomaly. As the lead satellite approached the mountain it was pulled toward it because of the gravitational force exerted by the mass anomaly. Thus, the distance between the two satellites increased. As the trailing satellite approached the mountain, it was pulled toward the anomalous mass while the lead satellite was held back by the same mass anomaly, and the distance between the satellites decreased. Finally, as the trailing satellite moved away from the mountain, it was held back, and their distance increased again. In a nutshell, measurement of the distance (range) and changes in distance (range-rate) between the two satellites are related to the gravitational potential of earth masses. A K-band microwave tracking system continuously measures changes in the distance between the satellites with a precision better than 1 μm (Tapley et al., 2004). The measurements are so precise that mass changes associated with ocean circulation, atmospheric circulation, and terrestrial water storage redistribution can be detected by GRACE. Nongravitational accelerations are monitored by onboard accelerometers, and the precise positions of the satellites are measured via global positioning system (GPS). The monthly analysis of the range and range-rate measurements between the two satellites provides the temporal variations (i.e., anomalies) of the gravity field.

The successor of GRACE is the GRACE follow-on (GRACE-FO) mission. GRACE-FO, launched on May 22, 2018, carries on the work of its predecessor while testing a new laser ranging system designed to improve the precision of the intersatellite distance measurement.

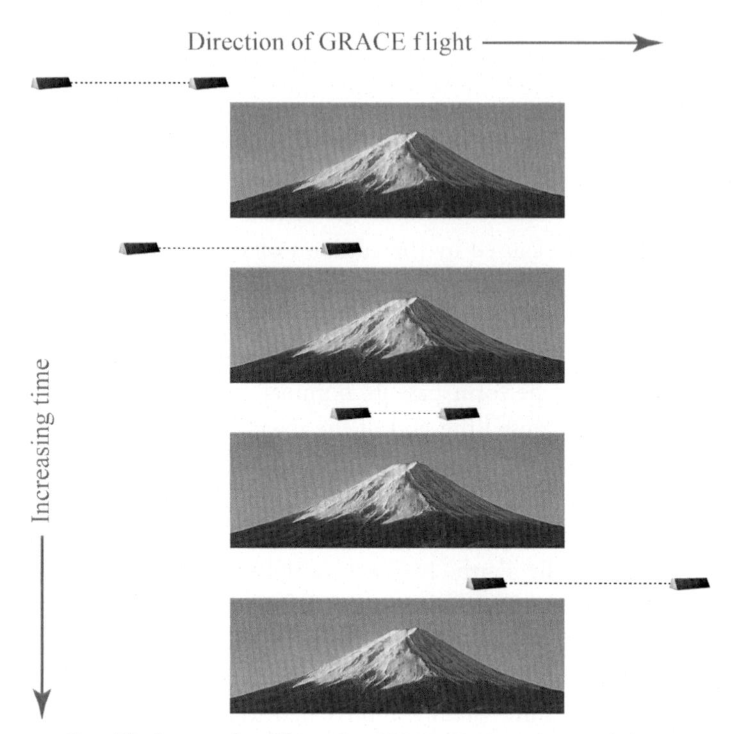

Figure 2.1 Simplified example of how the GRACE mission observed the time variable gravity field. The two GRACE satellites flew in tandem over the Earth (first panel). In this graphic the mountain represents a positive mass anomaly. As the lead satellite approached the mountain it was pulled toward it because of the gravitational force exerted by the mass anomaly. Thus, the distance between the two satellites increased (second panel). As the trailing satellite approached the mountain, it was pulled toward the anomalous mass while the lead satellite was held back by the same mass anomaly, and the distance between the satellites decreased (third panel). Finally, as the trailing satellite moved away from the mountain, it was held back, and their distance increased again (fourth panel).

2.3 Terrestrial water storage solutions

A monthly analysis of the range and range-rate measurements results in a map of temporal variations (i.e., anomalies) of the gravity fields caused by redistribution of earth masses. Masses can be static or time variable. Static masses include the total mass of the earth and mass heterogeneities that only vary on geologic timescales. This includes location of the continents, mountains, and depression in the crust. The static masses are "removed" when the long-term mean gravity field is saturated from a month solution to produce an anomaly map. The main drivers of temporal variations in the gravity field are oceanic and atmospheric circulations and redistribution of terrestrial water via the hydrological cycle. Atmosphere mass changes are removed during preprocessing using model based atmospheric pressure fields. An ocean model is used to remove high frequency (6-hourly to submonthly) wind and pressure-driven ocean motions that might otherwise alias into the monthly gravity solutions. What remains are anomalies in terrestrial water storage (TWS). Glacial isostatic adjustment must also be considered in certain regions such as Hudson's Bay in Canada, and a major earthquake can produce a significant gravitational anomaly, but the timescales of most solid earth processes are too long to be an issue.

Currently, the range-rate observations are processed using either spherical harmonics (Wahr et al., 1998) or the mass concentration (mascon, Rowlands et al., 2005) approach to produce the monthly gravity solutions.

The spherical harmonics solutions represent the gravity field via a set of coefficients (degree and order ≤ 120) of a given spherical harmonic expansion that describes the shape of the geoid (the surface of constant gravitational potential best matching the mean sea surface). The expansion coefficients can be manipulated using numerical devices such as Gaussian averaging functions in order to isolate mass anomalies (deviations from the baseline temporal mean) over regions of interest (e.g., Wahr et al., 1998). The spherical harmonic solutions have typically suffered from poor observability of east–west gradients, resulting in the so-called "stripes" that are conventionally removed via empirical smoothing and/or "destriping" algorithms (Swenson and Wahr, 2006). Although quite effective, especially for larger spatial scales, the destriping also removes some real geophysical signal along with the stripes, and the size, shape, and orientation of the signals strongly affects the effectiveness of destriping (Watkins et al., 2015). For this reason, errors in such estimates are inversely related to the size of the region, being as small as 1–2 cm equivalent height of water over continental-scale river basins, and being large enough to overwhelm the hydrology signal as the area drops below $\sim 150{,}000\ km^2$ (Rodell and Famiglietti, 1999; Wahr et al., 2006). Three centers are currently generating spherical harmonic fields: the Center for Space Research at the University of Texas; the GeoForschungsZentrum. Potsdam, Germany; and the Jet Propulsion Laboratory, California.

As an alternative to the spherical harmonic technique, the intersatellite ranging data can also be used directly to estimate regional "mass concentrations" (mascons) without the need to first derive a global gravity field (Rowlands et al., 2005). The primary advantage of using mascons as a basis function is that each mascon has a specific known geophysical location (unlike spherical harmonic coefficients which individually have no particular localization information). Mascons take advantage of this convenient property to specify a priori information (constraints) during the data inversion to internally remove the correlated error in the gravity solution. So, unlike the spherical harmonic solutions, the constrained mascon solutions typically do not need to be destriped or smoothed. The mascon approach also allows a better separation of land and ocean signals. Mascons solutions are currently freely available from different centers such as the Jet Propulsion Laboratory, California (Watkins et al., 2015); the Goddard Space and Flight Center, Maryland (Luthcke et al., 2013); and the Center for Space Research, Texas (Save et al., 2012).

2.4 Extremes in water storage

Extremes in water storage often are manifested as droughts and flooding events, driven by the surplus or deficit of water stored in a river basin or an aquifer. Understanding the spatial and temporal extent of these hydrologic extremes is critical for water resources management, hazard preparedness, and food security (Diffenbaugh and Scherer, 2011; D'Odorico et al., 2010). Few hydrologic observing networks yield sufficient data for comprehensive monitoring of changes in the total amount of water stored in a region (Rodell and Famiglietti, 2001). GRACE observations have helped to fill this gap. They have been used to characterize regional flood potential (e.g., Reager and Famiglietti, 2009) and to assess wetness and drought for the US Drought Monitor (e.g., Houborg et al., 2012). Furthermore, as an integrated measure of all surface and groundwater storage changes, GRACE data implicitly contains a record of seasonal to interannual water storage variations that can likely be exploited to lengthen early warning periods for regional flood and drought prediction (Thomas et al., 2014; Sun et al., 2017a,b). The next two subsections will provide a review of the extreme water storage events identified by the GRACE mission along with the potential for predictability.

2.4.1 Droughts

A meteorological drought is defined as an extended period of low precipitation, whereas hydrological droughts are generally defined as extended periods of water storage deficits leading to low streamflow. Satellite gravimetry is the only remote sensing technique able to monitor the water storage in a holistic manner. For this reason, GRACE data have been investigated alone, or in combination with land surface models, to analyze worldwide hydrological drought conditions and provide new methods to monitor droughts (Zhao et al., 2017). Moreover, GRACE can contribute to regional drought characterization by measuring water storage deficits in previously identified, drought-stricken areas. The duration and magnitude of the deficits can serve as metrics to help quantify hydrological drought severity (Thomas et al., 2014). This subsection provides a review of previous studies that used GRACE to quantify, analyze, and monitor droughts.

Recent studies have investigated the potential for using GRACE information to monitor droughts. In addition to Thomas et al. (2014) and Houborg et al. (2012), Cao et al. (2015) introduced the total storage deficit index, derived from the GRACE-recovered terrestrial water storage changes, to analyze drought characteristics in arid northwestern China. Yirdaw et al. (2008) employed the total storage deficit index, to characterize the 2002/2003 drought episode and generate a pictorial representation of the long-term dryness and wetness within the Saskatchewan River Basin in Canada. Yi and Wen (2016) established a GRACE-based hydrological drought index for drought monitoring in the continental United States from 2003 to 2012. Kusche et al. (2016) mapped the probabilities of extreme continental water storage changes form space gravimetry. They derived hot spot regions of high probability of peak anomalous water storage. Using the GRACE-derived TWS anomaly index, Wang et al. (2014) compared the precipitation anomaly index and the vegetation anomaly index to analyze drought events in the Haihe River Basin, China, from January 2003 to January 2013. Thomas et al. (2014) described a quantitative approach (based on water storage deficits) to measure the occurrence, severity, frequency, magnitude, and duration of hydrological droughts based on the GRACE-derived TWS (Sun et al., 2018).

Combined satellite/model drought monitoring tools are also becoming more common. Data assimilation systems merge observations with physically based models, using the model to provide spatially and temporally complete estimates of all drought-relevant hydrologic variables and the observation record to correct for model errors (Anderson et al., 2012). In particular, the application of the GRACE data assimilation approach for drought monitoring has recently been explored by NASA-funded investigations (Houborg et al., 2012; Rodell, 2012). These result in surface and root-zone soil moisture and groundwater drought indicators based on GRACE data assimilation results, and to evaluation of those indicators as inputs to the USDM products. Houborg et al. (2012) and Li et al. (2012) used GRACE data assimilation to evaluate its potential for developing new drought indicator products to serve as baselines for the Drought Monitor maps. The assimilation downscaled (in space and time) and disaggregated GRACE data into finer scale TWS components, which exhibited significant changes in their dryness rankings relative to those without data assimilation, suggesting that GRACE data assimilation could have a substantial impact on drought monitoring (Li et al., 2012; Houborg et al., 2012). The weekly indicators are available on the National Drought Mitigation Center's website (http://drought.unl.edu).

2.4.1.1 Major world droughts

Thomas et al. (2014) presented gravity-based measurements of water storage anomalies during recent droughts. They used GRACE to explore how hydrological droughts may be better characterized while estimating the associated regional water storage deficit. The average storage deficit during drought events can be identified using (1) the storage deficit as a negative departure from the seasonal cycle and (2) the drought duration as the number of months with continuous deficits (Thomas et al., 2014). In essence, the average storage deficit is an arithmetic mean of the storage deficit observed during a given drought event and it is used as a measure of drought intensity. Based on these definitions of storage deficit and drought duration, Humphrey et al. (2016) provided a review of all drought events identified in the GRACE period from 2002 to 2015. Fig. 2.2 shows the maximum average storage deficit ever observed for all drought events identified in April 2002–January 2017 by the GRACE record as derived by Humphrey et al. (2016). The year corresponding to this maximum is depicted in Fig. 2.2B. Here, we group these major drought events by continents. We discuss the potential for using GRACE data and models (via data assimilation) to improve predictability of these droughts events in the next section.

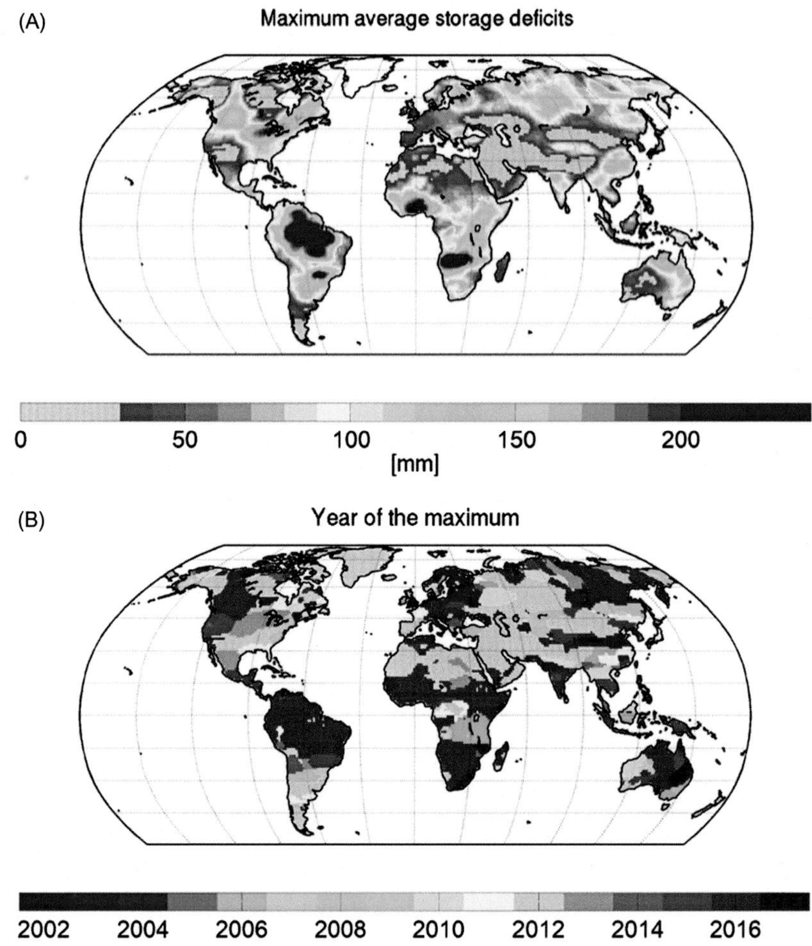

Figure 2.2 (A) Maximum value of the average storage deficit observed in the period April 2002–August 2015, expressed in mm of equivalent water height. (B) Year corresponding to the maximum value of the average storage deficit, showing only regions with a deficit larger than 30 mm. *Adapted from Humphrey, V., Gudmundsson, L., Seneviratne, S.I., 2016. Assessing global water storage variability from GRACE: Trends, seasonal cycle, subseasonal anomalies and extremes. Surveys Geophys., 37(2), 357–395.*

South America

The Amazon River water storage during year 2016 was one of the lowest during the GRACE record (Fig. 2.2). Other droughts, not shown, are identified for this region, such as the 2004–08, and 2010 dry-period reported in the works of Humphrey et al. (2016), Frappart et al. (2012), and Thomas et al. (2014). Getirana (2016) uses GRACE to identify a water loss rate of 26.1 cm/year for southeastern Brazil over the period from 2012 to 2015. Drought events in the Amazon River basin were shown to be related to precipitation deficits and to the El Niño Southern Oscillation (ENSO, Davidson et al., 2012; Frappart et al., 2012). The Amazon basin extreme droughts cause ineffective hydropower production. In fact, water storage droughts in this region cause massive agricultural losses, water supply restrictions, and energy rationing.

The 2008–12 La Plata basin (Argentina) drought is also seen in Fig. 2.2. Chen et al. (2010) identify the onset of this drought using GRACE. In their work, GRACE revealed details of the temporal and spatial evolution of this drought, with onset around austral spring 2008 in the lower La Plata basin, which spread to the entire La Plata basin. Wetness and drought in this region is correlated with ENSO, with dry and wet seasons corresponding to ENSO events, respectively (Chen et al., 2010; Abelen et al., 2015; Sordo-Ward et al., 2017).

North America

The 2012–15 drought in the Central Valley of California is also shown in Fig. 2.2. This drought reduced natural recharge to the aquifers and caused low annual snow accumulations in the Sierra Nevada Mountains (Margulis et al., 2016). Groundwater acts as the key strategic reserve in times of drought and the Central Valley is the most productive agricultural region for fresh produce in the United States (Faunt et al., 2009). Excessive groundwater pumping during a period of meteorological drought can cause unrecoverable losses at aquifer storage capacity (Faunt et al., 2009; Famiglietti et al., 2011; Famiglietti, 2014; Scanlon et al., 2012a,b). For example, Famiglietti et al. (2011) estimate a depletion in groundwater that totaled 20 km^3 of water during 2012–15. Continued groundwater depletion at this rate may well be unsustainable, with potentially dire consequences for the economic and food security of the United States (Famiglietti et al., 2011). Groundwater depletion can cause significant land subsidence (Galloway and Riley, 1999). The analysis of TWS for the entire GRACE record also indicates dry episodes for previous years over the southwestern United States. For example, Scanlon et al. (2012a) and Famiglietti et al. (2011) report that groundwater depletion was observed in the Central Valley during the drought period from April 2006 through September 2009.

Notably, most of the US experienced droughts in 2012 (Fig. 2.2). The US Drought Monitor (Schmugge et al., 2002) estimated that over three-quarters of the contiguous US experienced at least abnormally dry conditions by the summer's end with nearly half of the region, especially the Great Plains, experiencing severe drought (Hoerling et al., 2014). This period lacked the usual abundance of slow-soaking precipitation-bearing systems and evening thunderstorms along the Great Plains, and as a result, surface moisture conditions greatly deteriorated. GRACE data was used to identify this massive event (Humphrey et al., 2016). Most importantly, GRACE TWS data helped identify the recovery time. In March 2013, heavy winter rains broke a 3-year pattern of meteorological drought in much of the southeastern United States, while drought conditions still plagued the Great Plains and other parts of the United States. Chew and Small (2014) found that standard drought indices misrepresent the recovery time from the drought, putting it much earlier than the one obtained with the auxiliary information of GRACE TWS. Accurate understanding of drought onset and recovery time is critical because drought impacts grain and wheat prices (Boyer et al., 2013). A large disruption to production of these crops can have a substantial impact on international grain markets. Also visible in Fig. 2.2 are the droughts during the years 2007–09 over the eastern United States. Houborg et al. (2012) demonstrated the capabilities of using GRACE data assimilation for characterizing the hydrological drought (more discussion in Section 2.4.1.2). Furthermore, the 2010–13 drought in Texas is shown in Fig. 2.2 and was described by Long et al. (2014).

Africa

Fig. 2.2 identifies the 2006–07 dry conditions over an extended area around the Lake Victoria (Swenson and Wahr, 2009). The 2006–08 Zambezi basin (Thomas et al., 2014) and Lake Victoria droughts are captured as one large-scale and spatially contiguous event. Droughts in East Africa are recurring phenomena with significant humanitarian impacts. GRACE total water storage estimates reveal that water storage declined in much of East Africa by up to 60 mm/year (Swenson and Wahr, 2009) during the 2006–07 drought. The complex and highly variant nature of many hydrometeorological drivers such as ENSO, sea surface temperature (SST), and land–atmosphere feedback adds to the daunting challenge of drought monitoring and forecasting in Africa. The 2010–11 drought of the Horn of Africa is not shown in Fig. 2.2 but it was identified using GRACE in previous studies (e.g., Anderson et al., 2012) to characterize its temporal and spatial evolution.

Eurasia

Hydrological droughts are common in Europe, and several episodes of severe droughts, including the 2003 drought (spreading across western and central Europe) and the 2007–08 droughts (affecting southern and southwestern Europe) were detected by GRACE TWS (Li et al., 2012). In particular, the 2003 event was associated with the 2003 European heat wave (Rebetez et al., 2006). For most of Europe, TWS values were relatively high at the beginning of 2003 due to heavy precipitation in the summer and autumn of 2002, but they decreased very rapidly in the period February–August, resulting in very dry conditions during the entire summer of 2003. The soil drying in 2003 exceeded the long-term average by far (García-Herrera et al., 2010). Following the 2003 drought and heat wave, forest fires burned a large area of Siberia (García-Herrera et al., 2010). Drought conditions can also be found in northern India for the period 2009–10 (Fig. 2.2). The year 2009 was the driest year of the decade for this region in terms of precipitation (Chen et al., 2014) and resulted in higher groundwater abstraction rates (Rodell et al., 2009).

Fig. 2.2 identifies the 2004 drought in southeast China, around the Yangtze River basin. The Yangtze River is experiencing an increasing trend in terms of the frequency of extreme events (Dai et al., 2008), which often leads to severe reductions in crop outputs, along with other related social and economic losses (Chao et al., 2016). Other severe drought events in 2006 and 2011 were identified (Sun et al., 2018). Previous works based on the analysis of GRACE data quantify a terrestrial water storage deficit of −621, −617, and −192 mm for the 2004, 2006, and 2011 droughts, respectively (Chao et al., 2016; Sun et al., 2018).

The Sumatra region exhibits a minimum in 2004 (Fig. 2.2). However, this is an artifact caused by the 2004 earthquake, with GRACE observing changes in the gravity field due to crustal dilatation and vertical displacement of Earth's layered density structure (Han et al., 2006).

Australia

In Australia, multiyear droughts have been related to precipitation deficits (García-García et al., 2011). Van Dijk et al. (2013) explained drivers of the drought and its impacts. These were analyzed using climate, water, economic, and remote sensing data combined with biophysical modeling. ENSO explained two-thirds of the rainfall deficit, and a contribution of global climate change remains plausible. The millennium drought (2001–09) was the worst drought on record for southeast Australia

(Van Dijk et al., 2013). It has led to the almost complete drying of surface water resources, which account for most of the water used for irrigation and domestic purposes. Using GRACE data, Leblanc et al. (2009) showed the propagation of the water deficit through the hydrological cycle and the rise of different types of drought.

2.4.1.2 The potential for using GRACE in monitoring and predicting droughts

Currently, standard drought quantitation methods and products including the US Drought Monitor (USDM) rely heavily on in situ observations of precipitation, streamflow, snowpack data, and subjective human-based judgments (Rodell, 2012). USDM drought maps are published on a weekly basis by a team of authors and widely considered to be the premier US drought products available for use by governments, farmers and other stakeholders, and the public, despite limited groundwater and soil moisture data as direct inputs. Although meteorological drought is defined as an extended period with deficient precipitation, hydrological and agricultural droughts are also influenced by precursor conditions of groundwater and soil moisture. Surface moisture conditions can fluctuate quickly with the weather, while the deeper components of TWS (e.g., groundwater) are well suited to drought quantification, particularly hydrologic droughts, because they integrate meteorological conditions over timescales of weeks to years (Rodell, 2012). Prior to USDM use of GRACE data assimilation-based drought indices (Houborg et al., 2012), drought products did not incorporate systematic observations of soil moisture, groundwater, or total TWS. Groundwater storage and shallow and deep soil moisture are still major gaps in current drought monitoring capabilities and monitoring of droughts has suffered from the lack of reliable information on the water stored below the uppermost soil layer. Remote sensing-based drought indices have opened a new path forward in drought monitoring and detection (Niemeyer, 2008), allowing accurate spatial information to be obtained with global or regional coverage with high reliability and a high repetition rate (Sun et al., 2018). In particular, since GRACE measures the water storage changes in the entire profile, it provides valuable information on drought conditions beyond what can be seen at the surface (Li et al., 2012). Furthermore, given that GRACE offers information on the total water storage deficit it can be used to estimate the amount of water (precipitation) needed to recover from drought events (Thomas et al., 2014; AghaKouchak et al., 2014).

2.4.2 Floods

Flood events are projected to become more frequent as global warming amplifies the atmosphere's water holding capacity, increasing the occurrence of extreme precipitation events (Slater and Villarini, 2016; Groisman et al., 2012; Wang et al., 2017a,b). Many countries have developed flood alert systems (Molodtsova et al., 2016), such as the European Flood Alert System (Bartholmes et al., 2009) and the US National Weather Service Automated Flood Warning System (Scawthorn, 1999). While most of these warning systems rely on a dense network of gauging stations, a significant portion of the economic losses caused by floods occur in developing countries where ground flood monitoring and management programs are still inefficient (Molodtsova et al., 2016). To augment ground-based observations, flood monitoring has increasingly relied on products based on space-borne sensor observations. Among the remote sensing products that have been used for flood monitoring, data from the GRACE mission are unique because the total amount of terrestrial water can be directly measured (Molodtsova et al., 2016). The terrestrial water storage signal relates to the ability of the land surface to absorb and process water and accounts for the water in plants, groundwater, soil moisture, and snow (Reager and Famiglietti, 2009). Over the course of a year, a region can transition between its minimum and maximum TWS due to the annual cycle of precipitation. When the region is near the maximum, it can store and process only a finite amount of water before the saturated ground will force additional precipitation to run off (Reager and Famiglietti, 2009). Generally, the land surface modulates the connection between precipitation and runoff generation through two basic mechanisms: infiltration limitation and saturation excess. During floods, one of these two mechanisms is typically identified as a driver based on the preflood conditions and the nature of the precipitation event (Reager et al., 2015). That is, the prerainfall event wetness of a watershed can determine its response to rainfall (Sayama et al., 2011; Brutsaert, 2008; Kirchner, 2009), leading to variability in flood generation (Li and Simonovic, 2002). Slater and Villarini (2016) show that flood patterns are dependent on the overall wetness and potential water storage, with fundamental implications for water resources management, agriculture, insurance, navigation, ecology, and populations living in flood-affected areas.

2.4.2.1 Flood potential from GRACE

Reager and Famiglietti (2009) introduced the concept of "flood potential" to highlight the information relevant for regional flooding contained in the GRACE extreme (wet) TWS values. TWS that is near its maximum can be a harbinger of future flood events at several months lead time (Reager and Famiglietti, 2009). The concept of flood potential has been used in several studies (Section 2.4.2.2) to analyze flood risk in different watersheds around the world.

Reager and Famiglietti (2009) assume that the regional storage capacity can be approximated by the historic record maximum in the GRACE TWS storage anomaly time series to give a quantitative estimate (in cm) of the saturation point of the regional land surface. In a nutshell, GRACE time series can be used to calculate an effective maximum storage capacity in each region. Reager and Famiglietti (2009) isolate, temporally and spatially, those months in which a high percentage of the storage capacity is achieved and high precipitation continues. Finally, they normalized this new dataset to remove regional heterogeneity and create a global flood potential index. The resulting flood potential amount is the quantity of incoming water that exceeds the storage capacity for the current month based on the regionally observed storage anomaly maxima. Similar to a traditional "bucket model," when this quantity exceeds zero, flooding is likely to occur.

Fig. 2.3A shows a global map of the flood index for the year 2007 as derived by Reager and Famiglietti (2009). This is compared to the Dartmouth Flood Observatory map of reported floods (Fig. 2.3B). The Dartmouth Flood Observatory catalogues historic floods based on satellite observations, weather service, and news reports (https://www.dartmouth.edu/~floods/). The red areas in Fig. 2.3B indicate observed floods during 2007. While the GRACE derived flood index (Fig. 2.3A) misses some flooded regions in Africa and Asia, it is generally successful at predicting large-scale flood-affected areas. The index captures the general patterns of flooding on most continents, and captures some rather high spatial resolution structure, including events on the southernmost tip of Africa and the unique "T" shape of Mississippi river flooding (Reager and Famiglietti, 2009). Furthermore, Reager et al. (2014) demonstrated the ability of the flood potential concept to derive longer lead times (as much as 5–11 months in advance) in flood warnings at a regional scale (e.g., for the Missouri, 2011, the Columbia River, 2011, and Indus River, 2010 flood events).

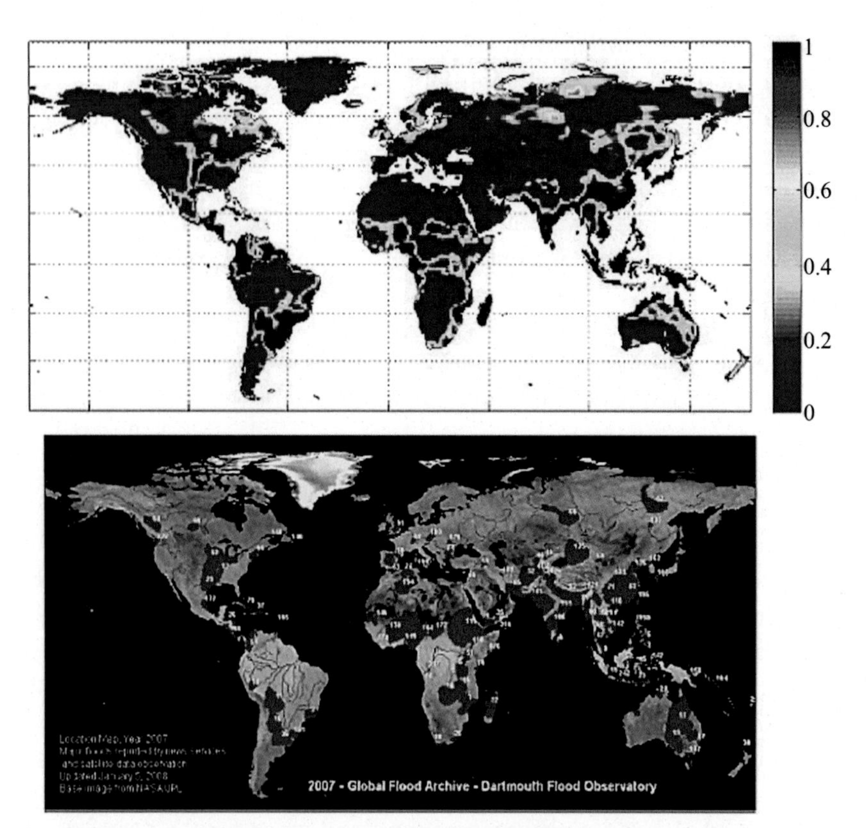

Figure 2.3 A comparison of the (top) 2007 flood index maxima and (bottom) 2007 Dartmouth Flood Observatory reported floods. *Reager, J.T., Famiglietti, J.S., 2009. Global terrestrial water storage capacity and flood potential using GRACE. Geophys. Res. Lett., 36(23):L23402.*

However, several limitations still exist for using satellite gravimetry in an operational flood prediction mode (Reager et al., 2015). The prevailing limitations are: (1) coarse spatial resolution ($\sim$150,000 km^2, Rowlands et al., 2005; Swenson and Wahr, 2006); (2) the aggregated observation of multiple water storage components (i.e., snow, soil moisture, and groundwater) as a single integrated value for each grid cell; and (3) latency in data product processing and release (2–4 months lag for GRACE and at least 10 days for GRACE-FO). These obstacles make many applications difficult, as water resource management tends to occur at small watershed scales and on daily to weekly timescales. For this reason, similarly to the drought monitoring work by Houborg et al. (2012), data assimilation can be used as a tool to partially overcome these limitations. In their work, Reager et al. (2015) assimilate GRACE data within the catchment land surface model and evaluate the assimilation scheme performance under flood conditions. The assimilation results made the model wetter in the months preceding flood, thus increasing the model capability to predict floods.

2.4.2.2 Historical flood events characterized by GRACE

The investigation of GRACE data to characterize flooding events has not been exploited as much as the characterization of drought events. This section lists historical flooding events that have been investigated with the use of GRACE data.

dutt Vishwakarma et al. (2013) discussed two major flooding events in India, one being the 2005 monsoon flooding in Mumbai and nearby states and other being the flood experienced by Bihar in 2008. GRACE detected these two flooding events at a large spatial scale. Long et al. (2014) used the flood potential index derived by Reager and Famiglietti (2009) to characterize the 2008 severe flood in southwest China. Chinnasamy (2017) developed a regression model between GRACE data and observed discharge flow. Their model was tested to predict the 2008 Koshi floods, India. Their results indicate that the saturation of water storage units in the basin play a vital role in the prediction of peak floods and their lead times. Chen et al. (2010) used GRACE to identify the Amazon River 2009 flood event, which caused many casualties, and its connection to ENSO. Sun et al. (2017a,b) used GRACE data to evaluate the flood potential index (Reager and Famiglietti, 2009) in the Yangtze River Basin of China. Zhou et al. (2017) also investigated the potentials for using GRACE data for identifying and predicting floods in the Yangtze River Basin. In particular, the 2010 flood was identified as the most serious disaster during the study period, with discharge and precipitation values 37.95% and 19.44% higher, respectively, than multiyear average values for the same period. Zhou et al. (2017) highlight that while the GRACE derived flood potential index can identify extremes for large floods, it is currently not suitable for smaller and/or short-term flood events (Sun et al., 2017a,b). Tangdamrongsub et al. (2016) used GRACE data to quantify flooding of the Tonlé ap basin, Cambodia between 2002 and 2014. For this region, the 2011 and 2013 flood events were clearly identified by GRACE. The basin-averaged TWS values for these two events were 42 cm (40% higher than the long-term mean peak value) and 36 cm (34% higher) equivalent water height, respectively. In May and June of

2011, a large flood crest moved down the Mississippi from a combination of snowmelt and heavy spring rains in the Missouri river basin. Flows in the Missouri were characterized as a 500-year event. Red River Landing, near the mouth of the Mississippi, went above flood stage on 19 March and did not fall below flood stage until 25 June. Reager et al. (2014) and Wang et al. (2014) investigated this flood event using GRACE data and they indicate that analysis of TWS observations has the potential for providing spatially distributed warning information. Similar conclusions were derived for the Columbia River flood event in May–June 2011, and the Indus River flood in 2010.

2.5 Conclusions

Extremes in water storage are often manifested as drought and flooding events, driven by the deficit or surplus of water stored in a river basin or an aquifer. These events are projected to become more frequent as a result of global warming (Pachauri et al., 2014). The existing in situ observing networks are not sufficient to detect and monitor these extremes in the total amount of water stored in a region. GRACE observations have helped to fill this information gap because GRACE has provided an unprecedented, holistic view of terrestrial water storage variations. GRACE Follow On, launched in May 2018, will extend that important data record.

GRACE has contributed to regional drought characterization by measuring water storage deficits in drought-stricken areas. It also has provided information useful for predicting regional flood potential. The concepts of flood potential and water storage deficit have been used in several studies to characterize the extremes observed by GRACE. This book chapter has reviewed many of the most significant studies.

Limitations still exist for using GRACE and GRACE Follow On data in an operational drought or flood prediction mode. The major limitations are related to the relatively coarse spatial and temporal resolutions of the GRACE observations and the fact that TWS is an aggregated observation of multiple water storage components (i.e., snow, soil moisture, and groundwater). Latency in the delivery of GRACE data is also a serious issue, but that is expected to be mitigated by the availability of low latency data products derived from GRACE Follow On (e.g., Sakumura et al., 2016). These limitations can be addressed through the combined used of observed TWS and models, that is, via data assimilation (e.g., Zaitchik et al., 2008; Girotto et al., 2016).

References

Abelen, S., Seitz, F., Abarca-del-Rio, R., Güntner, A., 2015. Droughts and floods in the La Plata basin in soil moisture data and GRACE. Remote Sensing 7 (6), 7324–7349.

AghaKouchak, A., Cheng, L., Mazdiyasni, O., Farahmand, A., 2014. Global warming and changes in risk of concurrent climate extremes: insights from the 2014 California drought. Geophys. Res. Lett. 41 (24), 8847–8852.

Anderson, R.G., Lo, M.H., Famiglietti, J.S., 2012. Assessing surface water consumption using remotely-sensed groundwater, evapotranspiration, and precipitation. Geophys. Res. Lett. 39 (16).

Bartholmes, J.C., Thielen, J., Ramos, M.H., Gentilini, S., 2009. The european flood alert system EFAS–Part 2: statistical skill assessment of probabilistic and deterministic operational forecasts. Hydrol. Earth Syst. Sci. 13 (2), 141–153.

Boyer, J.S., Byrne, P., Cassman, K.G., Cooper, M., Delmer, D., Greene, T., et al., 2013. The US drought of 2012 in perspective: a call to action. Global Food Sec. 2 (3), 139–143.

Brutsaert, W., 2008. Long-term groundwater storage trends estimated from streamflow records: climatic perspective. Wat. Resour. Res. 44, W02409.

Cao, Y., Nan, Z., Cheng, G., 2015. GRACE gravity satellite observations of terrestrial water storage changes for drought characterization in the arid land of northwestern China. Remote Sensing 7 (1), 1021–1047.

Chao, N., Wang, Z., Jiang, W., Chao, D., 2016. A quantitative approach for hydrological drought characterization in southwestern China using. Hydrogeol. J. 24 (4), 893–903.

Chen, J.L., Wilson, C.R., Tapley, B.D., Longuevergne, L., Yang, Z.L., Scanlon, B.R., 2010. Recent La Plata basin drought conditions observed by satellite gravimetry. J. Geophys. Res. Atmosph. 115 (D22).

Chen, Y., Deng, H., Li, B., Li, Z., Xu, C., 2014. Abrupt change of temperature and precipitation extremes in the arid region of Northwest China. Quatern. Int. 336, 35–43.

Chew, C.C., Small, E.E., 2014. Terrestrial water storage response to the 2012 drought estimated from GPS vertical position anomalies. Geophys. Res. Lett. 41 (17), 6145–6151.

Chinnasamy, P., 2017. Inference of basin flood potential using nonlinear hysteresis effect of basin water storage: case study of the Koshi basin. Hydrol. Res. 48 (6), 1554–1565. Available from: https://doi.org/10.2166/nh.2016.268.

Crow, W.T., Bindlish, R., Jackson, T.J., 2005. The added value of spaceborne passive microwave soil moisture retrievals for forecasting rainfall-runoff partitioning. Geophys. Res. Lett. 32, 18.

Dai, Z., Du, J., Li, J., Li, W., Chen, J., 2008. Runoff characteristics of the Changjiang River during 2006: effect of extreme drought and the impounding of the Three Gorges Dam. Geophys. Res. Lett. 35 (7).

Davidson, E.A., de Araújo, A.C., Artaxo, P., Balch, J.K., Brown, I.F., Bustamante, M.M., et al., 2012. The Amazon basin in transition. Nature 481 (7381), 321.

Diffenbaugh, N.S., Scherer, M., 2011. Observational and model evidence of global emergence of permanent, unprecedented heat in the 20th and 21st centuries. Clim. Change 107, 615–624.

Dijk, A.I., Beck, H.E., Crosbie, R.S., Jeu, R.A., Liu, Y.Y., Podger, G.M., et al., 2013. The Millennium Drought in southeast Australia (2001–2009): natural and human causes and implications for water resources, ecosystems, economy, and society. Water Resour. Res. 49 (2), 1040–1057.

D'Odorico, P., Laio, F., Ridolfi, L., 2010. Does globalization of water reduce societal resilience to drought? Geophys. Res. Lett. 37.

Entekhabi, D., Rodriguez-Iturbe, I., Castelli, F., 1996. Mutual interaction of soil moisture state and atmospheric process. J. Hydrol. 184, 3–17.

Famiglietti, J.S., 2004. Remote sensing of terrestrial water storage, soil moisture and surface waters. State Planet: Front. Challen. geophys. 197–207.

Famiglietti, J.S., 2014. The global groundwater crisis. Nat. Clim. Change 4 (11), 945.

Famiglietti, J.S., Lo, M., Ho, S.L., Bethune, J., Anderson, K.J., Syed, T.H., et al., 2011. Satellites measure recent rates of groundwater depletion in California's Central Valley. Geophys. Res. Lett. 38 (3).

Faunt, C.C., Hanson, R.T., Belitz, K., 2009. Chapter A. Introduction, Overview of Hydrogeology, and Textural Model of California's Central Valley. US Geological Survey professional paper, 1766.

Frappart, F., Papa, F., da Silva, J.S., Ramillien, G., Prigent, C., Seyler, F., et al., 2012. Surface freshwater storage and dynamics in the Amazon basin during the 2005 exceptional drought. Environ. Res. Lett. 7 (4), 044010.

Galloway, D., Riley, F.S., 1999. San Joaquin Valley, California. Land subsidence in the United States: US Geological Survey Circular 1182, 23–34.

García-García, D., Ummenhofer, C.C., Zlotnicki, V., 2011. Australian water mass variations from GRACE data linked to Indo-Pacific climate variability. Remote Sensing Environ. 115 (9), 2175–2183.

García-Herrera, R., Díaz, J., Trigo, R.M., Luterbacher, J., Fischer, E.M., 2010. A review of the European summer heat wave of 2003. Crit. Rev. Environ. Sci. Technol. 40 (4), 267–306.

Getirana, A., 2016. Extreme water deficit in Brazil detected from space. J. Hydrometeorol. 17 (2), 591–599.

Getirana, A., Kumar, S., Girotto, M., Rodell, M., 2017. Rivers and floodplains as key components of global terrestrial water storage variability. Geophys. Res. Lett. 44 (20).

Girotto, M., Margulis, S.A., Durand, M., 2014. Probabilistic SWE reanalysis as a generalization of deterministic SWE reconstruction techniques. Hydrol. Processes 28 (12), 3875–3895.

Girotto, M., De Lannoy, G.J., Reichle, R.H., Rodell, M., 2016. Assimilation of gridded terrestrial water storage observations from GRACE into a land surface model. Water Resour. Res. 52 (5), 4164–4183.

Girotto, M., De Lannoy, G.J., Reichle, R.H., Rodell, M., Draper, C., Bhanja, S.N., et al., 2017. Benefits and pitfalls of GRACE data assimilation: a case study of terrestrial water storage depletion in India. Geophys. Res. Lett. 44 (9), 4107–4115.

Gleeson, T., Wada, Y., Bierkens, M.F., van Beek, L.P., 2012. Water balance of global aquifers revealed by groundwater footprint. Nature 488 (7410), 197.

Groisman, P.Y., Knight, R.W., Karl, T.R., 2012. Changes in intense precipitation over the central United States. J. Hydrometeorol. 13 (1), 47–66.

Hall, D.K., Chang, A.T., Siddalingaiah, H., 1988. Reflectances of glaciers as calculated using Landsat-5 Thematic Mapper data. Remote Sensing Environ. 25 (3), 311–321.

Han, S.C., Shum, C.K., Bevis, M., Ji, C., Kuo, C.Y., 2006. Crustal dilatation observed by GRACE a er the 2004 Sumatra-Andaman earthquake. Science 313, 658–662. Available from: https://doi.org/10.1126/science.1128661.

Hoerling, M., Eischeid, J., Kumar, A., Leung, R., Mariotti, A., Mo, K., et al., 2014. Causes and predictability of the 2012 Great Plains drought. Bull. Am. Meteorol. Soc. 95 (2), 269–282.

Houborg, R., Rodell, M., Li, B., Reichle, R., Zaitchik, B.F., 2012. Drought indicators based on model-assimilated Gravity Recovery and Climate Experiment (GRACE) terrestrial water storage observations. Water Resour. Res. 48 (7).

Humphrey, V., Gudmundsson, L., Seneviratne, S.I., 2016. Assessing global water storage variability from GRACE: trends, seasonal cycle, subseasonal anomalies and extremes. Surveys Geophys. 37 (2), 357–395.

Kirchner, J.W., 2009. Catchments as simple dynamical systems: catchment characterization, rainfall-runoff modeling, and doing hydrology backwards. Wat. Resour. Res. 45, W02429.

Kusche, J., Eicker, A., Forootan, E., Springer, A., Longuevergne, L., 2016. Mapping probabilities of extreme continental water storage changes from space gravimetry. Geophys. Res. Lett. 43 (15), 8026–8034.

Leblanc, M.J., Tregoning, P., Ramillien, G., Tweed, S.O., Fakes, A., 2009. Basin-scale, integrated observations of the early 21st century multiyear drought in southeast Australia. Water Resour. Res. 45 (4).

Li, B., Rodell, M., Zaitchik, B.F., Reichle, R.H., Koster, R.D., van Dam, T.M., 2012. Assimilation of GRACE terrestrial water storage into a land surface model: evaluation and potential value for drought monitoring in western and central Europe. J. Hydrol. 446, 103–115.

Li, B., Rodell, M., Famiglietti, J.S., 2015. Groundwater variability across temporal and spatial scales in the central and northeastern U.S. J. Hydrol. 525, 769–780. Available from: https://doi.org/10.1016/j.jhydrol.2015.04.033.

Li, L., Simonovic, S.P., 2002. System dynamics model for predicting floods from snowmelt in North American prairie watersheds. Hydrol. Process. 16, 2645–2666.

Long, D., Shen, Y., Sun, A., Hong, Y., Longuevergne, L., Yang, Y., et al., 2014. Drought and flood monitoring for a large karst plateau in Southwest China using extended GRACE data. Remote Sensing Environ. 155, 145–160.

Luthcke, S.B., Sabaka, T.J., Loomis, B.D., et al., 2013. Antarctica, Greenland and Gulf of Alaska land ice evolution from an iterated GRACE global mascon solution. J. Glac 59 (216), 613–631. Available from: https://doi.org/10.3189/2013JoG12J147.

Margulis, S.A., Cortés, G., Girotto, M., Huning, L.S., Li, D., Durand, M., 2016. Characterizing the extreme 2015 snowpack deficit in the Sierra Nevada (USA) and the implications for drought recovery. Geophys. Res. Lett. 43 (12), 6341–6349.

Milly, P.C.D., Dunne, K.A., 1994. Sensitivity of the global water cycle to the water-holding capacity of the land. J. Clim. 7, 506–526.

Molodtsova, T., Molodtsov, S., Kirilenko, A., Zhang, X., VanLooy, J., 2016. Evaluating flood potential with GRACE in the United States. Nat. Hazards Earth Syst. Sci. 16 (4), 1011–1018.

Niemeyer, S., 2008. New drought indices. *Options Méditerranéennes*. Série A: Séminaires Méditerranéens 80, 267–274.

Pachauri, R.K., Allen, M.R., Barros, V.R., Broome, J., Cramer, W., Christ, R., et al., 2014. Climate change 2014: synthesis report. Contribution of Working Groups I, II and III to the fifth assessment report of the Intergovernmental Panel on Climate Change (p. 151). IPCC.

Reager, J.T., Famiglietti, J.S., 2009. Global terrestrial water storage capacity and flood potential using GRACE. Geophys. Res. Lett. 36 (23), L23402.

Reager, J.T., Thomas, B.F., Famiglietti, J.S., 2014. River basin flood potential inferred using GRACE gravity observations at several months lead time. Nat. Geosci. 7 (8), 588.

Reager, J.T., Thomas, A.C., Sproles, E.A., Rodell, M., Beaudoing, H.K., Li, B., et al., 2015. Assimilation of GRACE terrestrial water storage observations into a land surface model for the assessment of regional flood potential. Remote Sensing 7 (11), 14663–14679.

Rebetez, M., Mayer, H., Dupont, O., Schindler, D., Gartner, K., Kropp, J.P., et al., 2006. Heat and drought 2003 in Europe: a climate synthesis. Ann. Forest Sci. 63 (6), 569–577.
Richey, A.S., Thomas, B.F., Lo, M.-H., Swenson, S., Rodell, M., Famiglietti, J.S., 2015. Uncertainty in global groundwater storage estimates in a total groundwater stress framework. Wat. Resour. Res. 51, 5198–5216. Available from: https://doi.org/10.1002/2015WR017351.
Rodell, M., 2012. 11 Satellite Gravimetry Applied to Drought Monitoring. Remote Sensing Drought: Innovat. Monit. Approach. 261.
Rodell, M., Famiglietti, J.S., 1999. Detectability of variations in continental water storage from satellite observations of the time dependent gravity field. Water Resour. Res. 35 (9), 2705–2723.
Rodell, M., Famiglietti, J.S., 2001. An analysis of terrestrial water storage variations in Illinois with implications for the Gravity Recovery and Climate Experiment (GRACE). Water Resour. Res. 37 (5), 1327–1339.
Rodell, M., Chao, B.F., Au, A.Y., Kimball, J.S., McDonald, K.C., 2005. Global biomass variation and its geodynamic effects: 1982–98. Earth Interact. 9 (2), 1–19.
Rodell, M., Velicogna, I., Famiglietti, J.S., 2009. Satellite-based estimates of groundwater depletion in India. Nature 460 (7258), 999.
Rowlands, D.D., Luthcke, S.B., Klosko, S.M., Lemoine, F.G., Chinn, D.S., McCarthy, J. J., et al., 2005. Resolving mass flux at high spatial and temporal resolution using GRACE intersatellite measurements. Geophys. Res. Lett. 32 (4).
Sakumura, C., Bettadpur, S., Save, H., McCullough, C., 2016. High-frequency terrestrial water storage signal capture via a regularized sliding window mascon product from GRACE. J. Geophys. Res. Solid Earth 121 (5), 4014–4030.
Save, H., Bettadpur, S., Tapley, B.D., 2012. Reducing errors in the GRACE gravity solutions using regularization. J. Geod. Available from: https://doi.org/10.1007/s00190-012-0548-5.
Sayama, T., McDonnell, J.J., Dhakal, A., Sullivan, K., 2011. How much water can a watershed store? Hydrol. Process. 25, 3899–3908.
Scanlon, B.R., Longuevergne, L., Long, D., 2012a. Ground referencing GRACE satellite estimates of groundwater storage changes in the California Central Valley, USA. Water Resour. Res. 48 (4).
Scanlon, B.R., Faunt, C.C., Longuevergne, L., Reedy, R.C., Alley, W.M., McGuire, V. L., et al., 2012b. Groundwater depletion and sustainability of irrigation in the US High Plains and Central Valley. Proc. Nat. Acad. Sci. 109 (24), 9320–9325.
Scawthorn, C., 1999. Modeling flood events in the US. Proceedings of the EuroConference on Global Change and Catastrophe Risk Management, International Institute for Advanced Systems Analysis, Laxenburg, Austria.
Schmugge, T.J., Kustas, W.P., Ritchie, J.C., Jackson, T.J., Rango, A., 2002. Remote sensing in hydrology. Adv. Water Resour. 25 (8-12), 1367–1385.
Seneviratne, S., 2003. Terrestrial water storage: a critical variable for mid-latitude climate and climate change (Doctoral dissertation, ETH Zurich).
Seneviratne, S.I., Corti, T., Davin, E.L., Hirschi, M., Jaeger, E.B., Lehner, I., et al., 2010. Investigating soil moisture–climate interactions in a changing climate: a review. Earth-Sci. Rev. 99 (3-4), 125–161.
Slater, L.J., Villarini, G., 2016. Recent trends in US flood risk. Geophys. Res. Lett. 43 (24).
Sordo-Ward, A., Bejarano, M.D., Iglesias, A., Asenjo, V., Garrote, L., 2017. Analysis of current and future SPEI droughts in the La Plata Basin based on results from the Regional Eta Climate Model. Water 9 (11), 857.

Sun, A.Y., Scanlon, B.R., AghaKouchak, A., Zhang, Z., 2017a. Using GRACE satellite gravimetry for assessing large-scale hydrologic extremes. Remote Sensing 9 (12), 1287.
Sun, Z., Zhu, X., Pan, Y., Zhang, J., 2017b. Assessing terrestrialwater storage and flood potential using GRACE data in the Yangtze River Basin. China, Remote Sensing 9 (10). Available from: https://doi.org/10.3390/rs9101011.
Sun, Z., Zhu, X., Pan, Y., Zhang, J., Liu, X., 2018. Drought evaluation using the GRACE terrestrial water storage deficit over the Yangtze River Basin, China. Sci. Total Environ. Available from: https://doi.org/10.1016/j.scitotenv.2018.03.292.
Swenson, S., Wahr, J., 2006. Post-processing removal of correlated errors in GRACE data. Geophys. Res. Lett. 33 (8).
Swenson, S., Wahr, J., 2009. Monitoring the water balance of Lake Victoria, East Africa, from space. J. Hydrol. 370 (1-4), 163–176.
Syed, T.H., Famiglietti, J.S., Rodell, M., Chen, J., Wilson, C.R., 2008. Analysis of terrestrial water storage changes from GRACE and GLDAS. Water Resour. Res. 44 (2).
Tangdamrongsub, N., Ditmar, P.G., Steele-Dunne, S.C., Gunter, B.C., Sutanudjaja, E. H., 2016. Assessing total water storage and identifying flood events over Tonlé Sap basin in Cambodia using GRACE and MODIS satellite observations combined with hydrological models. Remote Sensing Environ. 181, 162–173.
Tapley, B.D., Bettadpur, S., Ries, J.C., Thompson, P.F., Watkins, M.M., 2004. GRACE measurements of mass variability in the Earth system. Science 305 (5683), 503–505.
Thomas, A.C., Reager, J.T., Famiglietti, J.S., Rodell, M., 2014. A GRACE-based water storage deficit approach for hydrological drought characterization. Geophys. Res. Lett. 41 (5), 1537–1545.
Wada, Y., van Beek, L.P., van Kempen, C.M., Reckman, J.W., Vasak, S., Bierkens, M. F., 2010. Global depletion of groundwater resources. Geophys. Res. Lett. 37 (20).
Wahr, J., Molenaar, M., Bryan, F., 1998. Time variability of the Earth's gravity field: hydrological and oceanic effects and their possible detection using GRACE. J. Geophys. Res. Solid Earth 103 (B12), 30205–30229.
Wahr, J., Swenson, S., Velicogna, I., 2006. Accuracy of GRACE mass estimates. Geophys. Res. Lett. 33 (6).
Wang, G., Wang, D., Trenberth, K.E., Erfanian, A., Yu, M., Bosilovich, M.G., et al., 2017a. The peak structure and future changes of the relationships between extreme precipitation and temperature. Nat. Clim. Change 7 (4), 268.
Wang, S., Zhou, F., Russell, H.A., 2017b. Estimating snow mass and peak river flows for the mackenzie river basin using grace satellite observations. Remote Sensing 9 (3), 256.
Wang, S.Y., Hakala, K., Gillies, R.R., Capehart, W.J., 2014. The Pacific quasi-decadal oscillation (QDO): an important precursor toward anticipating major flood events in the Missouri River Basin? Geophys. Res. Lett. 41 (3), 991–997.
Watkins, M.M., Wiese, D.N., Yuan, D.N., Boening, C., Landerer, F.W., 2015. Improved methods for observing Earth's time variable mass distribution with GRACE using spherical cap mascons. J. Geophys. Res. Solid Earth 120 (4), 2648–2671.
Yi, H., Wen, L., 2016. Satellite gravity measurement monitoring terrestrial water storage change and drought in the continental United States. Sci. Rep. 6, 19909.
Yirdaw, S.Z., Snelgrove, K.R., Agboma, C.O., 2008. GRACE satellite observations of terrestrial moisture changes for drought characterization in the Canadian Prairie. J. Hydrol. 356 (1-2), 84–92.
Zaitchik, B.F., Rodell, M., Reichle, R.H., 2008. Assimilation of GRACE terrestrial water storage data into a land surface model: results for the Mississippi River basin. J. Hydrometeorol. 9 (3), 535–548.

Zehe, E., Becker, R., Bárdossy, A., Plate, E., 2005. Uncertainty of simulated catchment runoff response in the presence of threshold processes: role of initial soil moisture and precipitation. J. Hydrol. 315 (1-4), 183–202.
Zhao, M., Velicogna, I., Kimball, J.S., 2017. Satellite observations of regional drought severity in the continental United States using GRACE-based terrestrial water storage changes. J. Clim. 30 (16), 6297–6308.
Zhou, H., Luo, Z., Tangdamrongsub, N., Wang, L., He, L., Xu, C., et al., 2017. Characterizing drought and flood events over the Yangtze River Basin using the HUST-Grace2016 solution and ancillary data. Remote Sensing 9 (11). Available from: https://dx.doi.org/10.3390/rs9111100.
dutt Vishwakarma, B., Jain, K., Sneeuw, N., Devaraju, B., 2013. Mumbai 2005, Bihar 2008 flood reflected in mass changes seen by GRACE satellites. J. Indian Soc. Remote Sensing 41 (3), 687–695.

Further reading

Chen, J., Famiglietti, J.S., Scanlon, B.R., Rodell, M., 2016. Groundwater storage changes: present status from GRACE observations. Surveys Geophys. 37 (2), 397–417.
Rippey, B.R., 2015. The US drought of 2012. Weather Clim. Extrem. 10, 57–64.
Svoboda, M., LeComte, D., Hayes, M., Heim, R., Gleason, K., Angel, J., et al., 2002. The drought monitor. Bull. Am. Meteorol. Soc. 83 (8), 1181–1190.
Wang, S., Russell, H.A., 2016. Forecasting snowmelt-induced flooding using GRACE satellite data: a case study for the Red River watershed. Can. J. Remote Sensing 42 (3), 203–213.

CHAPTER THREE

Utility of soil moisture data products for natural disaster applications

Wade T. Crow
USDA ARS Hydrology and Remote Sensing Laboratory, Beltsville, MD, United States

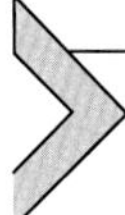

3.1 Sources of soil moisture information

3.1.1 Soil water balance modeling

Historically, soil moisture has been among the more difficult terrestrial variables to directly measure on a consistent and robust basis. Thus, the most common operational approach to estimating soil moisture has been, and remains, the use of modeling based on meteorological inputs (e.g., rainfall rates, near-surface air temperature, wind speed, and relative humidity) and various parameterizations of soil and vegetation water flux processes, to solve for soil moisture using a soil water balance approach.

These approaches vary widely in their complexity and spatial attributes; however, at their core, they are all based on a water balance formulation where soil moisture (θ) time dynamics ($d\theta/dt$) within a soil moisture depth of (Δz) are matched by the sum of fluxes of water due to precipitation (P), evapotranspiration (E_T), recharge (R), and surface runoff (Q):

$$d\theta/dt\Delta z = P - Q - E_T - R. \tag{3.1}$$

By applying constitutive relationships, Q, E_T, and R can all be written as a function of soil moisture and other (known) ancillary land surface properties. This allows Eq. (3.1) to be iteratively solved and increments of $d\theta/dt$ to be calculated over time. While complex variants of Eq. (3.1) also include subsurface lateral flow of water in response to topography (see, e.g., Hazenberg et al., 2015), most operational applications of Eq. (3.1) are one-dimensional in nature.

Extreme Hydroclimatic Events and Multivariate Hazards in a Changing Environment.
DOI: https://doi.org/10.1016/B978-0-12-814899-0.00003-1

As discussed below, variants of Eq. (3.1) appear in a wide range of hydrologic and land surface modeling systems aimed at the accurate representation of soil moisture for natural disaster applications. The horizontal resolutions of these systems vary widely, from as fine as 5–10 m for fine-resolution, hill-slope modeling (for, e.g., landslide risk assessment) to >10 km for regional-scale agricultural drought monitoring. The vertical depth (Δz) applied in balance calculations is typically on the order of 10–50 cm for drought applications aimed at characterizing soil moisture within a vertical depth containing most vegetation roots. However, finer vertical resolutions are common in rainfall-runoff applications where it is often necessary to accurately resolve the vertical propagation of precipitation wetting fronts within the soil column. In such cases, multiple vertical applications of Eq. (3.1) are integrated within a vertical stack of individual soil layers. The resulting system of coupled equations is then simultaneously solved using numerical techniques.

Critiques of Eq. (3.1) as a soil moisture estimate technique are plentiful in the literature. Some well-known deficiencies include: (1) the high degree of sensitivity in the relationship between θ and both R and $\mathrm{E_T}$ to poorly known soil hydraulic properties (Loosvelt et al., 2011), (2) difficulties in scaling up the relationship between θ and E_T for spatially heterogeneous landscapes (Crow and Wood, 2002), (3) the requirement for site-specific calibration to link θ and Q and (4) the large sensitivity of $d\theta/dt$ to errors in meteorological forcing variables (especially P) (Maggioni et al., 2012).

Despite these challenges, model-based soil moisture products remain the backbone of most operational disaster forecasting and monitoring systems, and, at a relatively coarse spatial resolution, operational examples of such products are widely available. For example, over the continental United States, near-real-time, hourly/0.125-degree soil moisture products (based on a range of land surface modeling approaches) are operationally produced by the North American Land Data Assimilation system and can be downloaded at http://www.emc.ncep.noaa.gov/mmb/nldas. Likewise, near-real-time, hourly/0.25-degree global land surface modeling products can be obtained from the Global Land Data Assimilation System at https://disc.gsfc.nasa.gov.

3.1.2 Ground observations

Due, in part, to known shortcomings in model-based soil moisture products (see above), there has been ongoing efforts to better develop and apply ground-based instrumentation within large-scale soil moisture networks. These efforts are challenged by the difficulty of directly measuring the volumetric fraction of the soil matrix filled by water (i.e., our benchmark definition of volumetric soil moisture). Therefore, ground-based soil moisture "measurements" are typically based on the actual measurement of proxy variables and the use of conversion functions to transform these measurements into true estimates of volumetric soil moisture. The gold standard of these proxy measurements are gravimetric soil moisture measurements obtained by taking the mass difference between wet and oven-dried soil and then normalizing this difference by the dry soil mass. Such measurements can then be transformed into volumetric soil moisture (suitable for comparison against model-based soil moisture estimates) via measurable values of bulk soil density.

However, in practice, the intensive labor and destructive nature of gravimetric soil moisture measurements make them difficult to sustain for long periods of time (and over large geographic regions). Instead, more intensive ground sampling typically relies on capacitance or thermal probes which measure the dissipation of an electric field or heat source within the soil medium and then relate this dissipation to the presence of water in the soil. Semipermanent soil moisture probes based on this concept represent the most common type of soil moisture instrumentation and comprise the backbone of most ground-based soil moisture networks. Ideally, these networks should be calibrated against gravimetric soil moisture observations (Cosh et al., 2006, 2008). There has also been substantial recent progress in the systematic collection, quality control, and dissemination of ground-based soil moisture datasets from these networks. Most notably through the establishment of the International Soil Moisture Network (Dorigo et al., 2011, 2013).

However, an inherent problem with any probe (or shovel-based) soil moisture measurement technique is that they sample only a very small volume of soil (typically on the order of $1-10\ cm^3$). Soil moisture commonly exhibits a large amount of unstructured spatial variability—even at very-fine linear spatial scales (<1 m). The result being that it is often difficult to upscale isolated, ground-based soil moisture measurements to coarser geographic extents. While soil moisture upscaling solutions have been proposed (Crow et al., 2012), the most robust approach appears to be the development of laterally dense ground-based soil moisture networks capable of spatially averaging across this extensive fine-scale variability (Colliander et al., 2017). Efforts to construct such networks has been aided by the development of new wireless downlink capabilities and reductions in the cost of individual data probes. In these networks, simple spatial averaging across multiple sensors has proven adequate for the validation of coarse-resolution ($\sim$30 km) soil moisture products estimates generated via remote sensing (Colliander et al., 2017).

Besides increasing the spatial density of ground-based measurements, a second possibility for effectively upscaling ground-based measurements is the use of alternative measurement approaches capable of averaging soil moisture conditions over a larger sampling volume. Examples of such technologies include the passive measurement of cosmic ray neutrons (Desilets et al., 2010), the use of ground-based instrumentation to passively measure surface reflectometry of signals from Global Position System (GPS) satellites (Larson et al., 2008) and the one-dimensional measurement of soil temperature gradients associated using either active or passive fiber optics (Sayde et al., 2010). While these techniques are capable of estimating mean soil moisture across a broader spatial area (often up to 10,000 m^2), they provide limited vertical resolution of soil moisture and, in the case of cosmic ray and GPS, are subject to temporal variations in the vertical support of their soil moisture estimates. Nevertheless, the cosmic ray and GPS approaches have become sufficiently mature such that they are currently being utilized in satellite soil moisture validation programs.

3.1.3 Remote sensing

The past two decades have seen rapid growth in the development and accessibility of soil moisture data products derived from satellite remote sensing. A wide variety of remote sensing approaches can potentially be applied to the estimation of soil moisture; however, the most advanced approaches are based on microwave remote sensing. Within the microwave region (1–10 GHz) of the electromagnetic spectrum, the emission properties of soil are sensitive to the presence (or absence) of water in the near surface. Wet soil is both more reflective and a less efficient greybody emitter of thermal terrestrial radiation. This physical principal forms the basis of approaches to measure soil moisture content using either remote passive and active microwave observations obtained from satellite-based sensors.

Passive microwave remote-sensing techniques are based on measuring the natural thermal microwave emission of the land surface and, via comparisons with estimates of physical land surface temperature, calculation of the effective land surface emissivity. Wet surfaces tend to have a lower emissivity, and thus via Kirkhoff's law of thermal radiation, higher reflectivity. Since the cold background sky emits very little microwave radiation relative to the (much warmer) terrestrial surface, a highly reflective land surface condition leads to a sharp reduction in microwave radiation detected by a downward-pointing microwave radiometer. Therefore, for a fixed physical temperature, reduced microwave radiation (or "brightness temperature") is associated with a relatively wetter surface. The relationship between soil moisture and surface temperature is also impacted by ancillary soil texture, surface roughness and vegetation characteristics. Passive soil moisture retrieval algorithms attempt to characterize, and therefore correct for, these ancillary characteristics and isolate the soil moisture component of the overall brightness temperature signal.

Extensive field and airborne instrumentation conducted in the 1980s and 1990s established that the L-band (1–2 GHz) microwave channel is optimal for soil moisture retrieval due to its relatively low sensitivity to both vegetation and atmospheric effects. Within the larger L-band, the 1.4 GHz subband has been specifically targeted due to its protected status for scientific research. However, relative to higher frequency channels commonly used for atmospheric and ocean surface research, the lower-frequency (and thus longer wavelength) L-band also poses a challenge with regards to lateral resolution. To date, the spatial resolution of operationally available, satellite-based soil moisture products has remained coarser than ~30 km. While not optimal for soil moisture estimates, (shorter-wavelength) X-band and C-band channels have also proven to be valuable for soil moisture retrieval and can be obtained at a slightly finer spatial resolution. Microwave-based soil moisture products also generally reflect soil moisture conditions within only the top few centimeters of the soil column. L-band retrievals, for example, are assumed to represent a vertically integrated measure of soil moisture content within the top 5 cm of the soil column. The vertical depth of X- and C-band retrievals is even shallower and assumed to capture only the top 1–3 cm.

Active microwave (i.e., radar) retrieval techniques share essentially the same physical basis as passive microwave techniques—with the key difference that the enhanced reflectivity of wet surfaces leads to higher levels of backscattering measurements for the satellite sensor. Active techniques also tend to be more sensitive to surface roughness and vegetation characteristics (since the radar signal must penetrate the canopy twice to complete its pathway to and from the remote sensor). However, even simple scatterometer-based systems have been successfully applied to soil moisture change detection (Naeimi et al., 2009). In addition, active techniques can be combined with Synthetic Aperture Radar (SAR) techniques, which use the motion of the sensor platform (i.e., the airplane or satellite) to effectively increase the size of the microwave antennae and thus improve the retrieval's spatial resolution. Using such techniques, satellite-based SAR radars can achieve resolutions on the order of tens of meters. However, such fine spatial resolution is generally combined with sparse or irregular temporal sampling of a single given site—limiting their value for real-time hazard monitoring.

A range of existing satellite-based products is currently available which utilize these properties to produce retrospective and near-real-time surface soil moisture products. The operational generation of satellite-based surface soil moisture products began in earnest in 2002 with the deployment of the Advanced Microwave Scanning Radiometer – Earth Observing System (AMSR-E) onboard the NASA Aqua satellite. AMSR-E utilized X- and C-band observations to retrieve a ~40–60 km soil moisture product using multiple algorithms. However, as noted above, the slightly higher frequencies available from AMSR-E (relative to the optimal L-band) have limited the performance of these products.

The L-band era of remote sensing observations began in earnest with the 2010 launch of the European Space Agency's Soil Moisture Ocean Salinity (SMOS) mission (Kerr et al., 2010). SMOS utilizes an array of antennae to passively measure L-band brightness temperature at a spatial resolution or ~45 km and a 2–3 day temporal repeat interval. Continuous SMOS soil moisture products are available since 2010 and currently posted at ~1-day latency (see, e.g., https://www.catds.fr). A second L-band mission concept, the NASA Soil Moisture Active/Passive (SMAP) mission was launched in 2015 (Entekhabi et al., 2010; Chan et al., 2018). As indicated by its name, the SMAP mission concept was based on the simultaneous acquisition of both active and passive L-band microwave observations. The SMAP radar failed in July 2015; however, the L-band portion of the mission continues to function well and has been producing an operational 36-km soil moisture product continuously since March 31, 2015. Initial validation results for SMAP soil moisture data products have been highly encouraging (Colliander et al., 2017; Chan et al., 2018; Chen et al., 2018). In particular, the 5-year temporal gap between the SMAP and SMOS launches gave SMAP mission developers the opportunity to address unexpected L-band radio frequency inference (RFI) found by SMOS over Europe and Asia (Mohammed et al., 2016). SMAP soil moisture and brightness temperature data products are available for download at https://nsidc.org/data/smap/smap-data.html.

In terms of active sensors, the C-band Advanced Scatterometer (ASCAT) instrument on board the EUMETSAT Metop-A satellite has been producing a continuous relative saturation (i.e., soil moisture normalized by porosity) product since 2008. Relative to C- and X-band retrievals products by the (passive) AMSR-E sensor, ASCAT soil moisture products have demonstrated unexpected precision in moderately-vegetated areas (Draper et al., 2013). Retrospective integration efforts for AMSR-E, ASCAT, and SMOS soil moisture products are ongoing as part of the ESA Soil Moisture Climate Change Initiative. Combined with (less accurate and frequent) soil moisture products produced in the between 1979 and 2001, a continuous soil moisture data product has been generated from the present back to 1979 and is available for download at http://www.esa-soilmoisture-cci.org/.

As noted above, the spatial resolution of satellite soil moisture products can be improved by several orders of magnitude via the use of SAR systems. However, the development of consistent, high-resolution soil moisture products using SAR-based soil moisture products has been slow. Existing L-band SAR systems (e.g., JAXA PALSAR) do not offer frequent enough temporal sampling for operational or forecasting applications. More regular imaging is available for X- and C-band systems (e.g., the Canadian RADARSAT series); however, these systems do not provide the ideal (i.e., L-band, multipolarimetric) class of measurements optimal for the soil moisture retrieval. Nevertheless, a wealth of (temporally sporadic) high-resolution soil moisture maps is available from existing aircraft and satellite platforms. As described below, such maps are potentially highly useful for flood and landslide risk mapping. In addition, while the temporal availability of SAR systems is not currently sufficient for regular monitoring and forecast applications, the next generation of planned SAR satellites (e.g., the Argentinean SAOCOM missions and the NASA/India SAR (NISAR) missions) have the potential to provide multipolarimetric, L-band observations at temporal repeat frequencies (i.e., on the order of 5–15 days) consistent with the need for ongoing, high-resolution monitoring of disaster prone regions.

3.1.4 Land data assimilation

A common theme discussed above is the inability of any single source of soil moisture information to meet the broad needs of the natural disaster community in terms of accuracy, temporal repeat, vertical sampling and horizontal resolution. It is therefore natural that considerable efforts have been focused on the development of data assimilation techniques capable of merging multiple, independent sources of soil moisture information into a single, unified analysis. The most common form for these techniques is to employ a continuous, soil water balance model (see Section 3.1.1) to provide background soil moisture predictions and then apply occasional updates to model-based state predictions (of, e.g., soil moisture and/or soil temperature) at points in time when remote sensing (or ground-based) observations become available. This process, referred to as sequential data assimilation, provides a means of correcting random errors in the background model due to, for example, random uncertainty in rainfall forcing data fed into the water balance surface model. Likewise, it allows for an efficient means of interpolating time and space gaps within satellite soil moisture products, and, when based on sophisticated (i.e., Kalman filter-based) approaches, a means for vertically extrapolating superficial (i.e., top 5-cm) soil moisture estimates deeper into the subsurface using vertical soil water processes represented in a multilayer soil water balance model.

These approaches have become refined enough to form the basis of several operational soil moisture products. For example, the SMAP mission produces Level 4 Surface and Root-zone Soil Moisture (SMAP_L4SM) products based on the assimilation of SMAP L-band brightness temperature observations into a 9-km, global implementation of the NASA Catchment land surface model. As such, the SMAP_L4SM soil moisture product provides global, 9-km, hourly analysis of both surface (0–5 cm) and root-zone (0–100 cm) at a data latency of 2–3 days (Reichle et al., 2017a,b). This product is available for download at https://nsidc.org/data/SPL4SMAU. Likewise, the European Center for Medium-range Weather Forecasting (ECMWF) and EUTMETSAT produces a separate 25-km, daily root-zone soil moisture analysis (with a data latency of <12 hours) based on the assimilation of ASCAT surface saturation retrievals into the ECMWF H-TESSEL model. This product (referred to "SM-DAS-2-H14") is available for viewing at http://hsaf.meteoam.it/soil-moisture.php.

In addition to these land data assimilation techniques, retrospective soil moisture datasets are commonly produced within atmospheric reanalysis systems where remotely sensed atmospheric observations are assimilated into coupled land-atmosphere models. Prominent examples of such systems include the NASA Modern Era Retrospective-analysis for Research and Applications (MERRA), the National Center for Environmental Prediction (NCEP) reanalysis, and the ECMWF Reanalysis (ERA). Efforts have been made to improve the land surface estimates of these systems by correcting model-predicted precipitation, surface radiation, and near-surface meteorology fields using actual surface observations and then using these corrected fields to force an off-line land surface model. The NASA Global Model Assimilation Office produces an operational "MERRA-Land" soil moisture product based on this principal (Reichle et al., 2011). However, it should be stressed that, unlike the SMAP Level 4 and EUMETSAT systems described above, these systems do not yet integrate remotely sensed soil moisture retrievals.

3.2 Soil moisture applications in natural disaster forecasting and monitoring

A large body of research is dedicated to exploiting the soil moisture products outlined above in Section 3.1 for natural disaster risk assessment, forecast, and monitoring applications. Below we will review this research and summarize the potential for these applications to reach operational maturity. We will focus on the following five natural disaster types: (1) floods (Section 3.2.1); (2) droughts (Section 3.2.2); (3) heat waves (Section 3.2.3); (4) forest fires (Section 3.2.4); and (5) landslides (Section 3.2.5).

3.2.1 Floods

Soil moisture data is potentially useful for both high-resolution risk assessment mapping and the operational monitoring of background soil moisture conditions for short-term flood forecasting. For example, prestorm soil moisture conditions play a key role in determining the efficiency with which incident rainfall is converted into runoff. Currently, the monitoring of prestorm soil moisture conditions is typically based on the application of soil water balance models (see Section 3.1.1 above) driven by observed rainfall and estimates of potential evapotranspiration (derived from observations of meteorological variables). High soil moisture conditions indicate a reduced soil capacity to infiltrate future rainfall accumulation. Therefore, flash flood risk assessment is based on overlaying spatially distributed quantitative prediction forecasts obtained from a numerical weather prediction (NWP) model with a spatial map of current soil moisture conditions obtained from soil water balance considerations. Areas exhibiting a combination of both elevated soil moisture conditions and large amounts of forecasted precipitation are assumed to be at risk of near-term flash flooding.

Recent work has verified that the assimilation of L-band soil moisture retrievals into these water balance approaches can significantly improve the ability of prestorm soil moisture estimates to predict the runoff efficiency found in the next storm event. For example, Fig. 3.1, taken from Crow et al. (2017), plots the observed rank coefficient of determination (R^2) sampled between prestorm soil moisture and storm-scale runoff efficiency (i.e., the ratio of precipitation volume converted into streamflow during a storm event).

Plotted results are based on the average R^2 values sampled within 16 medium-scale (2,000–10,000 km^2) basins in the United States Southern Great Plains region between April 2015 and May 2017. Results are plotted for a range of minimum daily rainfall thresholds used to define an individual storm event. The figure reveals a tendency for SMAP L4 soil moisture estimates (based on the assimilation of SMAP L-band observations into a land surface model) to correlate more strongly with storm-scale runoff efficiency than competing soil moisture products derived either from land surface modeling alone or older remote sensing products. This advantage grows slightly when larger storm events are considered. Crow et al. (2017) interpreted this enhanced correlation as being due to the improved precision of the SMAP L4 estimates relative to other surface soil moisture products. This underscores the operational potential of SMAP L4 products (and soil moisture data assimilation products in general) for improving our ability to monitor land surface conditions associated with increased flash flood risk.

Along the same lines, remotely sensed soil moisture has proven also useful for the direct specification of parameters for the popular curve number approach for partitioning rainfall into its infiltration and runoff components (Massari et al., 2014). These results suggest a role for

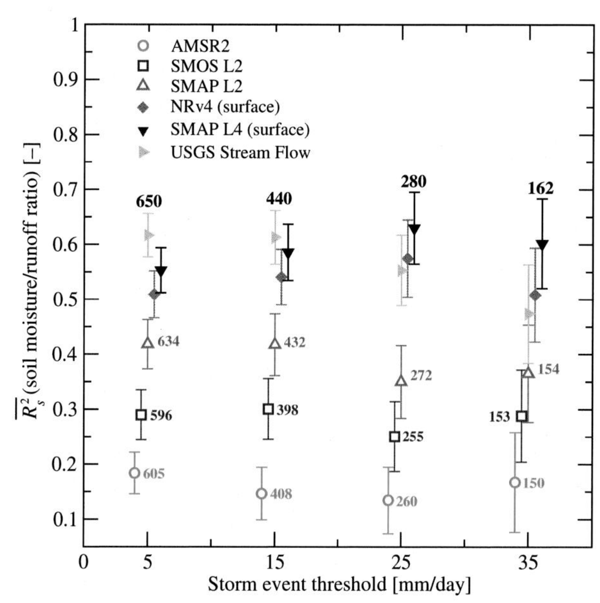

Figure 3.1 Spearman's rank coefficient of determination (R_s^2) between prestorm soil moisture and storm-scale runoff ratio (averaged across 16 basins in the United States Southern Great Plains) versus storm event precipitation accumulation threshold for a range of soil moisture products (plus antecedent streamflow observations). Error bars represent 1σ sampling uncertainty. SMOS and SMAP L2 represent remote sensing only products. NRv4 is a model-only soil moisture product, and SMAP L4 is based on the assimilation of SMAP observations into the NRv4 model. *Taken from Crow, W.T., Chen, F., Reichle, R.H., Liu, Q., 2017. L-band microwave remote sensing and land data assimilation improve the representation of pre-storm soil moisture conditions for hydrologic forecasting. Geophys. Res. Lett. 44. 10.1002/2017GL073642. 2017.*

remotely sensed soil moisture in providing improved antecedent soil moisture estimates for operational hydrologic modeling—particularly in poorly instrumented basins where it is challenging to force and/or calibrate a soil water balance model. Similar approaches have also been advocated for ground-based soil moisture observations (Section 3.1.2); however, their coverage tends to be restricted to heavily instrumented areas where water balance models already perform relatively well.

In large basins, high-quality rainfall and/or streamflow observations in upstream areas can be leveraged for improved flood forecasts in downstream areas. However, in cases where such observations are not operationally available from either rain or stream gauges, the inversion of soil moisture time series provides a means to improve the quality of rainfall information available for hydrologic routing (Román-Cascón et al., 2017;

Koster et al., 2018). Such approaches can be designed in tandem with a data assimilation approach aimed at improving antecedent soil moisture conditions—such that remotely sensed soil moisture is simultaneously leveraged to correct both prestorm soil moisture conditions and within-storm rainfall accumulations (Chen et al., 2014). However, important details regarding the temporal data latency at which these approaches can provide corrected rainfall have yet to be resolved. In addition, these approaches are subject to surface saturation effects that degrade their ability to detect very intense precipitation events of primary concern for flood forecast applications.

In smaller basins (and/or in larger basins at greater forecast lead times) flood forecast skill depends greatly on the quality of quantified precipitation estimates (QPF) products generated by numerical weather prediction. There is evidence that such products can be enhanced via the initialization of NWP models using improved soil moisture initial conditions. For example, the predictability of regional-scale flood events can potentially be improved if NWP forecasts are initialized using realistic soil moisture spatial patterns (Li et al., 2009). At finer spatial scales, there is anecdotal evidence that proper NWP representation of high-intensity precipitation in flash flood events requires the accurate representation of prestorm soil moisture spatial variability. Operational efforts to harness this predictability have primarily focused on the use of observed precipitation to drive soil water balance models—as opposed to water balance predictions driven by NWP precipitation forecasts. However, there are ongoing efforts at major NWP centers to also assimilate satellite-based soil moisture retrievals (or, more precisely, the L-band brightness temperature products from which these retrievals are derived) into the land surface model component of NWP forecast systems (see, e.g., Carrera et al., 2015).

In addition to flood forecasting, post-event mapping of inundated areas and pre-event mapping of flood plain risk zones is also critically important. The resolution requirements of such mapping is often difficult to meet. Indeed, the only existing viable soil moisture mapping tool for this purpose appears to be the use of satellite and airborne SAR imaging. For example, high-resolution SAR imagery is useful for mapping high-soil moisture areas along stream channels that are indicative of flood plains and thus at higher risk of future flooding. Likewise, SAR imaging is an excellent tool for the post-event mapping of saturated and inundated areas (Clement et al., 2018).

3.2.2 Droughts

Droughts come in a variety of types—each associated with its own definition and characteristics. Due to its close connection to soil moisture availability, we will focus here on agricultural drought, which is defined as the inability of soil water storage to meet minimal plant functional requirements. Therefore, agricultural drought is unique among the natural disaster types examined here in that soil moisture measurements *directly* describe the actual manifestation of the natural disaster. Estimates of vegetation health and biomass derived from visible/near-infrared (VIS/NIR) remote sensing also play an important role in agricultural drought monitoring. However, it is worth noting that the response of VIS/NIR vegetation indices to agricultural drought is typically lagged in time (on the order of weeks to months) with respect to the (detectable) onset of a low soil moisture anomaly (Bolten and Crow, 2012). Therefore, low soil moisture is a critical leading indicator of agricultural drought conditions and has a unique role in the design of drought early warning systems.

Such agricultural drought monitoring is most effective when soil moisture estimates can be placed in the context of a long-term historical data record. Such records allow for the calculation of soil moisture anomalies (i.e., differences relative to a climatological expectation for a particular day of the year), as opposed to absolute soil moisture values. Water use requirements for crops and agricultural management vary significantly throughout the growing season (and across crop types). Therefore, the calculation of climatological anomalies provides a means for end users to compare current conditions to typical conditions and facilitate improved decision support. In addition, long-term data records allow current conditions to be directly compared to past historical events (where the impact on agricultural production is already known). Such historical comparisons between existing drought conditions and reference conditions in the near-past represent an important analysis strategy in agricultural drought monitoring. This emphasis on the development of long-term products can be challenging for soil moisture product developers given the known tendency for different soil moisture estimation techniques (overviewed in Section 3.1) to produce contrasting soil moisture climatologies for the same location. As a consequence, absolute soil moisture observations obtained from a remote sensing source must typically be rescaled (via climatological considerations) before they can be directly compared to soil moisture products acquired from ground-based instrumentation or a land surface model. Likewise, remote sensing data records acquired from multiple sensors and/algorithms must be rescaled to some common climatology prior to their merger into a unified product. The functional form of such rescaling is relatively straightforward. However, deriving these corrections requires that each soil moisture product is available over a sufficiently long historical period to facilitate the robust sampling of a product-specific climatology. In practice, this data length requirement is often problematic and mutual rescaling is difficult for products which do not significantly overlap in time.

Model-based soil moisture anomaly products have been widely used both for both the retrospective characterization of historical droughts, as well as the real-time monitoring of existing droughts. For example, the Princeton University African Flood and Drought monitoring system (http://stream.princeton.edu/AWCM/WEBPAGE/interface.php?locale = en) utilizes long-term soil water balance modeling to calculate a normalized drought anomaly index which classifies current drought conditions relative to historical context. For the United States, the NLDAS-2 project generates daily soil moisture predictions using multiple land surface model and near-real-time forcing data observations. The availability of these modeling products back to 1979 provides important historical context. Research has also demonstrated the ability of these products to characterize root-zone soil moisture deficits associated with agricultural drought in data-poor areas. For example, within the developing world, the Famine Early Warning System Network (FEWS-NET) routinely generates a real-time soil moisture product using available ground- and satellite-based meteorological data.

Given that agricultural drought is defined as a lack of water within the vertical root-zone of crops, the limited vertical penetration depth of microwave-based satellite soil moisture products is commonly cited as a severe obstacle to their use in agricultural drought monitoring activities. However, the assimilation of surface soil moisture information into soil water balance models has been noted to significantly improve the ability of root-zone soil moisture estimates to characterize future variations in vegetation health and biomass (Bolten and Crow, 2012). For example, the USDA Foreign Agricultural Service (USDA FAS) provides its analysts with a global root-zone soil moisture product generated via the assimilation of SMOS and SMAP soil moisture retrievals into their global soil water balance model. Their analysts, in turn, consider this information when assessing global crop condition, growth stage and expected productivity. This data assimilation approach is particularly useful in areas of the world where water balance modeling is hampered by large errors in observed precipitation associated (associated with sparse ground-based rain gauge coverage). Since data-poor areas tend to overlap with food-insecure regions, the approach is potentially useful for famine relief efforts as well. Fig. 3.2, taken from Bolten and Crow (2012), describes the (nationally-averaged) correlation between this month's soil moisture anomaly and Normalized Difference Vegetation Index (NDVI) values several months on into the future. Results in the figure are shown for a range of food-insecure countries and for the cases of both root-zone soil moisture anomalies extracted from both a soil water balance model ("Model-only") and the assimilation of AMSR-E soil moisture retrievals into the same model ("Model + LPRM"). While AMSE-E does not

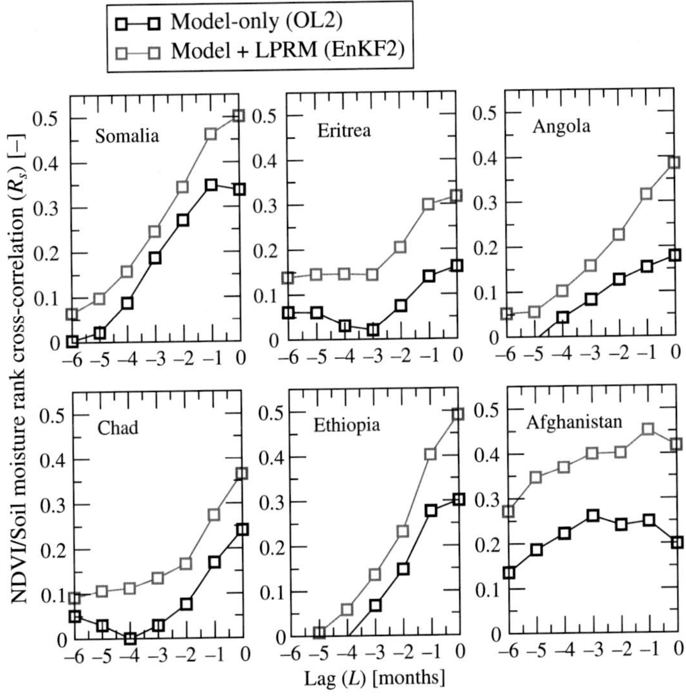

Figure 3.2 The impact of AMSR-E (LPRM) data assimilation on the lagged-rank correlation between soil moisture and NDVI anomalies (in the case where soil moisture precedes NDVI by L months). Listed countries are among the most food-insecure (and data-poor) in the world. *Taken from Bolten, J.D., Crow, W.T., 2012. Improved prediction of quasi-global vegetation conditions using remotely-sensed surface soil moisture. Geophys. Res. Lett. 39. L19406. 10.1029/2012GL053470.*

directly measure root-zone soil moisture anomalies, their assimilation into a water balance model significantly improves the ability of the resulting root-zone soil moisture estimates to anticipate future NDVI anomalies. This demonstrates that the assimilation of remotely sensed soil moisture retrievals adds skill to the forecasting of drought impacts within food-insecure countries.

Likewise, comparisons to operational county-level (~50–100 km) yield data in the United States, suggest that such land data assimilations can add modest predictive yield to the in-season forecasting of corn and soybean yield anomalies (Mladenova et al., 2017). The examples given above are all based on the assimilation of remotely sensed surface soil moisture into a water balance model with static vegetation considerations. Assimilation directly into dynamic crop growth models (where vegetation growth is estimated over time) is currently an active area of research (Ines et al., 2013). Eventual operational versions of these approaches will likely involve the simultaneous assimilation of both surface soil moisture and satellite-derived leaf area index estimates. However, preliminary results on the net impact of soil moisture data assimilation alone in such systems have been mixed.

It should be noted that soil moisture measurements can be used for drought forecasting without the implementation of a data assimilation system. For example, temporal changes in satellite-based soil moisture can help drought monitors track dynamic changes in drought intensity. Based on this principle, the United States Drought Monitoring system is currently ingesting SMAP surface soil moisture retrievals into their operational drought mapping system to better characterize the temporal evolution of existing drought conditions. In parallel to these remote sensing efforts, there have been extensive (local and state-level) efforts in the United States to deploy state-level "mesonet" networks which often include profile soil moisture observations. Past studies have established the value of these networks to characterize the extent and severity of agricultural drought (Ford et al., 2015). These efforts have been particularly notable for networks such as the Oklahoma Mesonet network that feature a well-developed, dense and mature network of profile soil moisture sensors. Given the generally large-scale nature of agricultural drought events, the spatial sampling density of many state-level mesonets are adequate to the capture regional drought events and efforts are ongoing to unify existing US-based state networks into a unified, national-level system. There have also been preliminary steps made towards the development of data assimilation systems capable of simultaneously integrating (spatially continuous, low-resolution and low accuracy) satellite-based products with (spatially sporadic, high-resolution and high-accuracy) ground-based observations into a unified root-zone soil moisture analysis (Gruber et al., 2018). Such systems arguably represent the future of efforts to fully leverage existing soil moisture observations for the real-time tracking of root-zone soil moisture anomalies.

As with positive precipitation anomalies associated with floods, the initialization of short-term and season NWP models has been shown to aid incrementally in the forecasting of dry/warm conditions associated with the development and intensification of agricultural drought (Koster et al., 2011). As noted above, this information is commonly provided by initializing the land surface component of NWP modeling systems with observed, as opposed to forecasted, precipitation data fields. However, efforts are underway to improve the seasonal forecasting of drought via the assimilation of surface soil moisture retrievals.

3.2.3 Heat waves

There is significant overlap between surface and lower atmospheric conditions (e.g., anticyclonic circulation) that are indicative of agricultural drought and extreme heat waves. However, it is generally assumed that soil moisture conditions have a stronger impact on future air temperature anomalies than on analogous precipitation anomalies (Koster et al., 2011). Consequently, soil moisture is considered to play a relatively larger role in the development and intensification of heat waves than drought. Recent work has shed light on the role of soil moisture in the local intensification of heat waves and evidence exists for a link between dry soil moisture and anomalously high near-surface air temperature (Miralles et al., 2014). To date, much of this evidence has been based on the calculation of soil moisture proxies (such as normalized precipitation indices) rather than actual soil moisture estimates. Indeed, it has been suggested that satellite-based soil moisture products lack sufficient vertical sampling depth to adequately describe the coupling between soil moisture and near-surface air temperature extremes (Hirschi et al., 2014). However, these conclusions need to be revisited for newer, more-accurate soil moisture products generated by the L-band SMOS and SMAP satellite missions.

3.2.4 Fire

Overall bulk surface (i.e., soil plus liter plus vegetation) dryness is a critical risk assessment variable for the short-term assessment of forest fire risk.

Agricultural drought assessment derived from in situ and/or water balance calculations can be used to assess overall surface dryness within forests and assess fire risk (Krueger et al., 2015). There has also been recent interest in the integration of remotely-sensed soil moisture products into these assessments (Bartsch et al., 2009). While some remote sensing approaches struggle to accurate separate microwave signals due to vegetation water content versus soil moisture, this distinction is less critical for fire risk applications—since the primary valuable of interest is the integrated dryness of the bulk surface (i.e., the combination of soil plus vegetative plus subcanopy liter). Soil moisture assessments can be paired with microwave-based assessments of vegetation water content (also available from microwave remote sensing) to provide this type of integrated water content assessment.

However, given the spatial patchiness of many forested landscapes, the coarse spatial resolution of satellite-based soil moisture products represents a more serious obstacle. Consequently, forest fire applications are often associated with attempts to downscale soil moisture estimates using finer-scale observations of surface temperature and vegetation density. These efforts have reached operational status using downscaled SMOS soil moisture data products in Spain (Chaparro et al., 2016). Within the United States, the deployment of ground-based instrumentation is increasingly moving into forested areas—with the expectation that in situ soil moisture (and related vegetation water content) measurements will eventually be used for operational forest fire risk assessment.

3.2.5 Landslides

The amount of soil water within the vertical soil column strongly impacts the ability of soil to withstand shear stress and is therefore a key variable for assessing landslide risk. These assessments are often based on observed antecedent rainfall and basic water balance considerations. In addition, despite the challenges inherent in retrieving and/or extrapolating soil moisture within complex topography, ground-based and remotely sensed soil moisture estimates have been shown to contribute to the assessment of regional landslide risk (Brocca et al., 2016). As with fire risk assessment, landslide applications of satellite soil moisture retrievals are frequently paired with downscaling techniques to better resolve individual hill slope units.

References

Bartsch, A., Baltzer, H., George, C., 2009. The influence of regional surface soil moisture anomalies on forest fires in Siberia observed from satellites. Environ. Res. Lett. 4, 45021-45021.

Bolten, J.D., Crow, W.T., 2012. Improved prediction of quasi-global vegetation conditions using remotely-sensed surface soil moisture. Geophys. Res. Lett. 39, L19406. Available from: https://doi.org/10.1029/2012GL053470.

Brocca, L., Ciabatta, L., Moramarco, T., Ponziani, F., Berni, N., Wagner, W., 2016. Use of satellite soil moisture products for the operational mitigation of landslides risk in Central Italy. In: Srivastava, P.K., Petropoulos, G.P., Kerr, Y.H. (Eds.), *Satellite Soil Moisture Retrieval*. Elsevier, pp. 231–247. Available from: http://dx.doi.org/10.1016/B978-0-12-803388-3.00012-7.

Carrera, M.L., Bélair, S., Bilodeau, B., 2015. The Canadian Land Data Assimilation System (CaLDAS): description and Synthetic Evaluation Study. J. Hydrometeor. 16, 1293–1314. Available from: https://doi.org/10.1175/JHM-D-14-0089.1.

Chan, S., et al., 2018. Development and assessment of the SMAP enhanced passive soil moisture product. Remote Sensing Environ. 204, 931–941. Available from: https://doi.org/10.1016/j.rse.2017.08.025.

Chaparro, D., Piles, M., Vall-llossera, M., 2016. Remotely sensed soil moisture as a key variable in wildfires prevention services: towards new prediction tools using SMOS and SMAP data. In: Srivastava, P.K., Petropoulos, G.P., Kerr, Y.H. (Eds.), Satellite Soil Moisture Retrieval. Elsevier, pp. 249–269. Available from: http://dx.doi.org/10.1016/B978-0-12-803388-3.00013.

Chen, F., Crow, W.T., Ryu, D., 2014. Dual forcing and state correction via soil moisture assimilation for improved rainfall-runoff modeling. J. Hydrometeorol. 15 (5), 1832–1848. Available from: https://doi.org/10.1175/JHM-D-14-0002.1.

Chen, F., Crow, W.T., Bindlish, R., Colliander, A., Burgin, M.S., Asanuma, J., et al., 2018. Global-scale evaluation of SMAP, SMOS and ASCAT soil moisture products using triple collocation. Remote Sensing Environ. 214, 1–13. Available from: https://doi.org/10.1016/j.rse.2018.05.008.

Clement, M., Kilsby, C., Moore, P., 2018. Multi-temporal synthetic aperture radar flood mapping using change detection. J. Flood Risk Manage. 11, 152–168. Available from: https://doi.org/10.1111/jfr3.12303.

Colliander, A., et al., 2017. Validation of SMAP surface soil moisture products with core validation sites. Remote Sensing Environ. 191, 215–231. Available from: https://doi.org/10.1016/j.rse.2017.01.021.

Cosh, M.H., Jackson, T.J., Starks, P.J., Heathman, G., 2006. Temporal stability of surface soil moisture in the Little Washita River Watershed and its applications in satellite soil moisture product validation. J. Hydrol. 323, 168–177.

Cosh, M.H., Jackson, T.J., Moran, S., Bindlish, R., 2008. Temporal persistence and stability of surface soil moisture in a semi-arid watershed. Remote Sensing Environ. 122, 304–313.

Crow, W.T., Wood, E.F., 2002. The value of coarse-scale soil moisture observations for regional surface energy balance modeling. J. Hydrometeorol. 3, 467–482.

Crow, W.T., Berg, A.A., Cosh, M.H., Loew, A., Mohanty, B.P., Panciera, R., et al., 2012. Upscaling sparse ground-based soil moisture observations for the validation of coarse-resolution satellite soil moisture products. Rev. Geophys. 50, RG2002. Available from: https://doi.org/10.1029/2011RG000372.

Crow, W.T., Chen, F., Reichle, R.H., Liu, Q., 2017. L-band microwave remote sensing and land data assimilation improve the representation of pre-storm soil moisture conditions for hydrologic forecasting. Geophys. Res. Lett. 44, 2017. Available from: https://doi.org/10.1002/2017GL073642.

Desilets, D., Zreda, M., Ferre, T., 2010. Nature's neutron probe: land surface hydrology at an elusive scale with cosmic rays. Water Resour. Res 46. Available from: https://doi.org/10.1029/2009WR008726W11505.

Dorigo, W.A., Wagner, W., Hohensinn, R., Hahn, S., Paulik, C., Xaver, A., et al., 2011. The International Soil Moisture Network: a data hosting facility for global in situ soil moisture measurements. Hydrol. Earth Syst. Sci. 15 (5), 1675–1698.

Dorigo, W.A., Xaver, A., Vreugdenhil, M., Gruber, A., Hegyiová, A., Sanchis-Dufau, A. D., et al., 2013. Global automated quality control of in situ soil moisture data from the International Soil Moisture Network. Vadose Zone J. Available from: https://doi.org/10.2136/vzj2012.0097.

Draper, C., Reichle, R., de Jeu, R., Naeimi, V., Parinussa, R., Wagner, W., 2013. Estimating root mean square errors in remotely sensed soil moisture over continental scale domains. Remote Sensing Environ. 137, 288–298. Available from: https://doi.org/10.1016/j.rse.2013.06.013.

Entekhabi, D., et al., 2010. The Soil Moisture Active and Passive (SMAP) Mission. Proc. IEEE 98 (5), 704–716. Available from: https://doi.org/10.1109/JPROC.2010.2043918.

Ford, T.W., McRoberts, D.B., Quiring, S.M., Hall, R.E., 2015. On the utility of in situ soil moisture observations for flash drought early warning in Oklahoma, USA. Geophys. Res. Lett. 42, 9790–9798. Available from: https://doi.org/10.1002/2015GL066600.

Gruber, A., Crow, W.T., Dorigo, W., 2018. Assimilation of spatially sparse in situ soil moisture networks into a continuous model domain. Water Resour. Res. 54, 1353–1367. Available from: https://doi.org/10.1002/2017WR021277.

Hazenberg, P., Fang, Y., Broxton, P., Gochis, D., Niu, G.-Y., Pelletier, J.D., et al., 2015. A hybrid-3D hillslope hydrological model for use in Earth system models. Water Resour. Res. 51, 8218–8239. Available from: https://doi.org/10.1002/2014WR016842.

Hirschi, M., Mueller, B., Dorigo, W., Seneviratne, S., 2014. Using remotely sensed soil moisture for land-atmosphere coupling diagnostics: the role of surface vs. root-zone soil moisture variability. Remote Sensing Environ. 154, 246–252.

Ines, A., Das, N., Hansen, J., Njoku, E., 2013. Assimilation of remotely sensed soil moisture and vegetation with a crop simulation model for maize yield prediction. Remote Sensing Environ. 138, 149–164. Available from: https://doi.org/10.1016/j.rse.2013.07.018.

Kerr, Y., et al., 2010. The SMOS mission: new tool for monitoring key elements of the global water cycle. Proc. IEEE 98 (5), 666–687.

Koster, K.D., Crow, W.T., Reichle, R.H., Mahanama, S.P., 2018. Estimating basin-scale water budgets with SMAP soil moisture data. Water Resour. Res. 54. Available from: https://doi.org/10.1029/2018WR022669.

Koster, R.D., et al., 2011. The second phase of the global land–atmosphere coupling experiment: soil moisture contributions to subseasonal forecast skill. J. Hydrometeor. 12, 805–822. Available from: https://doi.org/10.1175/2011JHM1365.1.

Krueger, E.S., Ochsner, T.E., Engle, D.M., Carlson, J.D., Twidwell, D.L., Fuhlendorf, S. D., 2015. Soil Moisture Affects Growing-Season Wildfire Size in the Southern Great Plains. Agronomy & Horticulture -- Faculty Publications, https://digitalcommons.unl.edu/agronomyfacpub/838.

Larson, K.M., Small, E.E., Gutmann, E.D., Bilich, A.L., Braun, J.J., Zavorotny, V.U., 2008. Use of GPS receivers as a soil moisture network for water cycle studies. Geophys. Res. Lett 35, L24405. Available from: https://doi.org/10.1029/2008GL036013.

Li, H., Luo, L., Wood, E., Schaake, J., 2009. The role of initial conditions and forcing uncertainties in seasonal hydrologic forecasting. J. Geophys. Res. 2009 (114), 1–10.

Loosvelt, L., Pauwels, V.R.N., Cornelis, W.M., De Lannoy, G.J.M., Verhoest, N.E.C., 2011. Impact of soil hydraulic parameter uncertainty on soil moisture modeling. Water Resour. Res. 47, W03505. Available from: https://doi.org/10.1029/2010WR009204.

Maggioni, V., Anagnostou, E.N., Reichle, R.H., 2012. The impact of model and rainfall forcing errors on characterizing soil moisture uncertainty in land surface modeling. Hydrol. Earth Syst. Sci. 16, 3499–3515. Available from: http://dx.doi.org/10.5194/hess-16-3499-2012. 2012.

Massari, C., Brocca, L., Barbetta, S., Papathanasiou, C., Mimikou, M., Moramarco, T., 2014. Using globally available soil moisture indicators for flood modelling in Mediterranean catchments. Hydrol. Earth Syst. Sci. 18, 839–853. Available from: https://doi.org/10.5194/hess-18-839-2014.

Miralles, D.G., Teuling, A.J., van Heerwaarden, C.C., Vilà-Guerau de Arellano, J., 2014. Mega-heatwave temperatures due to combined soil desiccation and atmospheric heat accumulation. Nat. Geosci. 7 (5), 345–349. Available from: https://doi.org/10.1038/ngeo2141.

Mladenova, I.E., Bolten, J.D., Crow, W.T., Anderson, M.C., Hain, C.R., Johnson, D.M., et al., 2017. Inter-comparison of soil moisture, evaporative stress and vegetation indices for estimating corn and soybean yields over the U.S. J. Selected Topics Appl. Earth Observat. Remote Sensing 10 (4), 1328–1343. Available from: https://doi.org/10.1109/JSTARS.2016.2639338.

Mohammed, P.N., Aksoy, M., Piepmeier, J.R., Johnson, J.T., Bringer, A., 2016. SMAP L-band microwave radiometer: RFI mitigation prelaunch analysis and first year on-orbit observations. IEEE Trans. Geosci. Remote Sensing 54 (10), 6035–6047. Available from: https://doi.org/10.1109/TGRS.2016.2580459.

Naeimi, V., Scipal, K., Bartalis, Z., Hasenauer, S., Wagner, W., 2009. An improved soil moisture retrieval algorithm for ERS and METOP scatterometer observations. IEEE Trans. Geosci. Remote Sens 47 (7), 1999–2013. Available from: https://doi.org/10.1109/TGRS.2008.2011617.

Reichle, R., et al., 2017a. Assessment of the SMAP level-4 surface and root-zone soil moisture product using in situ measurements. J. Hydrometeorol. 18 (10), 2621–2645. Available from: https://doi.org/10.1175/JHM-D-17-0063.1.

Reichle, R., et al., 2017b. Global assessment of the SMAP Level-4 surface and root-zone soil moisture product using assimilation diagnostics. J. Hydrometeorol. 18 (12), 3217–3237. Available from: https://doi.org/10.1175/JHM-D-17-0130.1.

Reichle, R.H., Koster, R.D., De Lannoy, G.J., Forman, B.A., Liu, Q., Mahanama, S.P., et al., 2011. Assessment and enhancement of MERRA land surface hydrology estimates. J. Clim. 24, 6322–6338. Available from: https://doi.org/10.1175/JCLI-D-10-05033.1.

Román-Cascón, C., Pellarin, T., Gibon, F., Brocca, L., Cosme, E., Crow, W.T., et al., 2017. Correcting satellite-based precipitation products through SMOS soil moisture data assimilation in two land-surface models of different complexity: API and SURFEX. Remote Sensing Environ. 200, 295–310. Available from: https://doi.org/10.1016/j.rse.2017.08.022.

Sayde, C., Gregory, C., Gil-Rodriguez, M., Tufillaro, N., Tyler, A., van de Giesen, N., et al., 2010. Feasibility of soil moisture monitoring with heated fiber Optics. Water Resour. Res. 46, W06201. Available from: https://doi.org/10.1029/2009WR007846.

CHAPTER FOUR

Water surface elevation in coastal and inland waters using satellite radar altimetry

Stefano Vignudelli[1], Andrea Scozzari[2], Ron Abileah[3], Jesús Gómez-Enri[4], Jérôme Benveniste[5] and Paolo Cipollini[6]

[1]Biophysics Institute of the National Research Council (CNR-IBF), Pisa, Italy
[2]Consiglio Nazionale delle Ricerche (CNR-ISTI), Pisa, Italy
[3]Jomegak, San Carlos, CA, United States
[4]University of Cadiz, Cadiz, Spain
[5]European Space Agency (ESA-ESRIN), Directorate of Earth Observation Programmes, EO Science, Applications and Climate Department, Frascati, Italy
[6]Telespazio VEGA UK Ltd. for ESA-ECSAT, Harwell, United Kingdom

4.1 Introduction and rationale

Changes in water surface elevation are traditionally measured in situ by single instruments at a fixed point, yielding a measurement relative to the land where the instrument is installed. Values are generally taken with frequent sampling (order of minutes or less). Such an instrument is called a *tide gauge* in the ocean environment and more generically a *water gauge* in lakes and reservoirs. In rivers, the instrument is usually called a *stream gauge*, because the water surface elevation is then used to derive discharges. Several types of installations are currently in use in hydrometric practice (Vuglinskiy et al., 2009; WMO, 2010), depending on the kind of water target.

Many databases around the world collect and make available water surface elevations. Concerning extreme events, there is a clear requirement for high-frequency (i.e., sampling at subhour intervals) information. The GESLA-2 (Global Extreme Sea Level Analysis version 2) database is an updated data set of high-frequency water surface elevations obtained from tide gauges operated by many agencies around the world (Woodworth et al., 2016). Fig. 4.1 shows the geographical distribution of the 1355 stations presently available in GESLA-2.

Extreme Hydroclimatic Events and Multivariate Hazards in a Changing Environment.
DOI: https://doi.org/10.1016/B978-0-12-814899-0.00004-3

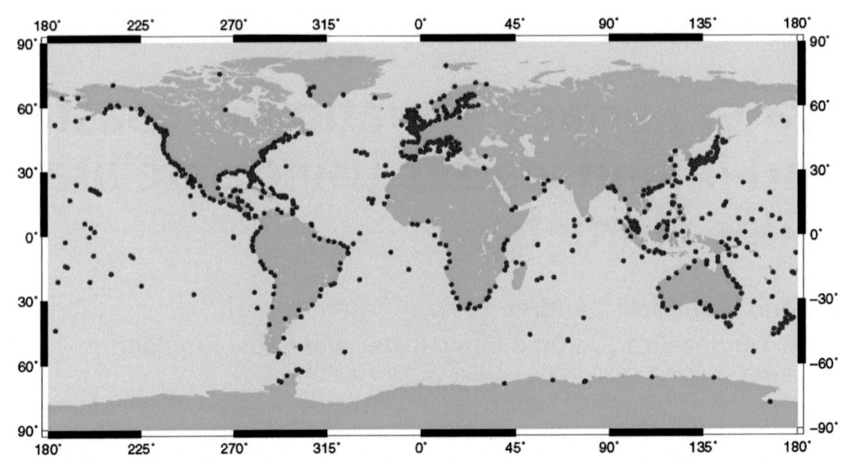

Figure 4.1 Map of tide gauge stations with water surface elevations in the GESLA-2 data set. *Data from Woodworth et al. (2016).*

Tide gauges are located in protected environments around the world's coastlines and some ocean islands and therefore are often not representative of offshore conditions. The distribution of tide gauge stations is particularly sparse in some regions (e.g., Africa). In other regions (e.g., India) GESLA-2 does not contain data, although such measurements are done. Some countries provide free access only to a limited number of tide gauges for commercial reasons or security policies. Recently, the EuroGOOS Tide Gauge working group launched a survey collecting responses from 40 organizations operating tide gauge networks across Europe and adjacent countries. The results indicated 674 tide gauge installations existing in 2016, however, 25% of these stations have problems of funding in some way for maintenance with respect to previous years (EuroGOOS, 2017).

A similar scenario is evident with reference to inland water surface elevations. Fig. 4.2 shows the distribution of stations produced by University of Texas at Austin, Center for Research in Water Resources (UT CRWR). The map contains locations of several data producers of inland water surface elevations. There are thousands of stations measuring water surface elevation shown in Fig. 4.2, at least half of which are in the United States. However, many stations have only historical data, some stations are quite old, others also contain recent data (GEOSS, 2014). The number of reporting stations reached the highest number in 1979 and decreased sharply thereafter (see Fig. 4.1 in Ruhi et al., 2018). There are

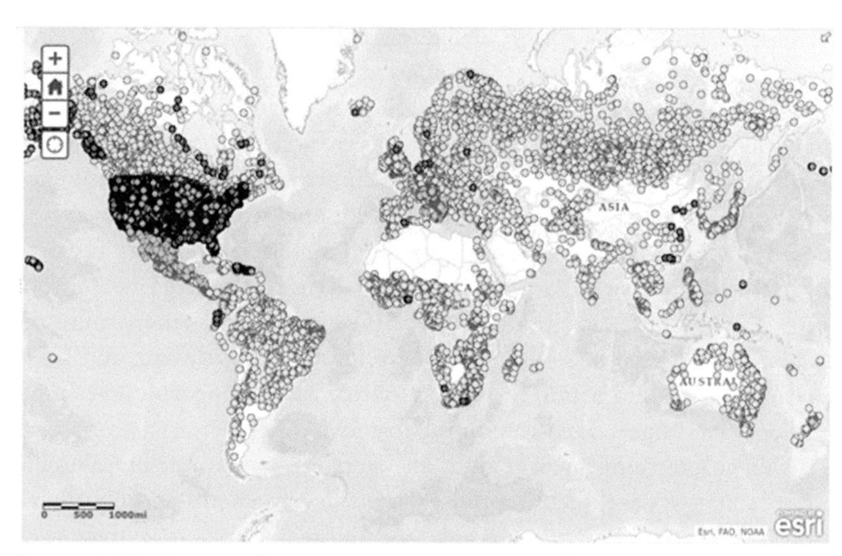

Figure 4.2 Distribution of stations measuring water surface elevations produced by University of Texas at Austin, Center for Research in Water Resources (UT CRWR). Data available at Esri ArcGIS Online web map viewer (http://bit.ly/19fUSPY) from different data producers of inland water surface elevations, for example, GRDC (yellow and violet) and USGS (blue) in US; CONAGUA in Mexico and ISPRA in Italy (green); HRC in New Zealand (cyan); INDRHI in Dominican Republic (red).

various reasons for this decrease, including the lack of funding to maintain the stations, impossibility to collect data, or restrictions on access to them for political reasons.

The advent of satellite altimetry (Fu and Cazenave, 2000) made it possible to obtain global surface elevation measurements in water targets with low temporal sampling (currently every 10 days or longer). The measurements are taken along the ground tracks, that is, the projection of the altimeter orbit on the Earth's surface. Radar altimetry has the particular advantage of providing measurements at locations where in situ data are not available due to remoteness or lack of investment. Satellite radar altimetry is also capable of characterizing, with a long-term observational data set, how water surface elevation variability evolves from the open ocean to the coastal zone and inland waters. Satellite altimetry was designed for open ocean, however, as shown later, more than a decade of Research & Development has permitted to extend the satellite-based water surface elevation records towards the coast (Vignudelli et al., 2011a,b; Cipollini et al., 2017a,b) and inland (Crétaux et al., 2017) with quality comparable to open ocean measurements.

One of the advantages of satellite radar altimetry is that measurements are collected globally in any weather conditions and regardless of the time of overpass. The coastal zone is affected by several local ocean processes, all impacting the sea surface elevation at different scales. Coastal dynamics has smaller spatial and temporal scales than open ocean and requires a monitoring at finer-scale that is difficult to satisfy with only sparse tide gauges.

The usage of altimeter data over inland waters was initially restricted to a limited number of large water targets, due to the large native footprint size of the radar altimeter. Over the recent years, the quality and quantity of data has been enhanced by refining the processing and therefore making the estimation of water surface elevations possible over much smaller water targets, as initiated by Berry et al. (2005).

This chapter aims at providing an overview of the satellite altimetry measurement system, focusing on the latest developments and possibilities offered by the various past, present, and future missions to extract water surface elevations around coasts and inland. The chapter also discusses the potential usage of these measurements in applications, which will be then explored further in the second part of the book. Section 4.2 provides an illustration of how the satellite altimetry system works. Section 4.3 shows how the water surface elevation is extracted in open ocean conditions. Section 4.4 outlines the past, present and future satellite radar altimetry missions. Section 4.5 illustrates the main processing aspects regarding the coastal zone and inland waters applications. Section 4.6 provides an overview of the data sets available for usage in the coastal zone and inland waters. The improvements in quality and quantity are shown in Section 4.7. Two case studies of extreme events are illustrated in Section 4.8 to demonstrate the benefits of using satellite-based water surface elevations. Finally, Section 4.9 summarizes the perspectives from upcoming satellite radar altimetry missions over coastal zone and inland waters and the future developments of radar altimetry in these challenging domains.

4.2 The concept of satellite radar altimetry

In satellite radar altimetry, a radar on board a satellite sends a sequence of short electromagnetic pulses to the surface of the earth by a nadir-pointed antenna. The echoes returned by the illuminated surface are collected by the antenna and forwarded to the internal processing. The carrier frequencies of the main pulses used for the measurement are in the Ku- or Ka-microwave bands, interspersed with less-frequent pulses in the C- or S-bands, which are used for atmospheric correction purposes. The pulse sequences can be continuous or organized in bursts, and the pulse repetition frequency (PRF) depends on the specifications of the adopted radar and the objectives of the mission. As an example, the RA-2 sensor (Radar Altimeter-2) on board the Envisat satellite had a PRF of about 1800 Hz, with a continuous transmission scheme. The received signal is sampled at regular intervals (called "gates" or "range bins"), with a regular sampling during a fixed-length time window. The satellite altimetry community usually calls "waveform" the sequence of samples that is collected.

The time the radar pulse takes to travel from the satellite to the illuminated surface and back again is deduced from the waveform and then converted into a range measurement, that is, the distance between the satellite antenna and the reflective surface. Over open ocean, the waveform has a characteristic shape that can be described analytically through the Brown model (Brown, 1977). The model assumes that the returned signal is a combination of elemental specular reflections uniformly distributed over the wave surface, and a wave height distribution given by the sea state.

The noisiness of each individual waveform is reduced by averaging multiple successive waveforms. Generally, until Sentinel-3 a number of waveforms were incoherently averaged on board the satellite, before being transmitted to the ground stations. With the advent of Sentinel-3, data are transmitted globally to the ground at full rate for the first time. Averaged measurements are delivered at reduced rates, typically in the range between 80 and 1 Hz, which corresponds to 7 km of flight (in the case of 1 Hz), down to less than 100 m (at 80 Hz rate) in the along-track direction.

In order to extract the relevant parameters (e.g., range, wave height, wind speed) a procedure called "retracking" is used. The procedure consists in fitting the theoretical Brown model to the collected waveforms and estimating the geophysical parameters by applying specific metrics (i.e., "estimators") during the fitting procedure (Gommenginger et al., 2011).

Fig. 4.3 shows an example of a real averaged waveform with a retracked waveform overlapped (red line). In particular, the range estimation is derived from the epoch (time delay) at mid-height of the fitted function, the signal strength is related to the backscatter coefficient, which in turn depends on the wind speed at sea surface, and the leading edge slope is affected by the significant wave height (SWH). As it will be seen later in this chapter, the retracking procedure based on the Brown model is effectively applicable only to open ocean conditions, while in the coastal zone and for inland waters other approaches are proposed, based on specific model functions, fitting estimators, and procedures.

The altimetric range estimated by retracking, needs to be further modified by several corrections, such as those needed to compensate for the variable delays introduced by the radiative transfer in the atmosphere. Range ($R_{altimeter}$) is usually corrected for the various effects with an equation of the form:

$$R_{corr} = R_{altimeter} + \sum R_{instr} + \sum R_{atmos} + \sum R_{surface}$$

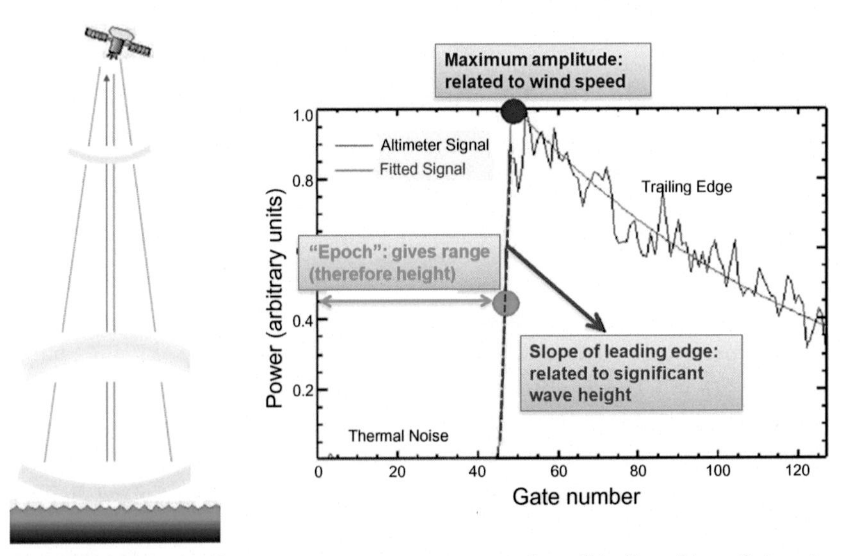

Figure 4.3 Example of 18 Hz Envisat RA-2 ocean waveform (black) and best-fitting signal after retracking (red). The figure highlights how the epoch, slope of leading edge, and maximum amplitude of the waveform are related to range (i.e., sea level height), significant wave height, and wind speed, respectively. *Adapted from from Gómez-Enri, J., Vignudelli, S., Quartly, G., Gommenginger, C., Benveniste, J., 2009. Bringing satellite radar altimetry closer to shore, SPIE Newsroom, 1-3, doi: 10.1117/2.1200908.1797.*

where:

$\sum R_{instr}$ is the correction of instrument dependent errors (e.g., biases, drifts, delays).

$\sum R_{atmos}$ accounts for the path delays due to the atmospheric refraction. In particular, there are three main components of the delay and attenuation due to the radiation transfer in the atmosphere: the first originates from dry gases, the main contributor being oxygen (O_2), the second one from water vapor (H_2O), and the third one from ionospheric electrons.

$\sum R_{surface}$ compensates for the error sources connected with the interaction between the electromagnetic signal and the reflective surface, for example, sea state bias.

4.3 Water surface elevation in the open ocean

The estimated range is not directly usable in applications, as users generally want to measure the elevation of the sea surface relative to a reference. Such quantity is called Sea Surface Height (SSH). The reference system adopted in the satellite altimetry context is usually the ellipsoid, which is a rough approximation of the Earth's surface. The computation of SSH requires independent measurements of the satellite orbital trajectory. The distance of the satellite above the ellipsoid, called altitude (H), is determined by the Precise Orbit Determination (POD), a procedure that enables the acquisition of the orbit ephemeris (i.e., the three-dimensional location of satellite's centre of mass).

Fig. 4.4 shows the satellite altimetry system. The SSH at a given instant is therefore calculated by subtracting the corrected range (R_{corr}) from the satellite altitude (H). The actual shape of the SSH results from different forcings. The Earth rotation, shape, and mass distribution contribute to what is called the geoid, that is, the equipotential gravity surface on which the water would relax if there were no other effects, which varies between -105 and $+85$ m with respect to a reference ellipsoid. External forcing from the atmosphere (winds, pressure, solar heating, precipitation) and gravitational effects from the moon and the sun (tides) generate motions of the ocean resulting in a SSH topography that departs a few m from the geoid. By measuring SSH oceanographers can observe

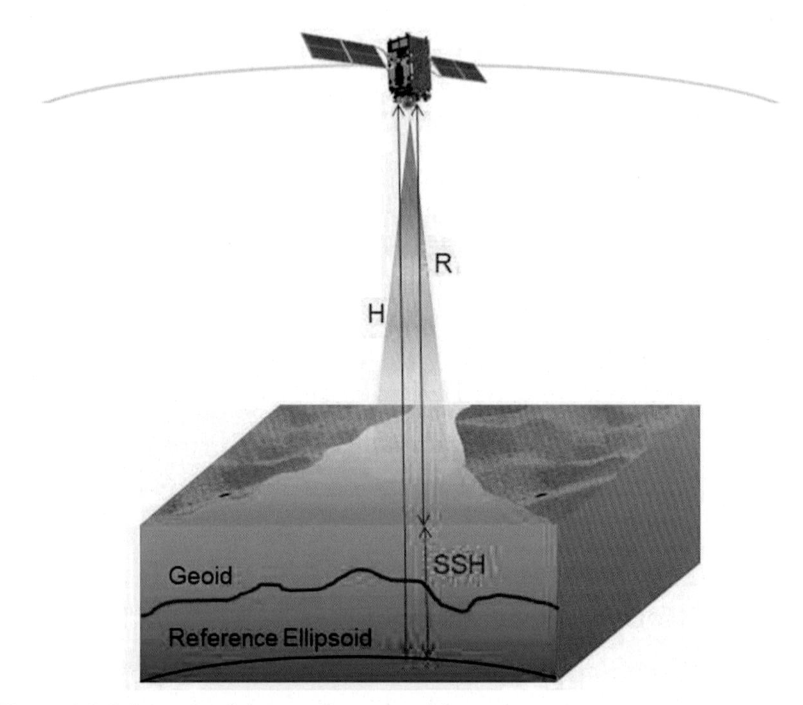

Figure 4.4 Schematic of the satellite radar altimetry system.

the signal due to these processes and in many cases isolate and monitor each process, as explained in detail in Fu and Cazenave (2000).

The satellite-based sea surface elevation differs from a point-wise elevation of the sea surface provided by a traditional shore-based tide gauge at a given location. Compared to in situ measurements of water surface elevations, the computation of satellite-based SSH therefore requires independent measurements of the satellite orbital trajectory and auxiliary information to correct for various effects (instrument errors, atmospheric refractions, perturbations caused by the surface interaction). Moreover, the SSH measurement is an average across the radar footprint (with a radius of 2–15 km according to sea state conditions). Some averaging is typical done in the along-track direction of the satellite in order to reduce the noise of individual waveforms.

When the ellipsoid reference is used, the larger signal captured in the sea surface elevation measurements is the geoid. However, for many applications the spatial resolution of the existing geoids is not sufficient. Nevertheless, the geoid is assumed to be constant at our timescales, suggesting focusing only on the variations of the sea surface elevation over time (i.e., the sea surface elevation anomalies). The SSH is therefore often given with reference to a Mean Sea Surface (MSS) and therefore expressed as a sea-level anomaly (SLA) or sea surface height anomaly (SSHA). The MSS can be computed by averaging SSH over several years at a fixed along-track ground point or using models that are based on SSH from all satellite altimetry missions over a given time period (e.g., Schaeffer et al, 2012; Andersen et al., 2015). Then, in practice the MSS is subtracted from SSH to compute SLA as $SLA = SSH - MSS$ with $SSH = H - R_{corr}$.

Some applications require additional geophysical corrections to remove the signature of other geophysical phenomena. These phenomena include: (1) tides (ocean, solid earth, polar, loading tides) and (2) the ocean's response to the atmospheric pressure at low frequency (inverse barometric correction) and to the wind field at high frequency (dynamic atmospheric correction). These corrections can be done through the use of models. However, in the case of extreme events (e.g., storm surges), the interest is in the actual level the sea reaches, therefore the signals mentioned above are an integral part of the water surface elevation which results from the combined effects of ocean dynamics, tidal effects, and atmospheric forcing.

4.4 Satellite radar altimetry missions

The development of satellite altimetry technology started in the 1970s. Fig. 4.5 shows the timeline of the historical altimeter missions, including those still operating at the time of writing.

A total of 10 altimeter missions have flown in the past. Generally, altimeter satellites are only operational for a limited number of years. The first altimeter was on-board the Skylab satellite in 1973 and was intended to measure the shape of the Earth (Marsh and Chang, 1976). Geos3 (1975) made measurements sufficiently accurate for geoid studies (Lerch et al., 1979). With Seasat (launched in 1978), the potential for ocean studies was shown (Tapley et al., 1982). Unfortunately, the mission failed after only 3 months. The next satellite altimetry mission, Geosat, was launched in 1985 and was the first one to work for several years and produce the first routine ocean product (McConathy and Kilgus, 1987). However, the modern age of satellite radar altimetry with uninterrupted global

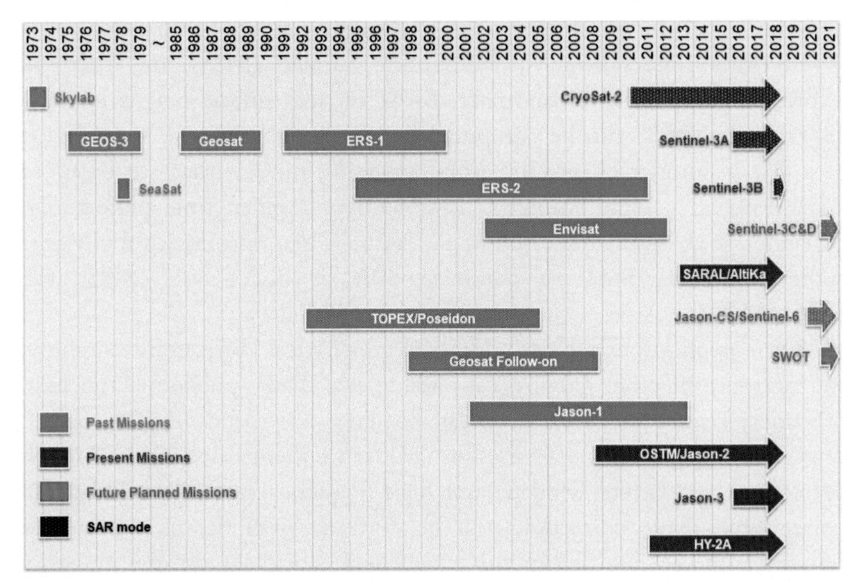

Figure 4.5 Global satellite altimetry missions: past, now, and then. *Courtesy: PODAAC.*

measurements started in 1991, with the first European satellite to carry a radar altimeter, ERS-1 (Andersen et al., 2013). This satellite was designed to have three different orbital configurations, including a 35-day repeat orbit lasting from April 1992 until December 1994. ERS-2 was launched on 1995 and operated simultaneously to ERS-1 shifted by 8 km, until ERS-1 was retired, in March 2000 (Andersen et al., 2013). The launch of TOPEX/Poseidon satellite mission in 1992 provided the greatest impetus for radar altimetry research in the 20th century (Fu et al., 1994). Its launch was followed by Jason-1 (2001), Jason-2 (2008), and Jason-3 (2016) over the same ground track pattern (Ménard et al, 2003; Lambin et al., 2010) known now as the "reference orbit." US Navy Geosat Follow-On (known as GFO) was launched in 1998 (Walker and Barry 1997). ESA follow-up satellites were Envisat, launched in 2002 (Resti et al., 1999; Benveniste et al., 2001), CryoSat-2, launched in 2010 (Parrinello et al, 2018) and Sentinel-3 A and B flying since 2016 and 2018, respectively (Donlon et al., 2012). HY-2 (HaiYang, meaning "ocean" in Chinese) is a marine remote sensing satellite series launched by China in 2011 (Jiang et al, 2012). AltiKa, the altimeter onboard SARAL mission (launched in 2013), is the first and so far the only one to operate in the Ka-band (Verron et al., 2015).

At time of the writing, there are seven altimetry satellites in service:

- two satellites, Jason-2 and Jason-3, with a relatively short repeat cycle (10 days), able to observe the same water target frequently but with relatively widely-spaced ground tracks (350 kilometres at the equator). Jason-3 is on the same "reference orbit" as their predecessors, TOPEX/Poseidon (1992–2005), Jason-1 and Jason-2, with Jason-2 having been shifted after the launch of Jason-3 onto a new orbit, mid-way between its previous tracks, with a 5-day time lag;
- one satellite, CryoSat-2, with an altimeter (SIRAL) capable of working with a Synthetic Aperture Radar (SAR) mode (also known as Delay-Doppler processing mode) to satisfy the scientific requirements for observing the poles and the ice sheets, but also some coastal and inland water regions with a drifting orbit (revising the same place after 369 days);
- two satellites, Sentinel-3A and -3B, with a longer repeat cycle (27 days) but tighter ground track spacing (around 100 kilometres at the equator) forming a constellation;
- one satellite, HY-2, which was for 2 years on a 14-day orbit with ground track spacing of 315 kilometres at the equator, and has then been moved from its nominal orbit to a drifting (i.e., nonrepeat) orbit; and
- one satellite, SARAL, with AltiKa altimeter that had originally covered the old ERS/EnviSat 35-day repeat for more than 4 years and was then moved to a drifting orbit.

The characteristics of past and present flying altimetric missions are given in Table 4.1. The combined global altimetry historical data set begins in the early 1970s. However, the older altimeter missions provided water surface elevations with low accuracy. The precise (and uninterrupted) era of altimetry started in 1991 with the launch of TOPEX/Poseidon and ERS-1. Thus, we have a potential situation today that there is one long-term record since 1992 thanks to TOPEX/Poseidon and the Jason 1/2 series that will continue in the same orbit configuration with the Jason-3 mission. At present, in addition to the Jason-3 mission, there are other six missions that provide complementary observations. Along-track resolution can be improved by utilizing new SAR techniques, as explained further in this chapter. SAR altimetry has been first tested with CryoSat-2 over selected portions of the ocean, and it is now definitely operational globally with Sentinel-3A and -3B.

Table 4.1 Main characteristics of satellite altimetry missions operating until now and planned for the future.

Mission	Agency	Time period	Orbit altitude (km)	Equatorial ground track distance (km)	Orbit revisiting (days)	Band	Pulse repetition frequency (Hz)
Skylab	NASA	1973–1974	440			Ku	
Geos-3	NASA	1975–1979	830			Ku	
Seasat	NASA	1978	800	800	17	Ku	1020
Geosat	US Navy	1985–1990	800	164	17	Ku	1020
ERS-1	ESA	1991–2000	785	20–80	3,35,168	Ku	1020
TOPEX/ Poseidon	NASA/CNES	1992–2006	1336	350	10	Ku/C	4500
ERS-2	ESA	1995–2011	785	80	35	Ku	1020
GFO	US Navy	1998–2008	800	170	17	Ku	1020
Jason-1	NASA/CNES	2001–2013		315	10	Ku/C	1800
Envisat	ESA	2002–2012	782	80	35	Ku/S	1800
Jason-2	NASA/CNES/ EUMETSAT/NOAA	2008-now	1336	350	10	Ku/C	1800
CryoSat-2	ESA	2010-now	725	7.5	369	Ku	1970/18181
HY-2A	CNSA/CNES	2011-now	971	315	14,168	Ku/C	2000

SARAL/AltiKa	ISRO/CNES	2013-now	785	80	35,	Ka	3800
Jason-3	NASA/CNES/ EUMETSAT/NOAA	2016-now	1336	350	10	Ku/C	2060
Sentinel-3A	ESA/EC/ EUMETSAT	2016-now	815	104	27	Ku/C	1924/17800
Sentinel-3B	ESA/EC/ EUMETSAT	2018-now	815	104	27	Ku/C	1924/17800
Sentinel-6	ESA/EC/ EUMETSAT/NASA/ CNES/NOAA	2020	1336	350	10	Ku/C	9000
Sentinel-3C/3D	ESA/EC/ EUMETSAT	2021	815	104	27	Ku/C	1924/17800
SWOT	NASA/CNES/CSA/BSA	2021	891	137	21	Ka	4420

Most of the satellite altimetry missions (e.g., ERS-1/2, Envisat, TOPEX/Poseidon, Jason-1/2/3, Sentinel-3A, -3B, -3C, and -3D) have been designed to fly in repeat orbits for most of the time, which means that the satellite revisits the same geographical location after a certain period. Some missions were moved into drifting orbits for geodesy objectives for some limited periods (e.g., Geosat, ERS-1, Envisat, Jason-2 SARAL/AltiKa). The repeat orbit duration dictates the distance between ground tracks and therefore the cross-track spatial resolution. Table 4.1 shows that CryoSat-2 has the best cross-track spatial resolution, 7.5 km but at the expense of a 369-day repeat cycle, while at the other end of the spectrum we find TOPEX/Poseidon and Jason series with a wide cross-track spacing (350 km), but much faster repeat (every ~10 days). This series of missions provides, during 10 days, a global coverage but its coarse coverage resolution does not sufficiently sample the coastal zone and also misses several inland water targets. However, a large lake could be crossed several times during a cycle.

An example is Lake Issykul in Kyrgystan. Envisat is passing every 16 days with two ascending and descending tracks. It increases the temporal resolution that is a requirement during flooding. With reference to rivers, for each track crossing the river a water virtual time series can be formed and used. With satellite missions in drifting orbits (e.g., CryoSat-2) it would take 369 days to fly over the same location, so the time series approach cannot be applied. Drifting orbits provide the possibility to observe the same river in more locations at different times, but the changing over time of the river dynamics has to be taken into account. This calls for different usage when in synergy with modeling tools compared from those used for virtual water station time series.

4.5 Altimeter processing in the coastal zone and inland waters

Satellite radar altimetry was originally designed for the open ocean domain. As on-board instrumentation and signal processing improved, new applications to coastal zones and to inland water targets became possible and continue to be researched, developed, and exploited today. Indeed, the application of satellite radar altimetry to inland waters has the potential to provide a strong contribution to the hydrological sector.

As anticipated in Section 4.2, the radar signal reflected by water targets with surrounding land does not adhere to the typical Brown model, which characterizes open ocean waveforms. The altimeter maintains the received signal inside a fixed-length analysis window by using an onboard pretracker. The pretracker continuously adjusts the acquisition delay of the received signals, trying to keep the leading edge of the collected waveform (corresponding to the earliest return from water) tied to a nominal tracking point within the time window. A deeper explanation of this aspect can be found in Gommenginger et al. (2011). The presence of nonwater targets (e.g., land, artificial structures, etc.) can introduce artifacts in the positioning and shape of the received waveforms. Thus, the signal contamination due to nonwater scatterers may happen in the coastal zone and inland waters, resulting in inaccurate range estimations as a consequence.

Classical models for the radar altimeter waveform assume a homogeneous wave field and constant sigma-0. When approaching the coastline it is common to observe the variable wave spectrum, variable wind, and surfactant streaks, all contributing to the variability of the ocean backscatter. Ignoring such variability leads to errors in range computation, which the retracking algorithms need to mitigate. In standard altimetry products, data in the coastal zone are flagged as bad and left unused. Several dedicated retrackers have been developed to improve altimetry measurements near coasts (e.g., see Gommenginger et al., 2011 for a review), including the latest promising subwaveform approach (Passaro et al., 2014; Roscher et al., 2017; Peng and Deng, 2018). Nevertheless modern altimetry using SAR mode already shows better coastal performance (Dinardo et al., 2018). Many studies also show improvements over inland water targets with the retracking of the waveforms from conventional altimeters (e.g., see recently Yuan et al., 2017; Huang et al., 2018) and from recent altimeters operating in SAR mode (e.g., see recently Villadsen et al., 2016; Moore et al., 2018).

Moreover, the corrections and auxiliary information are not optimized for the coastal zone, neither for inland water applications, and must be revisited in order to enhance accuracy. Another important point is that the combination of coastal morphology and local winds may produce calm water areas that appear to the radar as "bright targets," as demonstrated in (Gómez-Enri et al., 2010; Scozzari et al., 2012).

Some data sets are already available for usage in the coastal zone and inland waters, currently oriented to big lakes and rivers, with regards to the hydrological applications. Specialized processing chains, including the usage of adaptive retracking scenarios based on machine learning, the usage of dedicated retrackers robust to signal contamination due to land, and the merging of multiple missions, are some of the particular features of the various products and services developed for hydrological applications. Section 4.6 provides deeper illustration of these data sets. Like for oceans, hydrological applications imply calibration and validation requirements of needed reference measurements, as mentioned in Bonnefond et al. (2011) and Crétaux et al. (2011a,b).

The typical representation technique used to visually inspect the acquired data is called "radargram." Similarly to GPR (Ground Penetrating Radar) applications, the radargram for altimetry applications is a 2-D plot representing the reflection amplitudes versus time (or gate number) along the satellite track. In a radargram each waveform is projected onto a 2-D space where the x-axis represents the satellite position (along-track) and the y-axis the range bin (or gate); signal amplitudes are represented according to an assigned colormap.

Fig. 4.6 shows an example of radargram taken in a descending orbit of Envisat across the Ligurian Sea. A sequence of Brown-shaped returns can be identified between the two dashed red lines. The remainder of the radargram clearly shows the effect of land interference, even when the satellite track is over the sea but close to the coast, like in the proximity of Corsica Island. The ground projection of the satellite orbit is shown in the inset of Fig. 4.6.

Land interference as observed in the radargrams is also apparent by analyzing single waveforms. Fig. 4.7 shows two waveforms derived from

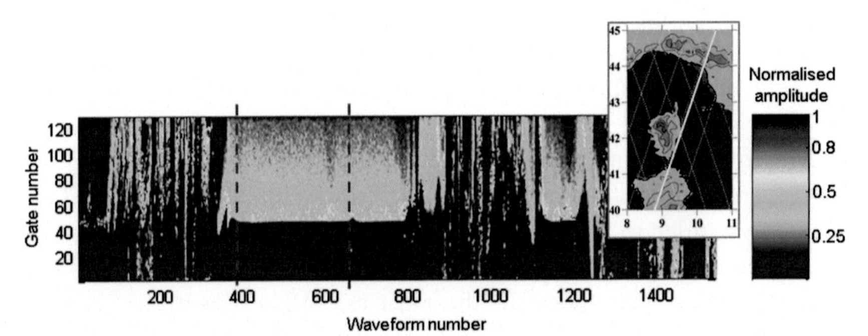

Figure 4.6 Example of radargram taken in a descending orbit of Envisat across the Ligurian Sea. A sequence of Brown-shaped returns can be identified between the two dashed red lines.

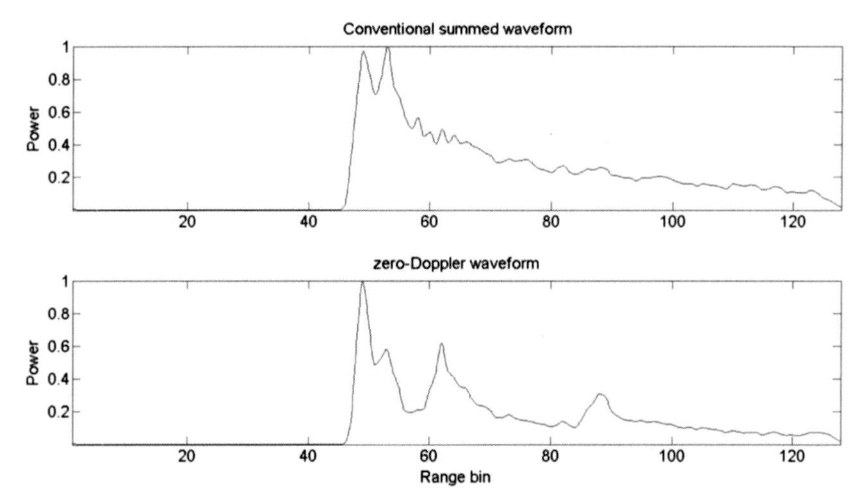

Figure 4.7 Example of radar echo impacted by presence of land.

the RA-2 altimeter on board Envisat satellite, demonstrating how the radar echoes may be impacted by the presence of land. In this example, the Envisat satellite track is almost perpendicular to the coastline of Make-jima, a small island in Japan (Abileah et al., 2013). The waveform shown in the upper panel was obtained by incoherent averaging of $N = 1984$ raw waveforms (about one second at Envisat's Pulse Repetition Frequency, PRF). It deviates from Brown's theoretical model, being characterized by a double-peaking feature due to land contamination to the signal.

An improvement is expected by coherent summing and exploitation of Doppler velocity for beam sharpening, like it happens for SAR techniques. In Abileah et al. (2013) the possibility to do coherent sums at relatively low PRF was demonstrated, and a simplified method was proposed, named "zero-Doppler," because it uses only a zero-delay, thus ignoring the off-nadir backscatter. The zero-Doppler velocity method produces a much-improved waveform by reducing the land interference (Abileah et al., 2013).

As mentioned in the previous section, CryoSat-2 carries the first radar altimeter (SIRAL) that uses the SAR technique. The SAR altimeter sends a burst of coherent pulses, which are combined (i.e., coherently summed) and organized in Doppler beams by partitioning in Doppler frequency bins. In this way, the surface footprint is split into narrow across-track strips from which sharper returns (waveforms) are obtained. As a result, a higher along-track resolution (~300 m) is achieved (Raney, 1998), leading to the ability of resolving shorter-scale surface features with respect to conventional radar altimetry. In order to ensure coherence of the returned signals, SIRAL operates at a PRF of 17.8 kHz. Following the pathway opened by CryoSat-2, the current Sentinel-3 constellation carries a SAR altimeter (SRAL); also the future Sentinel-6/Jason-CS mission due for launch after 2020 will be equipped with a SAR altimeter.

The RA-2 instrument offers a global archive of Individual Echoes (IEs), that is, raw waveforms collected at the native sampling rate of 1795 Hz (which is the instrument's PRF). RA-2 worked with a continuous pulse transmission scheme, like that expected from the upcoming Sentinel-6 mission. In Abileah et al. (2017) it was demonstrated that most rivers and small-to-medium sized lakes exhibit specular echoes, which have a well-defined Doppler evolution and can be processed with the SAR technique for water level estimation.

The radargram in Fig. 4.8, obtained by a data set of Envisat IEs, shows the evidence of bright specular echoes when the satellite orbit is crossing a tiny river (about 50 m large) in three distinct points, demonstrating the possibility to detect and range even such relatively small water targets. Such investigation, made by using RA-2 data, supports very much the perspectives of SAR altimetry applications to inland water.

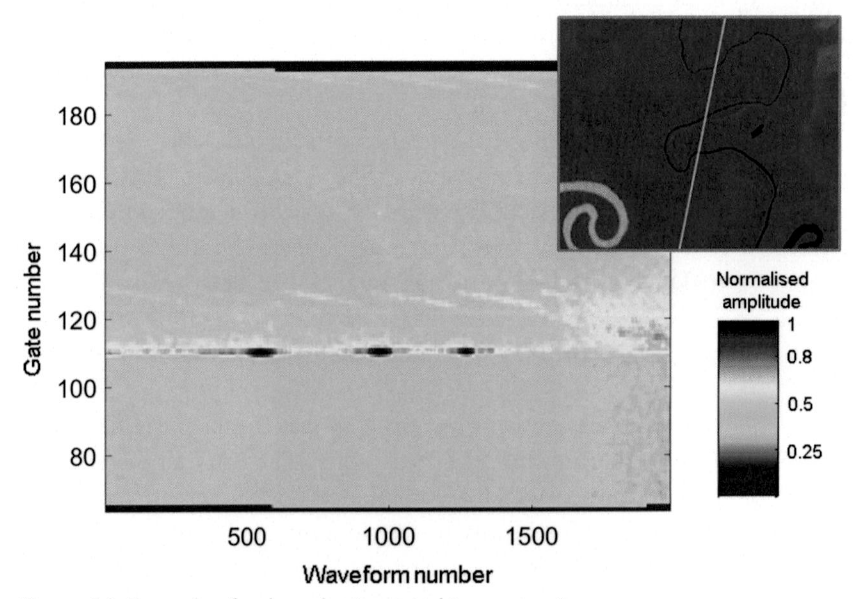

Figure 4.8 Example of radar echo impacted in narrow river.

Improvements may also be obtained by working in the Ka-band (shorter wavelength than Ku). The SARAL mission carries AltiKa (see Section 4.4), the first altimeter operating in such band. The usage of Ka-band enhances the spatial resolution, both in terms of smaller footprint and in terms of narrower range bins, thanks to the larger bandwidth with respect to Ku-band altimeters. These happen at the expense of a higher uncertainty due to the attenuation and propagation delays introduced by the water column in the atmosphere, to which the Ka-band is more sensitive than Ku.

4.6 Data sets available for usage in the coastal zone and inland waters

Data acquisition from the different altimetry satellites is performed by each space agency. There are four levels (from Level 0 to Level 4) of processed data depending on their processing stage. Space agencies generate a suite of altimeter products. Their availability depends on how long it takes to produce them. Three classes are usually considered: (1) products available to users within a few (e.g., 3) hours after data sensing; (2) products available to users after a few (e.g., 3) days; and (3) products available to users after weeks. The main output of Level 2 processing stage is the GDR (Geophysical Data Record) product. It usually exists in different versions, the main difference among them laying in their temporal availability and therefore in the different quality of the corrections (geophysical or engineering calibration) and orbits. The GDR product is normally generated at the rate of 1 Hz (i.e., a record every ~7 km). It usually contains all the altimeter data and connected corrections to compute the SSH or SLA. Some corrections come directly from the altimeter, some from the other on-board instruments, or other instruments; others from models, or a combination of all the former. The main output of Level 1 processing stage is the SGDR (Sensor Geophysical Data Record) that users require when they like performing their own retracking. The product is designed to supply both the GDR level output and the waveforms at the rate of 18/20 Hz (i.e., a waveform every ~0.35 km).

Every satellite mission has its own record format of products. Many "off the shelf" products, whether distributed on DVD or available online through FTP, become almost immediately out of date due to the continuous improvement in corrections and processing. The variety of sources of data components, together with improvements in processing, can change in time, giving rise to improved versions becoming available. AVISO for CNES and PO.DAAC for NASA are the official facilities to process and distribute the up-to-date suite of satellite altimetry products in netCDF format. Both services offer free access to these products in NRT and off-line for light (L4), advanced (L2, L3), and expert (L1) users.

The above dissemination services have been designed with the open ocean in mind. They remain the source especially for altimetry specialists. Most users may not be aware of, or have access to, the latest updates and most appropriate corrections to use for their application (Snaith et al., 2006). RADS (Radar Altimeter Database System) was the first service in which an user can select and combine altimeter data and models from various external sources "on the fly" (Scharroo et al., 2013) and recompute the updated SLA. RADS is maintained by Delft Technical University and NOAA. It provides a harmonized, validated, and cross-calibrated multimission L2/L3 database, including all up-to-date corrections, auxiliary information, in a consistent format, and a reference frame common for all altimeter missions. The multimission data set is accessible at http://rads.tudelft.nl/rads/rads.shtml.

The use of satellite altimetry in coastal zones and inland waters requires specialized processing that takes into account the typical problems of these regions. The PISTACH (Prototype Innovant de Système de Traitement pour l'Altimétrie Côtière et l'Hydrologie) project generated a product that includes all possible corrections and source data, including values from a number of different retracking scenarios (Mercier et al., 2010). This approach provides a wealth of detailed information for the technical specialist, but results in very large products, which are the domain of the expert. The COASTALT project had a similar approach, generating a product with a range of retracking scenarios and possible corrections, with data only generated for coastal areas (Vignudelli et al., 2011a,b). Only those corrections that can be calculated globally were included in the standard product and users were able to add regional specific corrections to provide a more tailored solution.

A decade of progress in improving processing made possible the availability of experimental and operational products dedicated to the monitoring of coastal and inland waters. Compared to standard products they include higher along-track resolution, new/improved retrackers, new/improved corrections, refined preprocessing and/or post-processing,

Table 4.2 (adopted from Cipollini et al., 2017a,b) summarizes the main characteristics of the available products for coastal zone. The following section provides more information on some specific products, that is, CTOH, ALES, COSTA, X-TRACK, PEACHI, GPOD SARvatore, highlighting their main characteristics. A data set with consistent coastal processing for all available missions is not yet available.

The CTOH is a French observation service dedicated to satellite altimetry studies (Birol et al., 2017). It maintains homogeneous altimetric databases (L2/L3) for the long-term monitoring of sea level. The reprocessing involves an ad hoc editing strategy of the data records and a careful extrapolation/interpolation of missing or imperfect corrections of the altimetric measurement in the coastal strip.

ALES is a data set retaining all the fields in the SGDR with the addition of new retracked fields (range, SWH, and backscatter) from the ALES algorithm (Passaro et al., 2014). It is available along-track at a resolution of around 350 m for Envisat, Jason-1, Jason-2, and AltiKa missions in the global coastal strip within 50 km from the coastline (Passaro et al., 2015). COSTA is a post-processed version for particular geographic location (Mediterranean Sea) in which raw data are already corrected and assembled on nominal tracks in the form of SLA time series (Passaro, 2017).

X-TRACK has been developed by CTOH for different altimetric missions and regions. It is essentially a standard product at an along-track resolution of 7 km, assembled on nominal tracks in the form of SLA time series, that is, extended to the coastal zone with improved editing and post-processing (Birol et al., 2017).

The recent G-POD service called SARvatore (SAR Versatile Altimetric Toolkit for Ocean Research & Exploitation) is a web platform that allows any scientist worldwide to process on-line, on-demand and with a user-selectable configuration both CryoSat-2 and Sentinel-3 data in SAR mode using a dedicated retracker for coastal zone and obtaining an along-track data set that is spaced approximately every 350 m (20 Hz) or 87 m (80 Hz) (Dinardo, 2014; Dinardo et al., 2015).

As stated previously, due to the processing complexity in inland waters, the know-how is restricted to a few expert groups that, however, sponsored by various agencies, now offer water height products derived from the satellite radar altimeters. Water heights are computed as time series over lakes and over rivers at the intersections with the satellite ground tracks (the so-called virtual stations).

Table 4.2 Available altimeter data sets for usage in the coastal zone.

ID	Produced by	Altimeter	Product level	Posting rate	Coverage	Download from	Comments
PISTACH	CLS CNES	j2	L2	20 Hz	Global	AVISO +	Experimental Jason-2 products for Hydrology and Coastal studies with specific processing. Will be discontinued at the end of 2016 in favor of PEACHI
PEACHI	CLS CNES	sa, (j2 to be added soon)	L2	40 Hz	Global	AVISO + / ODES	Experimental SARAL/AltiKa products including dedicated retracking and corrections leading to more accurate products for coastal zones, hydrology and ice. From 2017 expected to generate also j2 products
XTRACK	LEGOS-CTOH	tx, j1, j2, gfo, en (sa to be added soon)	L2, L3	1 Hz 20 Hz (test)	23 regions covering the whole coastal ocean	CTOH AVISO + / ODES	Specific processing using improved data screening and latest corrections available
ALES	NOC	j2, n1, (j1, j3 to be added soon)	L2	20 Hz	Global, <50 km from coast	PODAAC	Experimental products from the ALES processor included in SGDR-type files alongside the standard products and corrections.

SARvatore	ESA-ESRIN	c2 (SAR only)	L2	20 Hz	SAR mode regions	ESA GPOD	On-demand Processing service for the CryoSat-2 SAR mode data where the user can configure some processing parameters to meet specific requirements (for istance for the coastal zone)
COP	ESA	c2 (LRM/PLRM)	L2	20 Hz	Global	ESA	Global products for CryoSat-2 from an Ocean processor (output is in PLRM over the SAR mode regions) - but no specific coastal processing
COSTA	DGFI-TUM	e2, en (j1, j2, e1 to be added soon)	L3	1 Hz 20 Hz	Mediterranean and North Sea	PANGAEA	Dedicated coastal altimetry sea level measurements based on enhanced ALES retracker

e1: ERS-1; tx: TOPEX; e2: ERS-2; gfo: GEOSAT Follow-On-1; j1: Jason-1; n1: Envisat; j2: Jason-2; c2: CryoSat-2; sa: SARAL/ALtikA; j3: Jason-3. For CryoSat-2 (c2) LRM/PLRM are Low-Resolution Mode and Pseudo-LRM and SAR for Synthetic Aperture Radar mode. Product levels: L2: along-track data with corrections; L3: data projected to reference points along nominal satellite ground track.

Adapted from Cipollini, P., Calafat, F.M., Jevrejeva, S., Melet, A., Prandi, P., 2017a. Monitoring sea level in the coastal zone with satellite altimetry and tide gauges. Surveys Geophys., 38(1), 33-57; Cipollini, P., Benveniste, J., Birol, F., Fernandes, M.J., Obligis, E., Passaro, M., et al., 2017b. Satellite Altimetry in Coastal Regions, in Satellite Altimetry over Oceans and Land Surfaces, first ed. Cazenave A. & Stammer D., editors. Boca Raton, FL: CRC Press. Updated table with links at www.coastalt.eu.

The PISTACH product is also available for inland waters, with specific processing, developed by CLS (Satellite Location Collection), in which retracking is based on the classification of typical waveforms occurring in inland water bodies. A number of products have been developed to provide freely accessible water elevation time series for selected targets. There are some products now freely accessible (e.g., River & Lake, Hydroweb, DAHITI, HydroSat) providing water heights over rivers and lakes. There are two additional databases that provide water heights only over lakes (AltWater and G-REALM).

The Rivers and Lake project (http://tethys.eaprs.cse.dmu.ac.uk/RiverLake/shared/main) developed by ESA and Montfort University used mainly ERS-2/Envisat (2002–10) and Jason-2 (2009–11) satellite altimetry to provide historical and Near-Real Time water surface elevations data for a large number of virtual water stations around the world. The waveforms at resolution higher than 1 Hz are processed using the retracking expert system reported in Berry et al. (2005), based on different retracking algorithms that are adapted to changes in target type and waveform shape (Berry et al., 1997). Around 750 water targets with 35-day sampling and 57 with 10-day sampling are available (Berry and Wheeler, 2009).

HydroWeb was developed at Laboratoire d'Etudes en Géophysique et Océanographie Spatiales (LEGOS). The historical data set is accessible at http://ctoh.legos.obs-mip.fr/products/hydroweb. Based on historical and present altimeter data, Hydroweb provides time series over water levels of about 100 lakes and 250 sites on large rivers (Crétaux et al., 2011a,b). Hydroweb is now a fully operational service on the CNES/THEIA Platform at http://hydroweb.theia-land.fr/?lang = en&. The Hydroweb approach is to use standard products and combine different missions that fly over the water target to enhance precision and temporal sampling.

DAHITI (Database for Hydrological Time Series of Inland Waters) is developed by the Deutsches Geodätisches Forschungsinstitut der Technischen Universität München (DGFI-TUM) and provides virtual station time series from multiple missions over lakes, rivers, and reservoirs. The data set is accessible at http://dahiti.dgfi.tum.de/en/. DAHITI uses all the available altimetry missions, including data back to 1992 with TOPEX/Poseidon, with a processing strategy based on an extended outlier detection and a Kalman filtering (Schwatke et al., 2015).

HydroSat is developed by University of Stuttgart, Institute of Geodesy (GIS). It provides water height time series from satellite altimetry over some rivers along with other remotely sensed or derived variables, such as surface water extent and discharge (Tourian et al., 2016). The data set is accessible at http://hydrosat.gis.uni-stuttgart.de/php/index.php.

AltWater (Altimetry for inland Water) was developed at DTU space (National Space Institute, Technical University of Denmark) under the FP7 project Land and Ocean Take Up from Sentinel-3 (LOTUS). Currently, the database only contains water heights obtained from CryoSat-2 for a limited number of lakes (Nielsen et al., 2016), but other missions are planned to be added in the future. The main difference consists of the reprocessing of the CryoSat-2 waveforms, which is not performed in the processing chain of the other services. The processing methodology is described in Nielsen et al. (2015). The data set is accessible at http://altwater.dtu.space.

G-REALM (Global Reservoirs/Lakes) is developed by the United States Department of Agriculture's Foreign Agricultural Service (USDA-FAS), in cooperation with NASA and the University of Maryland (Birkett et al., 2017). It provides near-real time data from multiple missions that are utilized to routinely monitor lake and reservoir changes in water surface elevation. The data set is available at https://www.pecad.fas.usda.gov/cropexplorer/global_reservoir/. G-REALM is based on standard retracking algorithms provided in standard products at 18/20 Hz and provides indicators that help to assess quality of the measurements in challenging conditions (e.g., presence of islands in lakes; dry beds, presence of ice, etc.). It keeps the satellite missions separate.

The G-POD service also allows users to process, on line and on demand, low-level CryoSat-2 and Sentinel-3 Altimetry data products (FBR, Level 1 A) in SAR mode up to Level-2 geophysical products with self-customized options that are not available in the default processing of CryoSat-2 and Sentinel-3 Ground Segments (Dinardo, 2014; Dinardo et al., 2015). The service is open, free of charge, and accessible on line at https://gpod.eo.esa.int/services/CRYOSAT_SAR and https://gpod.eo.esa.int/services/Sentinel-3. It provides water surface elevations at the rates of 20 Hz and 80 Hz.

A detailed description of the data processing methods used to generate the altimeter-derived water elevation time series available in the previous databases can be found in the references. The above products may differ for processing approach and applied geophysical corrections.

4.7 Improvements in accuracy

Water surface elevations derived from satellite altimetry are required to be quantitatively self-consistent and have a certified accuracy and/or precision. Their validation is complicated by the fact that satellite radar altimetry relies on a combination of different types of measurements (range, orbit, and corrections). The traditional approach is to use a different (independent) data set, for example, comparison with different altimetry missions at crossover points or against in situ measurements at close sites. The method to compare satellite altimetry and in situ measurements is not unique and depends on the region and goals of the usage of altimeter data.

Comparisons with in situ water surface elevations can be done in a relative sense, that is, taking into account the variations over a certain time period. Satellite-based time series near the station are usually constructed and then compared to the time series obtained from the in situ instrument. Comparisons can be done using monthly averaged, daily, or near-simultaneous measurements. The frequent in situ observations are generally subsampled to the time of the altimeter overpass.

Extreme events occurring on the ocean and inland waters (e.g., storm surges and flooding) are short-term phenomena, therefore, the near-simultaneous comparison method between satellite altimetry and in situ observations is the more appropriate. The usage requires computing the accuracy of the satellite-based water surface elevations in order to discriminate the anomalous raised value with respect to the background. The relative total altimetric time series would be similar to those observed by the in situ instrument. Because the two time series are generally from spatially separated points, the residual time series will contain differences, which result from the two different locations where the in situ and altimetric data are made.

In fact, the satellite altimeter and in situ station can only overlap by chance. Therefore, the two systems do not necessarily observe the same area. The root-mean-square error (RMSE) is commonly used as an indicator of accuracy. It should decrease in points if near to the station and will indicate if the water surface elevations are within the acceptable threshold, for example, for assimilation in storm surge models.

Table 4.3 summarizes a list of recent papers aiming at the validation of coastal altimeter products near the coast. The authors generally use three statistical parameters for the comparison against the tide gauges: RMSE, Standard Deviation of the differences (SD), and Correlation Coefficient (R). Comparisons were made in coastal areas all around the world using

Table 4.3 Accuracy statistics from validation studies in coastal zone using satellite radar altimetry.

Altimeter mission	Region	RMS accuracy (cm)	Correlation (R)	References
Jason-2	Hong Kong (China)	SD: ~5		Xu et al. (2018)
SARAL/Jason-2/Jason -1/Envisat/ERS-2	Southern Bay of Biscay	[0–5] km coastal band (8–21)/(17–34)/(10–29)/(18–90)/(22–89)	[0–5] km coastal band: 0.99/(0.83–0.98)/(0.93–0.99)/(0.67–0.99)/(0.51–0.99)	Vu et al. (2018)
CryoSat-2/SARAL	Gulf of Cadiz (Iberian peninsula)	7.2/5.3		Gómez-Enri et al. (2017)
Jason-1/2	UK	3.8–5.8	0.75–0.45	Cipollini et al. (2017a,b)
Jason-2	Great Barrier Reef (Australia)	14–19	0.77–0.81	Idris et al. (2017)
CryoSat-2	213 selected PSMSL tide gauges around the world	SD: 5.6	0.85	Bouffard et al. (2017)
Envisat/SARAL	Strait of Gibraltar	RMS: [11.7–13.3]/~ 9.0		Gómez-Enri et al. (2016)

different satellite missions. RMSE values range between 3.8 cm (Jason-2) and 89 cm (ERS-2) with R between 0.45 (Jason-1) and 0.99 (SARAL/AltiKa, Jason-1, Envisat). The ocean variability at the location of the in situ instruments and their distance to the tracks justify some of the differences in the observed accuracy. The use of different retracking algorithms, editing strategies, and geophysical corrections also play their role.

The retrieval of water surface elevations of inland water targets (lakes, reservoirs, and rivers) has a much wider variability of measurement conditions than the coastal zone.

Table 4.4 provides a selection of papers regarding different sites, showing accuracy statistics of water surface elevation products, all generated by satellite radar altimetry. The RMSE estimates vary from a few centimeters (e.g., Lake Vättern) to about a meter (e.g., Zambezi River). The resulting accuracy is certainly influenced by the characteristics of the site. However, the inland water altimetry products are processed by different research groups, using different algorithms (e.g., retracking, corrections, editing), so the satellite-based water surface elevations from the various products might have different resolutions, accuracies, and errors. Some comparisons of different inland products against water surface elevation measurements in selected sites can be found in Schwatke et al. (2015) and Ricko et al. (2012).

As shown by Abileah et al. (2017), the behavior of water as a reflector is a function of the along-track size of the water target. In particular, the size of the water target determines the number of radar echoes returned from the water that can be usefully averaged to remove speckle. The aperture of the main reflection lobe originating from the illuminated surface determines the length of the satellite track in which the altimeter can collect a strong enough signal for range measurement. By applying basic electromagnetic theory, Abileah et al. (2017) demonstrated that the size of the main lobe first decreases with the increasing size of the water target, then, at some point, the relationship reverses and the lobe width increases with the size of the target, with a turning point at about 100 m in size. Other factors (e.g., water vapor, wind set-up, presence of ice, tides, waves) can also impact range and corrections.

4.8 Water surface elevation and extreme events

Extreme events due to meteorological phenomena can produce a rise in the water surface, which can be considerably larger than other

Table 4.4 Accuracy statistics from validation studies in inland waters using satellite radar altimetry.

Altimeter product and satellite missions	Region	Water target	Time period	RMS accuracy (cm)	References
PISTACH (Jason-2)	Australia	Lake Eildon	2008–2012	28	Jarihani et al. (2013)
River and Lake (Envisat)	Africa	Zambesi River	2002–2010	34–72	Michailovsky et al. (2012)
Hydroweb (Envisat, TOPEX/ Poseidon Jason-1/2, GFO)	Africa	Lake Thana	1992–2006	13.6	Duan and Bastiaanssen (2013)
DAHITI (Envisat and AltiKa)	Brasil	Rio Madeira	2002–2014	21.6	Schwatke et al. (2015)
HydroSat					
	Swedish	Lake Vättern		4.6	
ALTWater	USA	Lake Okeechobee	2010–2015	7.3	Göttl et al. (2016)
(CryoSat-2)	Irland	Lake Lough Neagh		6.5	
G-REALM	Afghanistan	Lake Kajaki	2008–2016	49	
(TOPEX/Poseidon	Iraq	Lake Mosul	2008–2016	41	Nguy-Robertson et al. (2018)
Jason-1/2)	Iraq	Lake Tharthar	2003–2011	40	
G-POD SARVATORE (CryoSat-2)	Africa	Mekong River	2011–2015	43.7–64.1	Moore et al. (2018)

background signals. Theoretically, water surface elevations observed by radar altimeters should reflect these transient signals. However, due to the low revisit time there is a low probability of flying over the event at a given location. In addition, abrupt height changes are usually assumed to be associated with erroneous data and are rejected by standard editing methods. Increasing the temporal and spatial coverage by altimetry would require more satellites flying together, with, optimally, coordinated orbital patterns. In the next sections, the potential of satellite altimetry in retrieving surface elevations is shown in the presence of strong river discharges and storm surges.

4.8.1 Strong river discharges as observed by satellite SAR altimetry

The effect of heavy freshwater discharges in river estuaries is of interest in the altimetry community. It represents the transition zone between inland waters and coastal zones. Fresh waters from rivers reduce the salinity levels in the ocean surface adjacent to the river mouth and hence might give a bulge in sea level (González-Ortegón and Drake, 2012; González Ortegón et al., 2010; Navarro et al., 2012). An example of the capabilities of CryoSat-2 SRAL (in SAR mode) to measure this bulge is presented here. The study area is located in the eastern continental shelf of the Gulf of Cadiz (south-western Iberian Peninsula) close to the estuary of the Guadalquivir River. The time period analyzed was from 7 December 2010 to 17 January 2011. During that period the river was in extreme conditions with daily averages of water discharges larger than 400 m^3/s (Diez-Minguito et al., 2012). The analysis of CryoSat-2 tracks available in the adjacent shelf of the estuary showed one descending track (absolute orbit #3884: 1 January 2011). The availability of optical RGB MODIS images (Terra) (from AERONET: http://lance-modis.eosdis.nasa.gov/imagery/subsets/?project = aeronet) gave only one cloud-free image (04-01-2011).

Fig. 4.9A shows the daily average of water discharges during the extreme event. Fig. 4.9B presents the RGB MODIS image showing the turbidity plume due to the heavy freshwater river discharge. The surrounding area is characterized by a diffuse plume due to the mixture of the discharged water with the surrounding waters. The profile of the along-track 20 Hz SLA is presented in Fig. 4.9C. SLA data were smoothed with a running mean of five 20 Hz measurements. A clear

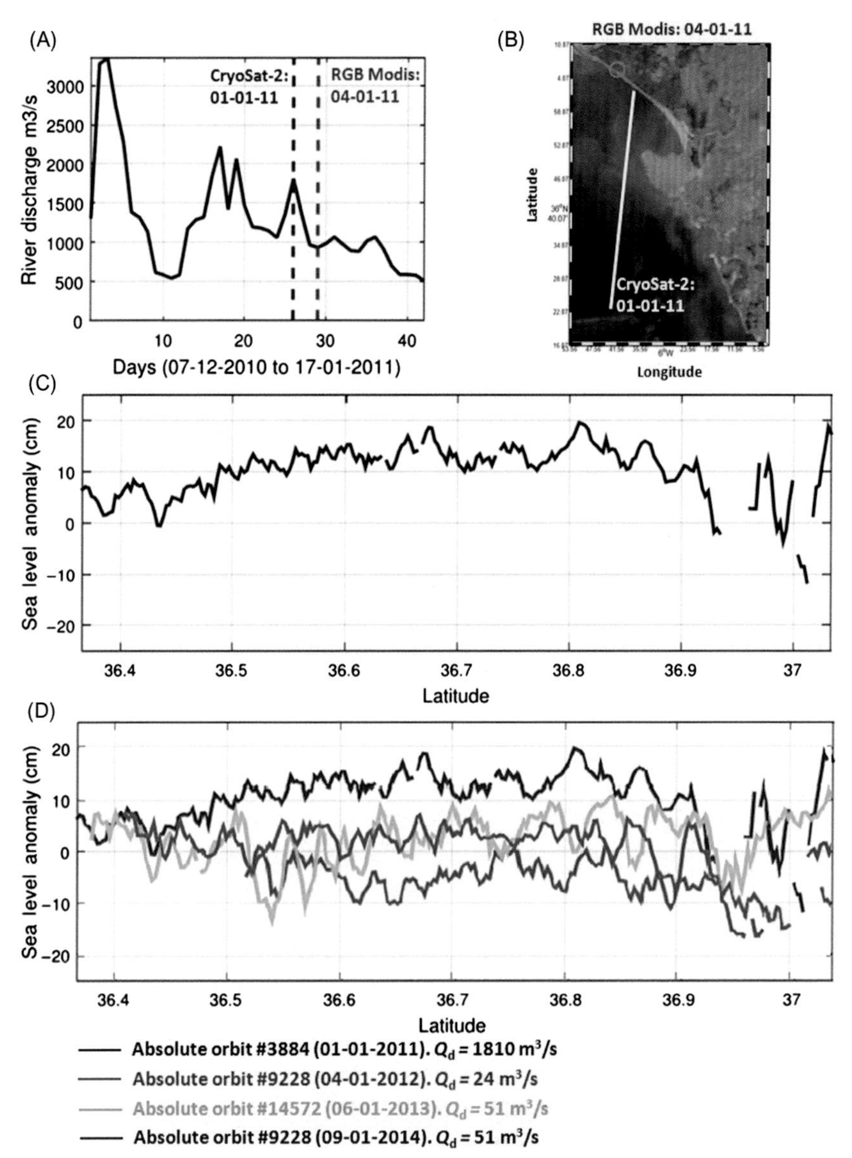

Figure 4.9 (A) Heavy river discharges in the estuary mouth of the Guadalquivir River (Spain); (B) RGB MODIS image (Terra) 4 January 2011 and example of CryoSat-2 track flying on 1 January 2011 near estuary; (C) Along-track water surface elevations relative to a Mean Sea Surface from CryoSat-2 passages around the peak discharge. *Adapted from Gómez-Enri, J., Vignudelli, S., Cipollini, P., Coca, J., Gonzalez, C.J., 2017. Validation of CryoSat-2 SIRAL sea level data in the Eastern continental shelf of the Gulf of Cadiz. Adv. Space Res., in press, doi: 10.1016/j.asr.2017.10.042.*

bulge of about 15 cm is clearly seen in the SLA profile over the continental shelf. The analysis of the same track at different 369-day cycles under no extreme conditions of water discharges is shown in Fig. 4.9A. The smoothed SLA profiles confirmed that the bulging shape was observed only during the event of heavy river discharge (January 1, 2011). More details of the response of CryoSat-2 SLA 20 Hz measurements during heavy river freshwater discharge can be found in Gómez-Enri et al. (2017). Daily gridded maps of SLA from multimission satellites were also used to analyze the sea level response to river discharge in Gómez-Enri et al. (2015).

4.8.2 Storm surges as observed by satellite altimetry

Satellite altimetry is not routinely used for storm surge services yet, although a possible inclusion was strongly recommended by Cipollini et al. (2009). A review of the available literature shows that there are open questions on how to use these data. Some studies demonstrated the capability of satellite altimetry of observing and studying storm surge features, for example, during Hurricane Katrina (Scharroo et al., 2005), Hurricane Sandy (Lillibridge et al., 2013), Hurricane Igor (Han et al, 2012), Cyclone Xaver (Fenoglio-Marc et al., 2015), Hurricane Isaac (Han et al., 2017), and Typhoon Seth (Li et al., 2018). Some assimilation experiments indicate that the improvement compared to tide gauges is rather limited as a consequence of getting good quality near-real-time data (e.g., Philippart and Gebrrada, 2002). Other investigations suggest that altimeter data could become beneficial if used in the initialization phase (e.g., Peng and Xie, 2006) or for validation of storm surge model outputs (e.g., Høyer and Bøvith, 2005). The European Space Agency (ESA) through the eSurge Venice Project supported the usage of satellite altimetry to improve storm surge simulations through data assimilation in the Adriatic Sea and around Venice (De Biasio et al., 2016; De Biasio et al., 2017).

An example is given in Fig. 4.10, which shows the satellite overpasses (Jason-1 and Envisat) crossing the Adriatic Sea around the storm surge of October 31, 2004. The signature of the surge is clearly evident in the Jason 1 overpass on October 30, 2004. This figure is just an example of how the other altimeter tracks can be eventually used to correct the model results (Bajo et al., 2017).

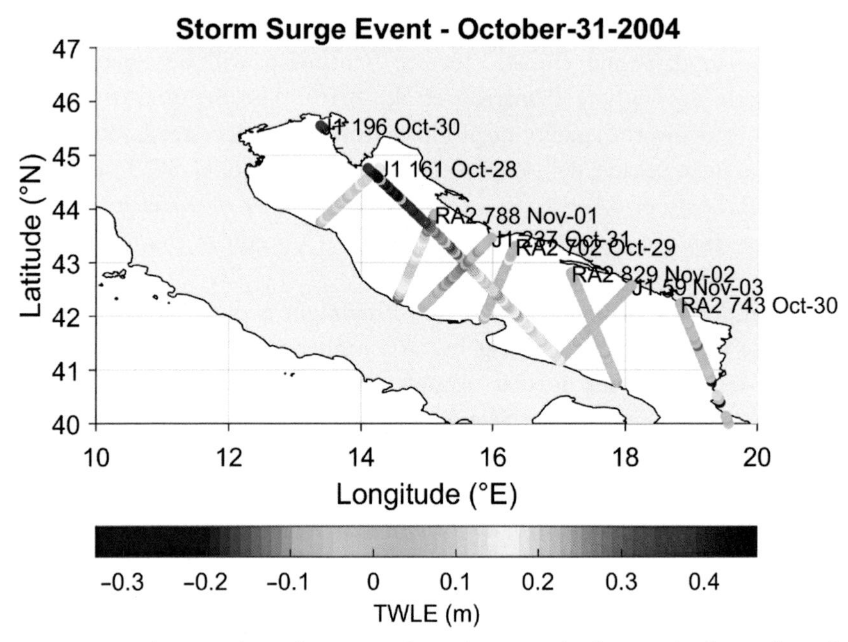

Figure 4.10 Along-track total water surface elevations (inclusive of tides and wind/pressure effects) relative to a Mean Sea Surface from satellite altimetry overpasses around the storm surge event of 31 October 2004. *Data from CTOH database at http://ctoh.legos.obs-mip.fr/.*

4.9 Future satellite radar altimetry missions in support of coastal zone and inland waters

With the advent of the along-track Delay-Doppler (SAR) processing, first implemented in CryoSat-2, and now in Sentinel-3A and -3B, more accurate surface elevations from satellite radar altimetry are anticipated for coastal and inland waters. The future satellite altimetry missions are expected to increase capabilities and coverage and represent a great opportunity to stimulate further exploitation of altimeters in coastal and inland water targets. A number of satellite missions are planned (Fig. 4.6), in particular: Sentinel-6 (to be launched 2020), Sentinel-3C and -3D (2021–23, to replace Sentinel-3A & -3B), SWOT (2021).

The Sentinel-6 mission program (also known as Jason Continuity of Service, Jason-CS) consists of two identical satellites (Sentinel-6A and Sentinel-6B), each one with a nominal lifetime of 5.5 years and a planned overlap of at least 6 months. The satellites will fly in the same orbit as its predecessors (TOPEX/Poseidon, Jason-1, Jason-2, and Jason-3) to continue the fundamental climatic record. Sentinel-6 will be operating in SAR mode everywhere (Scharroo et al., 2016). The Sentinel-6 missions will also maintain the quality of products from its predecessor Jason-3 mission. The next decade will see also the launch of Sentinel-3C/D that will follow-up Sentinel-3A/B to guarantee the continuity of water level measurements for the operational needs of the Copernicus program (ESA, 2017).

An important aspect of Sentinel-6 to highlight is the continuous high-rate pulse mode, like Envisat IEs but at a higher rate than its predecessor. The advantage of IEs is that complex waveforms can be coherently summed (Zero-Doppler, see Section 4.5) around the nadir crossing point of a narrow water body. Coherent averaging increases the accuracy of level estimation (Abileah et al., 2017).

CryoSat-2 and Sentinel-3 transmit pulses only one-third of the time and wait two-thirds of the time for pulses to be received (Scharroo et al., 2016). The Individual Echoes produced by the Envisat RA-2 platform (Resti et al., 1999) represent now the only currently available data set for the experimentation of continuous coherent processing on real data.

SWOT is a research mission under a joint effort between the US National Aeronautics and Space Administration (NASA), the French Space Agency (CNES), the Canadian Space Agency, and the United Kingdom Space Agency, that is, scheduled for launch in 2021 (Rodriguez et al., 2017; Rodriguez, 2016). The primary instrument on SWOT is a Ka-band radar interferometer (KaRIn) (Fjortoft et al., 2014). KaRIn will illuminate the surface with two 50 km swaths, separated by a 20 km nadir gap. Inside each swath, the intrinsic pixel resolution will vary from 10 to 60 m in the across-track direction and will be at best around 2 m along-track (Chevalier et al., 2017). The SWOT revisiting time is ~21 days, during which sites will be observed 2–4 times at irregular intervals, depending on the location latitude (Biancamaria et al., 2016). For the first time, SWOT brings together both hydrology and oceanography communities, and will extend the satellite-based record of water levels into coastal environments, at land-ocean interface (e.g., estuarine regions) and over inland water targets (e.g., rivers, lakes, reservoirs). While SWOT is not designed to monitor the fast temporal changes of extreme events (e.g., storm surges, flash floods, etc.), the swath coverage and its resolution will permit the better characterization of the spatial structure of their dynamics when they occur within the swath. SWOT data, regardless of latency, will be valuable for post-event reanalysis as well as model calibration and development (Hossain et al., 2017).

It is expected that advances in technology (e.g., reflectometry) or emerging concepts (CubeSats and miniaturization) will provide new complementary capabilities of collecting water surface elevations. For example, a novel approach that uses wideband signal of opportunity (SoOp) in a bistatic radar configuration is considered to address issues in coastal altimetry (Shah and Garrison, 2017). Small satellite designs were already proposed (Kilgus et al., 1989; Zheng, 1999; Richard et al., 2008). A CubeSat concept for ocean altimetry was proposed by Stacy (2012) and also considered by the US Navy (Mroczek and Jacobs, 2015). At present there are still technical challenges to measure water heights with the previous approaches, however, the potential benefits would be substantial.

Acknowledgments

We thank Alessandro Puntoni (CNR-IBF) for his technical support in preparing this manuscript.

References

Abileah, R., Gómez-Enri, J., Scozzari, A., Vignudelli, S., 2013. Coherent ranging with Envisat radar altimeter: a new perspective in analyzing altimeter data using Doppler processing. Remote Sensing Environ. 139, 271–276.

Abileah, R., Scozzari, A., Vignudelli, S., 2017. Envisat RA-2 individual echoes: a unique dataset for a better understanding of inland water altimetry potentialities. Remote Sensing 9 (6), 605.

Andersen, O., Knudsen, P., Stenseng, L., 2015. The DTU13 MSS (mean sea surface) and MDT (mean dynamic topography) from 20 years of satellite altimetry. In: Jin, S., Barzaghi, R. (Eds.), International Association of Geodesy Symposia 2014, vol. 144. Springer, Cham, pp. 111–121. Available from: http://dx.doi.org/10.1007/1345_2015_182.

Andersen, O.B., Knudsen, P., Berry, P., Smith, R., Rémy, F., Flament, T., et al., 2013. New views of Earth using ERS Satellite Altimetry. In: Fletcher, K. (Ed.), ERS Missions: 20 Years of Observing Earth. European Space Agency, Noordwijk, The Netherlands, pp. 225–254.

Bajo, M., De Biasio, F., Umgiesser, G., Vignudelli, S., Zecchetto, S., 2017. Impact of using scatterometer and altimeter data on storm surge forecasting. Ocean Model. 113, 85–94.

Benveniste, J., Roca, M., Levrini, G., Vincent, P., Baker, S., Zanife, O., et al., 2001. The Radar Altimetry Mission: RA-2, MWR, DORIS, and LRR. ESA Bulletin, No 106 (EnviSat Special Issue), pp. 67–76. Available at http://www.esa.int/esapub/bulletin/bullet106/bul106_5.pdf.

Berry, P.A.M., Wheeler, J.L., 2009. Jason2-ENVISAT exploitation, development of algorithms for the exploitation of Jason-2-ENVISATaltimetry for the generation of a river and lake product," Product Handbook, Internal Report DMU-RIVL-SPE-03-110, De Montfort University 3(5).

Berry, P.A.M., Jasper, A., Bracke, H., 1997. Retracking ERS-1 altimeter waveforms overland for topographic height determination: an expert system approach,". ESA Pub. SP414 1, 403–408.

Berry, P.A.M., Garlick, J.D., Freeman, J.A., Mathers, E.L., 2005. Global inland water monitoring from multi-mission altimetry. Geophys. Res. Lett. 32 (16), L16401.

Biancamaria, S., Lettenmaier, D.P., Pavelsky, T.M., 2016. The SWOT mission and its capabilities for land hydrology. Surveys Geophys. 37 (2), 307–337.

Birkett, C.M., Ricko, M., Beckley, B.D., Yang, X., Tetrault, R.L., 2017, December. G-REALM: a lake/reservoir monitoring tool for drought monitoring and water resources management. In AGU Fall Meeting Abstracts.

Birol, F., Fuller, N., Lyard, F., Cancet, M., Nino, F., Delebecque, C., et al., 2017. Coastal applications from nadir altimetry: example of the X-TRACK regional products. Adv. Space Res. 59 (4), 936–953.

Bonnefond, P., Haines, B.J., Watson, C., 2011. In situ absolute calibration and validation: a link from coastal to open-ocean altimetry. In: Vignudelli, S., Kostianoy, A.G., Cipollini, P., Benveniste, J. (Eds.), Coastal Altimetry. Springer-Verlag Berlin Heidelberg. Available from: http://dx.doi.org/10.1007/978-3-642-12796-0_11.

Bouffard, J., Naeije, M., Banks, C.J., Calafat, F.M., Cipollini, P., Snaith, H.M., et al., 2017. CryoSat ocean product quality status and future evolution. Advances in Space Research. Available from: http://dx.doi.org/10.1016/j.asr.2017.11.043 (in press).

Brown, G., 1977. The average impulse response of a rough surface and its applications. IEEE Trans. Anten. Propagat. 25 (1), 67–74.

Chevalier, L., Laignel, B., Turki, I., Lyard, F., Lion, C., 2017. Hydrological variability from gauging stations and simulated SWOT data, for Major French Rivers. J. Geosci. Environ. Protect. 5 (13), 54.

Cipollini, P., Benveniste, J., Bouffard, J., Emery, W., Gommenginger, C., Griffin, D., et al., 2010. The role of altimetry in Coastal Observing SystemsIn: Hall, J., Harrison, D.E., Stammer, D. (Eds.), Proceedings of OceanObs'09: Sustained Ocean Observations and Information for Society, Vol. 2. ESA Publication WPP-306, Venice, Italy, pp. 21–25. September 2009. Available from: http://dx.doi.org/10.5270/OceanObs09.cwp.16.

Cipollini, P., Calafat, F.M., Jevrejeva, S., Melet, A., Prandi, P., 2017a. Monitoring sea level in the coastal zone with satellite altimetry and tide gauges. Surveys Geophys. 38 (1), 33–57.

Cipollini, P., Benveniste, J., Birol, F., Fernandes, M.J., Obligis, E., Passaro, M., et al., 2017b. Satellite Altimetry in Coastal Regions. In: Cazenave, A., Stammer, D. (Eds.), Satellite Altimetry over Oceans and Land Surfaces, first ed CRC Press, Boca Raton, FL, USA.

Crétaux, J.F., Jelinski, W., Calmant, S., Kouraev, A., Vuglinski, V., Bergé-Nguyen, M., et al., 2011a. SOLS: a lake database to monitor in the Near Real Time water level and storage variations from remote sensing data. Adv. Space Res. 47 (9), 1497–1507.

Crétaux, J.F., Calmant, S., Romanovski, V., Perosanz, F., Tashbaeva, S., Bonnefond, P., et al., 2011b. Absolute calibration of Jason radar altimeters from GPS kinematic campaigns over Lake Issykkul. Marine Geodesy 34 (3-4), 291–318.

Crétaux, J.F., Nielsen, K., Frappart, F., Papa, F., Calmant, S., Benveniste, J., 2017. Hydrological applications of satellite altimetry: rivers, lakes, man-made reservoirs, inundated areas. In: Cazenave, A., Stammer, D. (Eds.), Satellite Altimetry Over Oceans and Land Surfaces, first ed CRC Press, Boca Raton, FL, USA.

De Biasio, F., Vignudelli, S., della Valle, A., Umgiesser, G., Bajo, M., Zecchetto, S., 2016. Exploiting the potential of satellite microwave remote sensing to hindcast the storm surge in the Gulf of Venice. IEEE J. Selected Topics Appl. Earth Observat. Remote Sensing 9 (11), 5089–5105.

De Biasio, F., Bajo, M., Vignudelli, S., Umgiesser, G., Zecchetto, S., 2017. Improvements of storm surge forecasting in the Gulf of Venice exploiting the potential of satellite data: the ESA DUE eSurge-Venice project. Europ. J. Remote Sensing 50 (1), 428–441.

Díez-Minguito, M., Baquerizo, A., Ortega-Sánchez, M., Navarro, G., Losada, M.A., 2012. Tide transformation in the Guadalquivir estuary (SW Spain) and process-based zonation. J. Geophys. Res. 117, 14. Available from: https://doi.org/10.1029/2011JC007344. C03019.

Dinardo, S., 2014. GPOD CryoSat-2 SARvatore Software Prototype User Manual. https://wiki.services.eoportal.org/tiki-index.php?page = GPOD + CryoSat-2 + SARvatore + Software + Prototype + User + Manual.

Dinardo, S., Restano, M., Ambrózio, A., Benveniste, J., 2015. SAR altimetry processing on demand service for CryoSat-2 and Sentinel-3 at ESA G-POD. In Sentinel-3 for Science Workshop 734, 16.

Dinardo, S., Fenoglio-Marc, L., Buchhaupt, C., Becker, M., Scharroo, R., Fernandes, M. J., et al., 2018. Coastal SAR and PLRM Altimetry in German Bight and West Baltic Sea. Advances in Space Research.

Donlon, C., Berruti, B., Buongiorno, A., Ferreira, M.H., Féménias, P., Frerick, J., et al., 2012. The global monitoring for environment and security (GMES) sentinel-3 mission. Remote Sensing Environ. 120, 37–57. Available from: https://doi.org/10.1016/j.rse.2011.07.024.

Duan, Z., Bastiaanssen, W.G.M., 2013. Estimating water volume variations in lakes and reservoirs from four operational satellite altimetry databases and satellite imagery data. Remote Sensing Environ. 134 (2013), 403–416.

ESA, European Space Agency, 2017. Sentinels for Copernicus, BR-319 (third ed.). ISBN 978-92-9221-108-0.

EuroGOOS, Sea level observation networks in and around Europe Challenges in monitoring increasing sea level hazards, Note to policymakers, report, 4 pp, April 2017. Available at www.eurogoos.eu.

Fenoglio-Marc, L., Scharroo, R., Annunziato, A., Mendoza, L., Becker, M., Lillibridge, J., 2015. Cyclone Xaver seen by geodetic observations. Geophys. Res. Lett. 42 (22), 9925–9932.

Fjortoft, R., Gaudin, J.M., Pourthie, N., Lalaurie, J.C., Mallet, A., Nouvel, J.F., et al., 2014. KaRIn on SWOT: characteristics of near-nadir Ka-band interferometric SAR imagery. IEEE Trans. Geosci. Remote Sensing 52 (4), 2172–2185.

Fu, L.L., Cazenave, A. (Eds.), 2000. Satellite Altimetry and Earth Sciences: A Handbook of Techniques and Applications, vol. 69. Elsevier.

Fu, L.L., Christensen, E.J., Yamarone, C.A., Lefebvre, M., Menard, Y., Dorrer, M., et al., 1994. TOPEX/POSEIDON mission overview. J. Geophys. Res. Oceans 99 (C12), 24369–24381.

GEOSS,GEOSS Water Services for Data and Maps Engineering Report GEOSS Architecture Implementation Pilot Phase 6 Version 0.6, report 31 pp, March 2014, doi:10.13140/2.1.2623.1364.

Gómez-Enri, J., Vignudelli, S., Quartly, G., Gommenginger, C., Benveniste, J., 2009. Bringing satellite radar altimetry closer to shore. SPIE Newsroom 1–3. Available from: https://doi.org/10.1117/2.1200908.1797.

Gómez-Enri, J., Vignudelli, S., Quartly, G.D., Gommenginger, C.P., Cipollini, P., Challenor, P.G., et al., 2010. Modeling ENVISAT RA-2 waveforms in the coastal zone: case study of calm water contamination. IEEE Geosci. Remote Sensing Lett. 7 (3), 474–478.

Gómez-Enri, J., Escudier, R., Pascual, A., Mañanes, R., 2015. Heavy Guadalquivir River discharge detection with satellite altimetry: the case of the eastern continental shelf of the Gulf of Cadiz (Iberian Peninsula). Adv. Space Res. 55, 1590–1603. Available from: https://doi.org/10.1016/j.asr.2014.12.039.

Gómez-Enri, J., Cipollini, P., Passaro, M., Vignudelli, S., Tejedor, B., Coca, J., 2016. Coastal Altimetry products in the Strait of Gibraltar. IEEE Trans. Geosci. Remote Sensing 54 (9), 5455–5466.

Gómez-Enri, J., Vignudelli, S., Cipollini, P., Coca, J., Gonzalez, C.J., 2017. Validation of CryoSat-2 SIRAL sea level data in the Eastern continental shelf of the Gulf of Cadiz. Adv. Space Res. Available from: https://doi.org/10.1016/j.asr.2017.10.042. (in press).

Gommenginger, C., Thibaut, P., Fenoglio-Marc, L., Quartly, G., Deng, X., Gómez-Enri, J., et al., 2011. Retracking altimeter waveforms near the coasts. In: Vignudelli, S., Kostianoy, A.G., Cipollini, P., Benveniste, J. (Eds.), Coastal Altimetry. Springer-Verlag Berlin Heidelberg. Available from: http://dx.doi.org/10.1007/978-3-642-12796-0_4.

González-Ortegón, E., Drake, P., 2012. Effects of freshwater inputs on the lower trophic levels of a temperate estuary: physical, physiological or trophic forcing? Aquatic Sci. 74, 455–469. Available from: https://doi.org/10.1007/S00027-011-0240-5.

González-Ortegón, E., Subida, M.D., Cuesta, J.A., Arias, A.M., Fernández-Delgado, C., Drake, P., 2010. The impact of extreme turbidity events on the nursery function of a temperate European estuary with regulated freshwater inflow. Estuarine Coastal Shelf Sci. 87, 311–324. Available from: https://doi.org/10.1016/J.ECSS.2010.01.013.

Göttl, F., Dettmering, D., Müller, F.L., Schwatke, C., 2016. Lake level estimation based on CryoSat-2 SAR altimetry and multi-looked waveform classification. Remote Sensing 8 (11), 885.

Han, G., Ma, Z., Chen, D., Chen, N., 2012. Observing storm surges from space: hurricane Igor off Newfoundland. Sci. Rep. 2, 1010.

Han, G., Ma, Z., Chen, N., Chen, N., Yang, J., Chen, D., 2017. Hurricane Isaac storm surges off Florida observed by Jason-1 and Jason-2 satellite altimeters. Remote Sensing Environ. 198, 244–253.

Hossain, F., Srinivasan, M., Peterson, C., Andral, A., et al., 2017. Engaging the User Community for advancing Societal Applications of The Surface Water Ocean Topography Mission, Report, 6 pp, American Meteorological Society, doi:10.1175/BAMS-D-17-0161.1

Huang, Q., Long, D., Du, M., Zeng, C., Li, X., Hou, A., et al., 2018. An improved approach to monitoring Brahmaputra River water levels using retracked altimetry data. Remote Sensing Environ. 211, 112–128.

Høyer, J.L., Bøvith, T., 2005. Validation and applications of near-real time altimetry data for operational oceanography in coastal and shelf seas, in Proceedings from the ENVISAT symposium, September 6-10, 2004, Salzburg, Austria, ESA SP-572.

Idris, N.H., Deng, X., Md Din, A.H., Idris, N.H., 2017. CAWRES: a waveform retracking fuzzy expert system for optimizing coastal sea levels from Jason-1 and Jason-2 Satellite Altimetry Data. Remote Sensing 9 (6), 603.

Jarihani, A.A., Callow, J.N., Johansen, K., Gouweleeuw, B., 2013. Evaluation of multiple satellite altimetry data for studying inland water bodies and river floods. J. Hydrol. 505, 78–90.

Jiang, X., Lin, M., Liu, J., Zhang, Y., Xie, X., Peng, H., et al., 2012. The HY-2 satellite and its preliminary assessment. Int. J. Digital Earth 5 (3), 266–281.

Kilgus, C., Hoffman, E., Frain, W., 1989. Monitoring the Ocean with NAVY Radar Altimeter Lightsats. Johns Hopkins University Applied Physics Laboratory, Laurel, MD.

Lambin, J., Morrow, R., Fu, L.-L., Willis, J., Hans, B., Lillibridge, J., et al., 2010. The OSTM/Jason-2 Mission. Marine Geodesy 33, 4–25. Available from: https://doi.org/10.1080/01490419.2010.491030.

Lerch, F.J., Klosko, S.M., Laubscher, R.E., Wagner, C.A., 1979. Gravity model improvement using Geos 3 (GEM 9 and 10). J. Geophys. Res. Solid Earth 84 (B8), 3897–3916.

Li, X., Han, G., Yang, J., Chen, D., Zheng, G., Chen, N., 2018. Using satellite altimetry to calibrate the simulation of typhoon seth storm surge off Southeast China. Remote Sensing 10 (4), 657.

Lillibridge, J., Lin, M., Shum, C.K., 2013. Hurricane Sandy storm surge measured by satellite altimetry. Oceanography 26 (2), 8–9.

Marsh, J.G., Chang, E.S., 1976. Detailed gravimetric geoid confirmation of sea surface topography detected by the Skylab S-193 altimeter in the Atlantic Ocean. J. Geodesy 50 (3), 291–299.

McConathy, D.R., Kilgus, C.C., 1987. The Navy GEOSAT mission: an overview. Johns Hopkins APL Technical Digest 8 (2), 170–175.

Ménard, Y., Fu, L.L., Escudier, P., Parisot, F., Perbos, J., Vincent, P., et al., 2003. The Jason-1 mission special issue: Jason-1 calibration/validation. Marine Geodesy 26 (3-4), 131–146.

Mercier, F., Rosmorduc, V., Carrere, L., Thibaut, P., 2010. Coastal and Hydrology Altimetry product (PISTACH) handbook. Centre National d'Études Spatiales (CNES). Paris, France, p. 4.

Michailovsky, C.I., McEnnis, S., Berry, P.A.M., Smith, R., Bauer-Gottwein, P., 2012. River monitoring from satellite radar altimetry in the Zambezi River basin. Hydrol. Earth Syst. Sci. 16, 2181–2192.

Moore, P., Birkinshaw, S.J., Ambrózio, A., Restano, M., Benveniste, J., 2018. CryoSat-2 Full Bit Rate Level 1A processing and validation for inland water applications. Adv. Space Res. 62 (6), 1497–1515. Available from: https://doi.org/10.1016/j.asr.2017.12.015.

Mroczek, A., Jacobs, G., 2015. Nanosats for radar altimetry. in 29th Annual AIAA/USU Conference on Small Satellites, Logan, Utah.

Navarro, G., Huertas, I.E., Costas, E., Flecha, S., Díez-Minguito, M., Caballero, I., et al., 2012. Use of a real-time remote monitoring Network (RTRM) to characterize the Guadalquivir estuary (Spain). Sensors 12, 1398–1421. Available from: https://doi.org/10.3390/S120201398.

Nguy-Robertson, A., May, J., Dartevelle, S., Birkett, C., Lucero, E., Russo, T., et al., 2018. Inferring elevation variation of lakes and reservoirs from areal extents: calibrating with altimeter and in situ data. vol. 9. Remote Sensing Applications: Society and Environment, January 2018, pp. 116–125.

Nielsen, K., Stensen, L., Andersen, O.B., Villadsen, H., Knudsen, P., 2015. Validation of CryoSat-2 SAR mode based lake levels. Remote Sens. Environ. 171, 162–170.

Nielsen, K., Stenseng L., Villadsen H., Andersen O.B., Knudsen, P., 2016. Altimetry for inland waters, Living Planet Symposium. In: L. Ouwehand (Ed.), Proceedings of the conference held 9–13 May 2016 in Prague, Czech Republic. ESA-SP Volume 740, ISBN: 978-92-9221-305-3, p. 132.

Parrinello, T., Shepherd, A., Bouffard, J., Badessi, S., Casal, T., Davidson, M., et al., 2018. CryoSat: ESA's ice mission-eight years in space. Adv. Space Res. Available from: https://doi.org/10.1016/j.asr.2018.04.014.

Passaro, M., 2017. User Manual - COSTA v1.0 - DGFI-TUM Along Track Sea Level Product for ERS-2 and Envisat (1996-2010) in the Mediterranean Sea and in the North Sea. Deutsches Geodätisches Forschungsinstitut der Technischen Universität München (DGFI-TUM), München, 7 pp, hdl:10013/epic.50369.d001.

Passaro, M., Cipollini, P., Vignudelli, S., Quartly, G.D., Snaith, H.M., 2014. ALES: a multi-mission adaptive subwaveform retracker for coastal and open ocean altimetry. Remote Sensing Environ. 145, 173–189.

Passaro, M., Cipollini, P., Vignudelli, S., Quartly, G.D., Snaith, H.M., 2015. ALES Jason-2 Coastal Altimetry Version 1. Ver. 1. PO.DAAC, CA, USA. Dataset accessed at https://podaac.jpl.nasa.gov.

Peng, F., Deng, X., 2018. A new retracking technique for Brown peaky altimetric waveforms. Marine Geodesy 41 (2), 99–125.

Peng, S.Q., Xie, L., 2006. Effect of determining initial conditions by four-dimensional variational data assimilation on storm surge forecasting. Ocean Modelling 14 (1-2), 1–18. Available from: https://doi.org/10.1016/j.ocemod.2006.03.005.

Philippart M.E., Gebraada A., 2002 Assimilating satellite altimeter data in operational sea level and storm surge forecasting, In: Proceeding of the Second International Conference on EuroGOOS. Elseviers Oceanography Series, vol. 66, pp. 469-479. doi:10.1016/S0422-9894(02)80053-8.

Raney, R.K., 1998. The delay/doppler radar altimeter. IEEE Trans. Geosci. Rem. Sens. 36 (5), 1578–1588. Available from: https://doi.org/10.1109/36.718861.

Resti, A., Benveniste, J., Roca, M., Levrini, G., Johannessen, J., 1999. The Envisat radar altimeter system (RA-2). ESA Bull. 98 (8).

Richard, J., Enjolras, V., Rys, L., Vallon, J., Nann, I., Escudier, P., 2008, July. Space altimetry from nano-satellites: payload feasibility, missions and system performances. In: Geoscience and Remote Sensing Symposium, 2008. IGARSS 2008. IEEE International (Vol. 3, pp. III-71). IEEE.

Ricko, M., Carton, J.A., Birkett, C.M., Crétaux, J.F., 2012. Intercomparison and validation of continental water level products derived from satellite radar altimetry. J. Appl. Remote Sensing 6 (1), 061710.

Rodríguez, E., 2016. Surface Water and Ocean Topography Mission Project Science Requirements Document., Jet PropulsionLaboratory, California Institute of Technology, JPL D-61923; March 2016; 28 p. Available online: https://swot.jpl.nasa.gov/documents.htm.

Rodriguez, E., Fernandez, D.E., Peral, E., Chen, C.W., De Bleser, J.W., Williams, B., 2017. In: Cazenave, A., Stammer, D. (Eds.), Wide-Swath Altimetry A Review, in Satellite Altimetry over Oceans and Land Surfaces, first ed CRC Press, Boca Raton, FL, USA.

Roscher, R., Uebbing, B., Kusche, J., 2017. STAR: spatio-temporal altimeter waveform retracking using sparse representation and conditional random fields. Remote Sensing Environ. 201, 148–164.

Ruhi, A., Messager, M.L., Olden, J.D., 2018. Tracking the pulse of the Earth's fresh waters. Nat. Sustain. 1 (4), 198.

Schaeffer, P., Faugere, Y., Legeais, J.F., Ollivier, A., Guinle, T., Picot, N., 2012. The CNES_CLS11 global mean sea surface computed from 16 years of satellite altimeter data. Marine Geodesy 35 (Suppl. 1), 3–19.

Scharroo, R., Bonekamp, H., Ponsard, C., Parisot, F., von Engeln, A., Tahtadjiev, M., et al., 2016. Jason continuity of services: continuing the Jason altimeter data records as Copernicus Sentinel-6. Ocean Sci. 12 (2), 471–479.

Scharroo, R., Leuliette, E.W., Lillibridge, J.L., Byrne, D., Naeije, M.C., Mitchum, G.T., 2013. RADS: consistent multi-mission products. In Proceedings of the Symposium on 20 Years of Progress in Radar Altimetry, Venice, 20–28 September 2012. European Space Agency Special Publication, ESA SP-710, 1–4.

Scharroo, R., Smith, W.H., Lillibridge, J.L., 2005. Satellite altimetry and the intensification of Hurricane Katrina. Eos, Trans. Am. Geophys. Union 86 (40), 366.

Schwatke, C., Dettmering, D., Bosch, W., Seitz, F., 2015. DAHITI—an innovative approach for estimating water level time series over inland waters using multi-mission satellite altimetry. Hydrol. Earth System Sci. 19 (10), 4345.

Scozzari, A., Gómez-Enri, J., Vignudelli, S., Soldovieri, F., 2012. Understanding target-like signals in coastal altimetry: experimentation of a tomographic imaging technique. Geophys. Res. Lett. 39 (2).

Shah, R., Garrison, J.L., 2017. Precision of Ku-band reflected signals of opportunity altimetry. IEEE Geosci. Remote Sensing Lett. 14 (10), 1840–1844.

Snaith, H., Scharroo, R., Naeije, M., 2006. Just-in-time altimetry: international collaboration in provision of altimetry datasets, Proceedings of the Symposium on 15 Years of Progress in Radar Altimetry, 13-18 March 2006, Venice, Italy (ESA SP-614, July 2006).

Stacy, N., 2012. 6U radar altimeter concept. 6U Cubesat Low Cost Space Missions Workshop. Australian Centre for Space Engineering Research, Canberra Australia.

Tapley, B.D., Born, G.H., Parke, M.E., 1982. The Seasat altimeter data and its accuracy assessment. J. Geophys. Res. Oceans 87 (C5), 3179–3188.

Tourian, M.J., Elmi, O., Shafaghi, Y., Sneeuw, N., 2016. HydroSat: a repository of global water cycle products from spaceborne geodetic sensors. Poster. OSTS meeting 2016, La Rochelle, France (01.11.–04.11.).

Verron, J., Sengenes, P., Lambin, J., Noubel, J., Steunou, N., Guillot, A., et al., 2015. The SARAL/AltiKa altimetry satellite mission. Marine Geodesy 38 (Suppl. 1), 2–21.

Vignudelli, S., Cipollini, P., Gommenginger, C., Gleason, S., Snaith, H.M., Coelho, H., et al., 2011a. Satellite altimetry: sailing closer to the coast. Remote Sensing of the Changing Oceans. Springer, Berlin, Heidelberg, pp. 217–238.

Vignudelli, S., Kostianoy, A.G., Cipollini, P., Benveniste, J. (Eds.), 2011b. Coastal Altimetry. Springer-Verlag, Berlin, Heidelberg. Available from: http://dx.doi.org/10.1007/978-3-642-12796-0, 578 pp.

Villadsen, H., Deng, X., Andersen, O.B., Stenseng, L., Nielsen, K., Knudsen, P., 2016. Improved inland water levels from SAR altimetry using novel empirical and physical retrackers. J. Hydrol. 537, 234–247.

Vu, P.L., Frappart, F., Darrozes, J., Marieu, V., Blarel, F., Ramillien, G., et al., 2018. Multi-satellite altimeter validation along the French Atlantic Coast in the Southern Bay of Biscay from ERS-2 to SARAL. Remote Sensing 10 (1), 93.

Vuglinskiy, V., Monteduro, M., Sessa, R., Gronskaya, T., Crétaux, J.F., 2009. Water level in lakes and reservoirs, water storage. Global Terrestrial Observing System (GTOS) Publication 59, 28.

WMO, 2010. Manual on Stream Gauging, vol. 1, n. 1044, World Meteorological Organization, Geneva, 245 pp.

Walker, D., Barry, R., 1997. The Navy GEOSAT follow-on altimeter-a true dual use technology. IEEE Aerospace Conference, Snowmass at Aspen, CO, 1997, vol.3, pp. 71–82. doi: 10.1109/AERO.1997.574853.

Woodworth, P.L., Hunter, J.R., Marcos, M., Caldwell, P., Menendez, M., Haigh, I., 2016. Towards a global higher-frequency sea level dataset. Geosci. Data J. 3 (2), 50–59.

Xu, X.Y., Birol, F., Cazenave, A., 2018. Evaluation of coastal sea level offshore hong kong from jason-2 altimetry. Remote Sensing 10 (2), 282.

Yuan, C., Gong, P., Zhang, H., Guo, H., Pan, B., 2017. Monitoring water level changes from retracked Jason-2 altimetry data: a case study in the Yangtze River. China. Remote Sensing Lett. 8 (5), 399–408.

Zheng, Y., 1999. The GANDER microsatellite radar altimeter constellation for global sea state monitoring. In: Paper presented at 13th AIAA/USU Conference on Small Satellite. Logan, Utah: AIAA, 1999.

Further reading

Rio, M.H., Guinehut, S., Larnicol, G., 2011. New CNES-CLS09 global mean dynamic topography computed from the combination of GRACE data, altimetry, and in situ measurements. J. Geophys. Res. Oceans 116 (C7).

CHAPTER FIVE

Remote sensing techniques for estimating evaporation

Thomas R.H. Holmes
Hydrological Science Lab, NASA Goddard Space Flight Center, Greenbelt, MD, United States

5.1 Introduction

In most general terms, evaporation is the process by which a substance is transformed from its liquid state into vapor. To a hydrologist, evaporation (E) refers specifically to the volume of water that evaporates in a given time-period from the Earth's surface into the atmosphere. Evaporation of water completes the hydrological cycle for ~60% of the precipitation that falls on the land areas of the Earth (e.g., Oki and Kanae, 2006). This exchange of water vapor between surface and atmosphere is associated with a large transfer of energy in the form of *latent heat*, the conversion of thermal energy into the molecular formation of water vapor from liquid. Latent energy is the dominant source of atmospheric heating when it is liberated through *condensation*. The process of evaporation and condensation transfers more than half of the annual solar energy received by the Earth's land masses to the atmosphere. Evaporation also accompanies the exchange of carbon dioxide and oxygen between growing vegetation and the atmosphere.

Despite this central role for evaporation as a link between the energy, water, and carbon cycles it is one of the least constrained components of the hydrological cycle in land surface models (LSM). Evaporation is difficult to measure remotely because it lacks a direct electromagnetic fingerprint that can be exploited by satellite retrievals. Even if the bulk portion of available energy that is diverted to latent heat can be estimated as a residual of the surface energy balance, the attribution to its source water reservoirs often depends on physical model assumptions. In order to improve estimates of overall evaporation and gain process understanding of the coupled carbon and water cycle, it is important to accurately

Extreme Hydroclimatic Events and Multivariate Hazards in a Changing Environment.
DOI: https://doi.org/10.1016/B978-0-12-814899-0.00005-5

estimate the partitioning of the bulk water flux into soil and vegetation source components. This is because the immediate source of liquid water that evaporates into the atmosphere determines the relative importance of meteorological, biophysiological, and hydrological controls on evaporation. The pathway of the water molecule also affects the circulation rate of the hydrological cycle, with implications for the prediction of available renewable freshwater resources (Oki and Kanae, 2006).

Passive sources of water for evaporation respond predominantly to meteorological conditions in proportion with water availability and surface texture. There is no biological control on these processes. If the immediate source is the moisture in the soil it is called *soil evaporation*. Globally this constitutes 20%–40% of the total evaporation. Similar to this is evaporation from open water bodies like lakes, rivers, but also temporary pools of rainwater, snow, and ice. A special case of a temporary source of water are pools or drops of water in the canopy that develop when rainwater is intercepted by the leaves. When this water evaporates before it reaches the soil it is referred to as *evaporation of intercepted water* (or interception for short) and can account for 10%–35% of the incident precipitation in forests (Miralles et al., 2011). Finally, some rainwater (or water from sprinkler irrigation (Cavero et al., 2009) evaporates before it reaches the ground, but this source of latent heat is typically neglected in land surface models.

In contrast to these passive sources of water for evaporation, the water contained in vegetation tissue is subject to biophysical regulation which can act to moderate the influence of evaporative demand. Plants are also connected to a larger soil reservoir through the root network which may sustain the water supply through periods of drought. When leaf-water evaporates it is called *transpiration* (T), which reflects its role as the main leaf-cooling process and can affect regional temperatures (e.g., Mueller et al., 2016). Transpiration is the dominant pathway for the total *Evapotranspiration* (ET) and is estimated to account for two-thirds of global land ET based on flux tower measurements (Jasechko et al., 2013; Schlesinger and Jasechko, 2014). While transpiration helps to keep plants cool, it is also a necessary side-effect of the plants need to breath in CO_2 for photosynthesis and sustain their growth. The biomass yield per unit of water use (crop-per-drop) is an important indicator of agricultural efficiency in water limited regions. Estimating crop water requirements and comparing it to antecedent precipitation is also a straightforward way to assess irrigation requirements (Allen et al., 1998). If the rate of water loss cannot be matched by water uptake from the roots, the leaf water potential drops, and ultimately the leaves will wilt. Different species of plants may have different strategies for optimizing their carbon gain while limiting the associated water loss through transpiration and this complicates the modeling of ecosystem response to drought (Konings and Gentine, 2017; Konings et al., 2017). Combined with a longer-term (climatological) baseline, remote sensing estimates of ET are used for (agricultural) drought monitoring (Anderson et al., 2007; Otkin et al., 2016).

Techniques for direct measurement of evaporation include the eddy-covariance (EC) method, Bowen ratio energy balance, and measurement of water loss with lysimeters or mass-balance methods, see for example, Allen et al. (2011) for details. Gas exchange measurements through the EC flux system is now a standard component of the experimental set-up of flux towers, many of which are organized in a global network that includes over 200 towers (Fluxnet; Baldocchi et al., 2001). EC systems have also been mounted on aircraft to measure gas exchange in the boundary layer over larger areal domains (e.g., Anderson et al., 2008; Wolfe et al., 2018). These in situ measurements are invaluable for the development and validation of process descriptions due to their continuous diurnal sampling and wealth of collocated instrumentation. However, to achieve regular monitoring of areal evaporation over land these sparse tower and aircraft measurements need to be combined with sustained satellite remote sensing. This chapter gives a broad overview of the satellite data products (Section 5.2) and the types of observation-based methodologies (Section 5.3) employed in the remote estimation of terrestrial evaporation at diverse spatial domains. Two specific examples of contrasting remote-sensing strategies for estimating evaporation over land are detailed in Sections 5.3.1 and 5.3.2. A brief discussion of more recent developments is included in Section 5.4.

5.2 Satellite measurements for evaporation retrievals

There is no single frequency-band that gives a unique fingerprint of evaporation of water into the atmosphere. Remote-sensing approaches for estimating evaporation from space typically rely on an assortment of more readily measurable meteorological and biophysical variables. This makes remote-sensing of evaporation reliant on several unique parts of the electromagnetic spectrum, from the visible, through infrared and microwave parts of the spectrum.

Many aspects of the vegetation state can be deduced from specific spectral features in the visible to shortwave infrared part of the spectrum. Such information is traditionally summarized in vegetation indices (VI) like leaf area index (LAI, see also Chapter 1.6 for further details) and normalized difference vegetation index (NDVI), and used for land classification. Simple relationships have been developed to estimate areal vegetation fraction from LAI (Carlson and Ripley, 1997). VI in combination with precipitation and air temperature has some explanatory information on monthly ET measurements. This was utilized by Jung et al. (2010) who trained a machine learning method on global flux tower observations of ET (Fluxnet) to predict global (monthly) land evaporation for 1982–2008 from satellite records of NDVI. Meteorological information like cloud properties and the resulting shortwave and longwave radiative budget of the land surface are dependent on measurements in the visible to infrared parts of the spectrum (e.g., CERES). Several authors have demonstrated the ability to retrieve more detailed plant functional traits and biophysical parameters from the spectral reflectance (e.g., Serbin et al., 2014; Frankenberg et al., 2011), but these have not yet been incorporated in global ET measurement approaches.

The microwave part of the spectrum has wavelengths of 1 mm to 1 m. These longer waves can penetrate soil, vegetation, and clouds to varying degrees depending on exact wavelength. Microwaves convey information on water content and temperature from the intervening soil, vegetation, and atmospheric layers. Since 1979, space-based passive-microwave (PMW) radiometers measure the emitted radiative energy at frequencies chosen for a variety of remote sensing applications. Their low spatial resolution (10–40 km) is appropriate for continental or global approaches. Radars are used to measure the microwave reflectance at somewhat higher resolution and are important for precipitation measurement (Smith et al., 2007; Skofronick-Jackson et al., 2017), and surface humidity (Jackson et al., 2009; Tomita et al., 2018). PMW land surface measurements with particular relevance to ET estimation are long-term soil moisture records (e.g., Dorigo et al., 2017). PMW observations also convey information on vegetation optical depth (Owe et al., 2008) which can be used to estimate vegetation water content or biomass

(Liu et al., 2013; Momen et al., 2017), and land surface temperature (LST) (Prigent et al., 2016; Holmes et al., 2016).

The most important part of the spectrum for the remote measurement of ET is the thermal infrared (TIR) region from 1 to 16 μm. Because the phase change of water from liquid to gas represents such a large sink in the surface energy balance, the thermal fingerprint is the most direct diagnostic of latent heat available to remote sensing. This is exploited by surface energy balance approaches (see below) that can leverage the high spatial resolution of TIR-based land surface temperature (LST). Satellites provide thermal radiance with spatial resolutions down to 30 m from low Earth orbit (e.g., Landsat-8 with a bi-weekly revisit time). More moderate spatial resolution but almost daily sampling is afforded by large-swath imagers like VIIRS (375 m) and MODIS (1 km). Multiband thermal infrared radiometers are also available from satellites in geostationary orbit, resulting in diurnal sampling at lower spatial resolution, e.g., 5-minute temporal and 4-km spatial sampling for the GOES Advanced Baseline Imager (ABI), and MSG-SEVIRI (3 km, 15 minutes). The accuracy of the LST retrieval depends on the ability to simultaneously estimate spectral emissivity, which depends on the number of thermal bands. Clouds are almost completely opaque to TIR emission which prevents the retrieval of TIR-based LST from cloud-covered surfaces. It can also impact the analysis of surrounding clear-sky surfaces if clouds are not adequately screened. The efficacy of the cloud screening is an important factor in the precision of LST product and is aided by additional thermal channels.

5.3 Evaporation retrieval approaches

ET retrieval approaches combine observable drivers within statistical or process-based methodologies. These process descriptions make use of the concept of *potential evaporation* (*Ep*), which is the rate of evaporation that a large area with growing vegetation would sustain if there is no limit on water availability. It is used in models to represent the atmospheric demand for water, and depends on meteorological conditions like surface humidity, net radiation, wind speed, and near-surface temperature gradients. Only in humid areas does the *actual evaporation* approach (or surpass) the potential for a large part of the year. The challenge is in estimating the actual evaporation from space, because it also depends on the surface hydrology and biophysical state of the surface.

There are two main categories of satellite-based methodologies to estimate the actual evaporation of water from the land surface. The first category includes methods that combine top-down *Ep* estimates with a bottom-up estimate of *evaporative stress*, or reduction from evaporative demand, that is, based on statistical parameterizations and observation-informed surface states. The second category of methodologies takes an entirely top-down approach and solves for actual evaporation as a residual of the surface energy-balance (EB). Although *Ep* is used to give context to the estimated *E*, it is not a driving dataset in energy-balance solutions.

There are several approaches to parameterize Ep and evaporative stress that have been adapted to available satellite data sets. The first set of algorithms implements the Penman-Monteith (P-M) formulation (Monteith, 1965) (e.g., Cleugh et al., 2007; Mu et al., 2011). The P-M formulation accounts for energy limitations, aerodynamic resistance, and stomatal conductance in its estimate of Ep, and parameterizes surface resistance based on VIs and surface humidity. The meteorological input requirements can be demanding for global implementation. The Priestley and Taylor (1972) (P-T) formulation is derived from the Penman-Monteith equations for a scenario with plentiful water so that the stomatal resistance is zero. P-T estimates of Ep are only based on insolation and temperature, which makes them more readily applicable to satellite measurements than the P-M methods. P-T applications must be combined with a way to estimate the reduction from the potential evaporation rate to account for hydrological or biophysical restrictions in water availability. Fisher et al. (2008) used estimates of water vapor pressure to parameterize water availability for soil evaporation, and VIs and air temperature to parameterize vegetation stress. This approach made use of the long satellite record of NDVI and net radiation to produce monthly ET estimates from 1984 to 2006 (PT-JPL). Another distinct approach to estimate evaporative stress is to combine P-T with a running-water-balance with inputs of precipitation and assimilation of soil moisture measurements (Miralles et al., 2011; Martens et al., 2017). The evaporative stress is subsequently parameterized based on prognostic model states, similar to methods employed by land surface models but with a more direct use of land surface remote sensing, see detailed description below (Section 5.3.1).

In contrast to these bottom-up approaches to estimating evaporative stress, energy balance approaches estimate actual evaporation directly from the thermal fingerprint of the latent heat flux. This is the reason that EB approaches are regarded as more purely diagnostic in comparison to approaches that include prognostic information. EB approaches have a long legacy and have found wide application in agricultural studies that benefit from the high spatial resolution afforded by TIR-based imagers. All EB approaches solve for E as the residual of the surface energy balance (net radiation—ground heat flux—sensible heat flux). The first group of larger scale EB approaches treat evaporation as a single bulk flux that includes soil and vegetation sources, for example, SEBAL (Bastiaanssen et al., 1998), SEBS (Su, 2002), METRIC (Allen et al., 2007), and SSEB (Senay et al., 2016). They evaluate the energy balance at "dry" and "wet" extremes and estimate ET between these extremes based on the spatial variation of internally calibrated temperature within the scene of the satellite image. These EB implementations rely on accurate estimation of the temperature difference between surface and air for the estimation of sensible heat. This is challenging to apply to larger domains due to the inherent biases in the independent estimation of air and surface temperatures.

Two-source EB approaches consider soil and vegetation as separate "sources" for heat and water exchange. They partition LST and net radiation between soil and canopy components and use these to solve a set of physical equations that represent the turbulent flux exchanges between the soil, canopy, and atmosphere (Kustas and Norman, 1999; Norman et al., 1995). These early applications face the same challenge in estimating consistent surface and air temperatures over large areas as the one-source approaches. Regional applications of EB approaches are enabled when the impact of errors in the LST and air temperature boundary conditions are reduced. Anderson et al. (2007) achieved this by basing the physical retrieval of the two-source EB approach on the rate of change in surface temperature during the morning. This reduces the impact of errors in the absolute (instantaneous) LST and air temperature retrievals and forms the basis of a multiscale integrated approach to estimating ET that is detailed in Section 5.3.2.

5.3.1 GLEAM, a water balance approach to estimating evaporative stress

An example of a bottom-up approach to estimating evaporative stress is GLEAM: Global Land Evaporation, Amsterdam Methodology (Miralles et al., 2011; Martens et al., 2017). It contains an observation-driven LSM specifically tailored to estimate global ET for long-term climatological analysis (e.g., Miralles et al., 2014). The LSM consists of a multilayer running-water-balance to estimate root zone water. Inputs to the water-balance are observed precipitation, and soil moisture observations (from PMW sensors). Central to this methodology is the use of P-T formulations to model potential evaporation based on insolation and temperature inputs (see Section 5.1).

The actual evaporation and transpiration are calculated as a fraction of Ep using parameterizations of evaporative stress. The stress is partly based on the modeled water availability in the surface layers which accounts for past precipitation, evaporation, and drainage. Estimates of evaporative stress for vegetation further account for vegetation water content through PMW measurements of vegetation optical depth. GLEAM includes separate stress functions and α for three dynamic surface components that contribute to ET: bare soil, short vegetation, tall vegetation.

GLEAM features the first global implementation of the Gash analytic model for the estimation of evaporation of intercepted rain water (Gash, 1979; Valente et al., 1997). The volume of water that evaporates from the wet canopy during and immediately after a rain storm is estimated as a fraction of daily rainfall. The model parameters further account for canopy cover fraction, canopy storage, mean rainfall rates, and evaporation rates during wet canopy conditions (Miralles et al., 2010). A novel feature is the use of observed lightning frequency (Cecil et al., 2014) to distinguish synoptic from convective precipitation to account for the associated differences in rain rates.

Snow depth estimates from PMW observations are used to divert precipitation into a snowpack that is subject to sublimation before the eventual melt and entry into the soil water reservoir. The contribution of lakes and rivers is not modeled so that the total evaporation estimated by GLEAM only refers to the land fraction of the total surface area.

GLEAM is designed to be implemented for the entire duration of the modern satellite record and has been used to create daily ET records at 0.25-degree resolution from 1980 to present. These long-term ET records have been used in a series of recent hydrological and climate studies to study the impact of climate change and El Nino Southern oscillations on the water cycle, land–atmosphere feedbacks, hydrometeorological extremes, benchmarking and evaluating climate models (Zhang et al., 2016; Miralles et al., 2014).

5.3.2 ALEXI: an integrated framework for two-source EB estimates of evaporation

The Atmosphere-Land Exchange Inverse (ALEXI) model (Anderson et al., 2007; Mecikalski et al., 1999) combines two-source energy balance models (Kustas and Norman, 1999; Norman et al., 1995) with atmospheric boundary layer (ABL) formulations for regional ET estimation. Model outputs include total ET, and estimates of soil evaporation and transpiration separately, with partitioning guided in part by the local vegetation cover fraction. ALEXI enables regional application of the two-source EB by basing the physical retrieval on the rate of change in surface temperature during the morning, reducing the impact of errors in the absolute (instantaneous) LST retrievals and in the air temperature boundary conditions.

The available energy is based on top-of-atmosphere shortwave and longwave radiances, albedo estimates, and longwave radiation calculated from LST according to Stefan-Boltzman's law. Ground heat flux is estimated as a fixed percent of net radiation. LAI is used for partitioning incoming radiation (Anderson et al., 2007). A set of equations describes the energy balance for soil to atmosphere fluxes and vegetation to atmosphere.

The ABL model simulates the changes in air temperature between the time of the morning LST observation (ideally 1 hour after sunrise) and the midday observation (ideally 1 hour before solar noon). The ABL-modeled air temperature at a reference height above the canopy provides a boundary condition that is consistent with the surface fluxes generated by the EB model. P-T Ep serves as an initial estimate for canopy evaporation and is iteratively reduced until the closed energy balance produces soil evaporation that is nonnegative. This procedure is based on the assumption that no condensation occurs during clear-sky daytime hours (Kustas and Norman, 1999, 2000).

Although ALEXI was developed for geostationary sensors (Anderson et al., 2007), global implementations with MODIS-LST (Hain and Anderson, 2017; Holmes et al., 2018) are now routinely available through NASA's Short-term Prediction Research and Transition (SPoRT). Combined with a long-term climatological record, the ratio of ALEXI ET over Ep is also used as an evaporative stress index to diagnose and monitor agricultural drought conditions (Anderson et al., 2011a). Because an EB-model like ALEXI does not model water availability it can be used to identify neglected sources and sinks of water in LSM's, such as groundwater depth, irrigation extent, and tile drainage density (Yilmaz et al., 2014; Hain et al., 2015).

Importantly, ALEXI is part of an integrated framework of multiscale ET estimation (Anderson et al., 2011b) that also includes a flux disaggregation scheme (disALEXI: Norman et al., 2003; Anderson et al., 2004). DisALEXI utilizes higher resolution LST measurements (e.g., Landsat, or airborne platforms) to disaggregate the continental scale physical ALEXI retrievals of daily ET. On the days with high-resolution thermal observations this approach compares well with eddy-covariance measurements from flux towers. In a final step to the multiscale ET analysis framework, the Spatial and Temporal Adaptive Reflectance Fusion Model (STARFM) (Gao et al., 2006) is applied to generate a time-continuous, daily ET product at the fine spatial resolution of the disALEXI output. This ET fusion approach has been successfully demonstrated over rain-fed and irrigated cotton, corn, and soybean fields (Cammalleri et al., 2014; Yang et al., 2017), irrigated vineyards (Semmens et al., 2016), as well as forested landscapes (Yang et al., 2017).

5.4 New developments

Intermodel comparisons of global evaporation products show that they are able to capture important aspects of seasonality and spatial distribution related to climate regimes (Jimenez et al., 2011; Mueller et al., 2013; Michel et al., 2016; Miralles et al., 2016). However, these studies also reveal a lack of agreement between models in terms of the relative contribution of soil evaporation, plant transpiration, and evaporation of the intercepted water to the total evaporation at a global scale. This reflects the large uncertainties of the contribution of transpiration (Wei et al., 2017), but also the treatment of interception as a distinct process. In general, current LSMs are found to underestimate the transpiration contribution to global E compared to estimates from in situ data, satellite products, and isotope measurements (Schlesinger and Jasechko, 2014). These disagreements in the ratio of T to total ET are a significant contributor to uncertainty in long-term predictions of changes in the coupled carbon and water cycle. New measurements that can be used to better estimate photosynthetic activity (Frankenberg et al., 2011), and thus water loss through transpiration, may allow for better observational constraints on the evaporation estimation.

Although surface energy balance approaches have a long legacy, long consistent records of global-scale ET estimates based on EB approach were not available until recently. This explains the lack of representation of EB approaches in the global evaporation comparisons discussed above, with SEBS as the lone exception. As was shown in Section 5.3.2, this situation is now changing and future assessments of global evaporation may be able to draw upon the full range of evaporation estimation approaches. Another development that will enhance the utility for EB approaches to climate science is in the application of more cloud-tolerant microwave observations to the estimation of LST (Holmes et al., 2009, 2016). This is intended to solve a central limitation of TIR-based LST, that no surface information penetrates clouds in the TIR frequency bands. This effectively limits (and biases) the temporal sampling of TIR-based approaches to clear conditions. The utility of microwave-based LST for EB estimates of evaporation has been demonstrated in the context of the ALEXI framework (Holmes et al., 2018). A full integration of PMW LST into the ALEXI framework will reduce the need for interpolation between days with clear-sky conditions and reduce uncertainty related to cloud filtering efficacy.

References

Allen, R.G., Pereira, L.S., Raes, D., Smith, M., 1998. Crop evapotranspiration-Guidelines for computing crop water requirements-FAO Irrigation and drainage paper 56. FAO, Rome, 300 (9), D05109.

Allen, R.G., Pereira, L.S., Howell, T.A., Jensen, M.E., 2011. Evapotranspiration information reporting: I. Factors governing measurement accuracy. Agric. Water Manage. 98 (6), 899–920.

Allen, R.G., Tasumi, M., Trezza, R., 2007. Satellite-based energy balance for mapping evapotranspiration with internalized calibration (METRIC)—model. J. Irrig. Drain. Eng. 133 (4), 380–394.

Anderson, M.C., Norman, J.M., Mecikalski, J.R., Torn, R.D., Kustas, W.P., Basara, J.B., 2004. A multiscale remote sensing model for disaggregating regional fluxes to micrometeorological scales. J. Hydrometeorol. 5 (2), 343–363.

Anderson, M.C., Norman, J.M., Mecikalski, J.R., Otkin, J.A., Kustas, W.P., 2007. A climatological study of evapotranspiration and moisture stress across the continental United States based on thermal remote sensing: 2. Surface moisture climatology. J. Geophys. Res. Atmosph. 112 (D11).

Anderson, M.C., Norman, J.M., Kustas, W.P., Houborg, R., Starks, P.J., Agam, N., 2008. A thermal-based remote sensing technique for routine mapping of land-surface carbon, water and energy fluxes from field to regional scales. Remote Sens. Environ. 112 (12), 4227–4241.

Anderson, M.C., Hain, C., Wardlow, B., Pimstein, A., Mecikalski, J.R., Kustas, W.P., 2011a. Evaluation of drought indices based on thermal remote sensing of evapotranspiration over the continental United States. J. Climate 24 (8), 2025–2044.

Anderson, M.C., Kustas, W.P., Norman, J.M., Hain, C.R., Mecikalski, J.R., Schultz, L., et al., 2011b. Mapping daily evapotranspiration at field to continental scales using geostationary and polar orbiting satellite imagery. Hydrol. Earth Syst. Sci. 15 (1), 223–239.

Baldocchi, D., Falge, E., Gu, L., Olson, R., Hollinger, D., Running, S., et al., 2001. FLUXNET: a new tool to study the temporal and spatial variability of ecosystem–scale carbon dioxide, water vapor, and energy flux densities. Bull. Am. Meteorol. Soc. 82 (11), 2415–2434.

Bastiaanssen, W.G., Menenti, M., Feddes, R.A., Holtslag, A.A.M., 1998. A remote sensing surface energy balance algorithm for land (SEBAL). 1. Formulation. J. Hydrol. 212, 198–212.

Cammalleri, C., Anderson, M.C., Gao, F., Hain, C.R., Kustas, W.P., 2014. Mapping daily evapotranspiration at field scales over rainfed and irrigated agricultural areas using remote sensing data fusion. Agric. Forest Meteorol. 186, 1–11.

Carlson, T.N., Ripley, D.A., 1997. On the relation between NDVI, fractional vegetation cover, and leaf area index. Remote Sensing Environ. 62 (3), 241–252.

Cavero, J., Medina, E.T., Puig, M., Martínez-Cob, A., 2009. Sprinkler irrigation changes maize canopy microclimate and crop water status, transpiration, and temperature. Agron. J. 101 (4), 854–864.

Cecil, D.J., Buechler, D.E., Blakeslee, R.J., 2014. Gridded lightning climatology from TRMM-LIS and OTD: dataset description. Atmosph. Res. 135, 404–414.

Cleugh, H.A., Leuning, R., Mu, Q., Running, S.W., 2007. Regional evaporation estimates from flux tower and MODIS satellite data. Remote Sens. Environ. 106 (3), 285–304.

Dorigo, W., Wagner, W., Albergel, C., Albrecht, F., Balsamo, G., Brocca, L., et al., 2017. ESA CCI Soil Moisture for improved Earth system understanding: state-of-the art and future directions. Remote Sensing Environ. 203, 185–215.

Fisher, J.B., Tu, K.P., Baldocchi, D.D., 2008. Global estimates of the land–atmosphere water flux based on monthly AVHRR and ISLSCP-II data, validated at 16 FLUXNET sites. Remote Sens. Environ. 112 (3), 901–919.

Frankenberg, C., Fisher, J.B., Worden, J., Badgley, G., Saatchi, S.S., Lee, J.E., et al., 2011. New global observations of the terrestrial carbon cycle from GOSAT: Patterns of plant fluorescence with gross primary productivity. Geophys. Res. Lett. 38 (17).

Gao, F., Masek, J., Schwaller, M., Hall, F., 2006. On the blending of the Landsat and MODIS surface reflectance: predicting daily Landsat surface reflectance. IEEE Trans. Geosci. Remote sens. 44 (8), 2207–2218.

Gash, J.H.C., 1979. An analytical model of rainfall interception by forests. Q J Roy Meteor. Soc. 105 (443), 43–55.

Hain, C.R., Anderson, M.C., 2017. Estimating morning change in land surface temperature from MODIS day/night observations: Applications for surface energy balance modeling. Geophys. Res. Lett. 44 (19), 9723–9733.

Hain, C.R., Crow, W.T., Anderson, M.C., Yilmaz, M.T., 2015. Diagnosing neglected soil moisture source–sink processes via a thermal infrared–based two-source energy balance model. J. Hydrometeorol. 16 (3), 1070–1086.

Holmes, T.R.H., De Jeu, R.A.M., Owe, M., Dolman, A.J., 2009. Land surface temperature from Ka band (37 GHz) passive microwave observations. J. Geophys. Res. Atmosph. 114 (D4).

Holmes, T.R., Hain, C.R., Anderson, M.C., Crow, W.T., 2016. Cloud tolerance of remote-sensing technologies to measure land surface temperature. Hydrol. Earth Syst. Sci. 20 (8), 3263.

Holmes, T.R.H., Hain, C.R., Crow, W.T., Anderson, M.C., Kustas, W.P., 2018. Microwave implementation of two-source energy balance approach for estimating evapotranspiration. Hydrol. Earth Syst. Sci. 22 (2), 1351.

Jackson, D.L., Wick, G.A., Robertson, F.R., 2009. Improved multisensor approach to satellite-retrieved near-surface specific humidity observations. J. Geophys. Res. Atmosph. 114 (D16).

Jasechko, S., Sharp, Z.D., Gibson, J.J., Birks, S.J., Yi, Y., Fawcett, P.J., 2013. Terrestrial water fluxes dominated by transpiration. Nature 49 (7445), 347.

Jimenez, C., Prigent, C., Mueller, B., Seneviratne, S.I., McCabe, M.F., Wood, E.F., et al., 2011. Global intercomparison of 12 land surface heat flux estimates. J. Geophys. Res. Atmosph. 116 (D2).

Jung, M., Reichstein, M., Ciais, P., Seneviratne, S.I., Sheffield, J., Goulden, M.L., et al., 2010. Recent decline in the global land evapotranspiration trend due to limited moisture supply. Nature 467 (7318), 951.

Konings, A.G., Gentine, P., 2017. Global variations in ecosystem-scale isohydricity. Global Change Biol. 23 (2), 891–905.

Konings, A.G., Williams, A.P., Gentine, P., 2017. Sensitivity of grassland productivity to aridity controlled by stomatal and xylem regulation. Nat. Geosci. 10 (4), 284.

Kustas, W.P., Norman, J.M., 1999. Evaluation of soil and vegetation heat flux predictions using a simple two-source model with radiometric temperatures for partial canopy cover. Agric. Forest Meteorol. 94 (1), 13–29.

Kustas, W.P., Norman, J.M., 2000. A two-source energy balance approach using directional radiometric temperature observations for sparse canopy covered surfaces. Agron. J. 92 (5), 847–854.

Liu, Y.Y., Dijk, A.I., McCabe, M.F., Evans, J.P., Jeu, R.A., 2013. Global vegetation biomass change (1988–2008) and attribution to environmental and human drivers. Global Ecol. Biogeogr. 22 (6), 692–705.

Martens, B., Gonzalez Miralles, D., Lievens, H., Van Der Schalie, R., De Jeu, R.A.M., Fernández-Prieto, D., et al., 2017. GLEAMv3: Satellite-based land evaporation and root-zone soil moisture. Geosci. Model Dev 10 (5), 1903–1925.

Mecikalski, J.R., Diak, G.R., Anderson, M.C., Norman, J.M., 1999. Estimating fluxes on continental scales using remotely sensed data in an atmospheric-land exchange model. J Appl. Meteorol. 38 (9), 1352–1369.

Michel, D., Jiménez, C., Miralles, D.G., Jung, M., Hirschi, M., Ershadi, A., et al., 2016. The WACMOS-ET project – part 1: tower-scale evaluation of four remote-sensing-based evapotranspiration algorithms. Hydrol. Earth Syst. Sci. 20 (2), 803–822. Available from: https://doi.org/10.5194/hess-20-803-2016.

Miralles, D.G., Gash, J.H., Holmes, T.R.H., de Jeu, R.A.M., Dolman, A.J., 2010. Global canopy interception from satellite observations. J. Geophys. Res. 115 (D16), D16122. Available from: https://doi.org/10.1029/2009JD013530.

Miralles, D.G., De Jeu, R.A.M., Gash, J.H., Holmes, T.R.H., Dolman, A.J., 2011. An application of GLEAM to estimating global evaporation. Hydrol. Earth Syst. Sci. Discuss. 8 (1).

Miralles, D.G., Van Den Berg, M.J., Gash, J.H., Parinussa, R.M., De Jeu, R.A.M., Beck, H.E., et al., 2014. El Niño–La Niña cycle and recent trends in continental evaporation. Nat. Clim. Change 4 (2), 122.

Miralles, D.G., Jiménez, C., Jung, M., Michel, D., Ershadi, A., McCabe, M.F., et al., 2016. The WACMOS-ET project, part 2: evaluation of global terrestrial evaporation data sets. Hydrol. Earth Syst. Sci. 20 (2), 823–842.

Momen, M., Wood, J.D., Novick, K.A., Pangle, R., Pockman, W.T., McDowell, N.G., et al., 2017. Interacting effects of leaf water potential and biomass on vegetation optical depth. J. Geophys. Res. Biogeo. 122 (11), 3031–3046.
Monteith, J.L., 1965. Evaporation and environment. Symp. Soc. Exp. Biol. 19 (205-23), 4.
Mu, Q., Zhao, M., Running, S.W., 2011. Improvements to a MODIS global terrestrial evapotranspiration algorithm. Remote Sens. Environ. 115 (8), 1781–1800.
Mueller, B., Hirschi, M., Jimenez, C., Ciais, P., Dirmeyer, P.A., Dolman, A.J., et al., 2013. Benchmark products for land evapotranspiration: LandFlux-EVAL multi-data set synthesis. Hydrol Earth Syst. Sci. Available from: http://repository.kaust.edu.sa/kaust/handle/10754/334520.
Mueller, N.D., Butler, E.E., McKinnon, K.A., Rhines, A., Tingley, M., Holbrook, N.M., Huybers, P., 2016. Cooling of US Midwest summer temperature extremes from cropland intensification. Nat. Clim. Change 6 (3), 317.
Norman, J.M., Kustas, W.P., Humes, K.S., 1995. Source approach for estimating soil and vegetation energy fluxes in observations of directional radiometric surface temperature. Agric. For. Meteorol. 77 (3–4), 263–293. Available from: https://doi.org/10.1016/0168-1923(95)02265-Y.
Norman, J.M., Anderson, M.C., Kustas, W.P., French, A.N., Mecikalski, J., Torn, R., et al., 2003. Remote sensing of surface energy fluxes at 101-m pixel resolutions. Water Resour. Res. 39 (8), 1221.
Oki, T., Kanae, S., 2006. Global hydrological cycles and world water resources. Science 313 (5790), 1068–1072.
Otkin, J.A., Anderson, M.C., Hain, C., Svoboda, M., Johnson, D., Mueller, R., et al., 2016. Assessing the evolution of soil moisture and vegetation conditions during the 2012 United States flash drought. Agric. Forest Meteorol. 218, 230–242.
Owe, M., de Jeu, R., Holmes, T., 2008. Multisensor historical climatology of satellite-derived global land surface moisture. J. Geophys. Res. Earth Surface 113 (F1).
Prigent, C., Jimenez, C., Aires, F., 2016. Toward "all weather," long record, and real-time land surface temperature retrievals from microwave satellite observations. J. Geophys. Res. Atmosph. 121 (10), 5699–5717.
Priestley, C.H.B., Taylor, R.J., 1972. On the assessment of surface heat flux and evaporation using large-scale parameters. Mon. Weather Rev. 100 (2), 81–92.
Schlesinger, W.H., Jasechko, S., 2014. Transpiration in the global water cycle. Agr. Forest Meteorol. 189, 115–117.
Semmens, K.A., Anderson, M.C., Kustas, W.P., Gao, F., Alfieri, J.G., McKee, L., et al., 2016. Monitoring daily evapotranspiration over two California vineyards using Landsat 8 in a multi-sensor data fusion approach. Remote Sens. Environ. 185, 155–170.
Senay, G.B., Friedrichs, M.K., Singh, R.K., Velpuri, N.M., 2016. Evaluating landsat 8 evapotranspiration for water use mapping in the Colorado river basin. Remote Sens. Environ. 185, 171–185.
Serbin, S.P., Singh, A., McNeil, B.E., Kingdon, C.C., Townsend, P.A., 2014. Spectroscopic determination of leaf morphological and biochemical traits for northern temperate and boreal tree species. Ecol. Appl. 24 (7), 1651–1669.
Skofronick-Jackson, G., Petersen, W.A., Berg, W., Kidd, C., Stocker, E.F., Kirschbaum, D.B., et al., 2017. The global precipitation measurement (GPM) mission for science and society. Bull. Am. Meteorol. Soc. 98 (8), 1679–1695.
Smith, E.A., Asrar, G., Furuhama, Y., Ginati, A., Mugnai, A., Nakamura, K., et al., 2007. International global precipitation measurement (GPM) program and mission: an overview. In Measuring Precipitation From Space. Springer, Dordrecht, pp. 611–653.
Su, Z., 2002. The Surface Energy Balance System (SEBS) for estimation of turbulent heat fluxes. Hydrol. Earth Syst. Sci. 6 (1), 85–100.

Tomita, H., Hihara, T., Kubota, M., 2018. Improved satellite estimation of near-surface humidity using vertical water vapor profile information. Geophys. Res. Lett. 45 (2), 899–906.
Valente, F., David, J.S., Gash, J.H.C., 1997. Modelling interception loss for two sparse eucalypt and pine forests in central portugal using reformulated rutter and gash analytical models. J. Hydrol. 190 (1–2), 141–162.
Wei, Z., Yoshimura, K., Wang, L., Miralles, D.G., Jasechko, S., Lee, X., 2017. Revisiting the contribution of transpiration to global terrestrial evapotranspiration. Geophys. Res. Lett. 44 (6), 2792–2801.
Wolfe, G.M., Kawa, S.R., Hanisco, T.F., Hannun, R.A., Newman, P.A., Swanson, A., et al., 2018. The NASA Carbon Airborne Flux Experiment (CARAFE): instrumentation and methodology. Atmosph. Measurement Tech. 11 (3), 1757.
Yang, Y., Anderson, M.C., Gao, F., Hain, C.R., Semmens, K.A., Kustas, W.P., et al., 2017. Daily landsat-scale evapotranspiration estimation over a forested landscape in North Carolina, USA, using multi-satellite data fusion. Hydrol. Earth Syst. Sci. 21 (2), 1017–1037.
Yilmaz, M.T., Anderson, M.C., Zaitchik, B., Hain, C.R., Crow, W.T., Ozdogan, M., et al., 2014. Comparison of prognostic and diagnostic surface flux modeling approaches over the Nile River basin. Water Resour. Res. 50 (1), 386–408.
Zhang, Y., Peña-Arancibia, J.L., McVicar, T.R., Chiew, F.H.S., Vaze, J., Liu, C., et al., 2016. Multi-decadal trends in global terrestrial evapotranspiration and its components. Sci. rep. 6, 19124.

CHAPTER SIX

Vegetation

Jean-Christophe Calvet[*]
CNRM (University of Toulouse, Meteo-France, CNRS), Toulouse, France

In the last four decades, data assimilation techniques have been implemented in weather numerical prediction models to monitor extreme weather conditions. At the same time, soil moisture observations and simulations have been used to properly initialize the bottom boundary of atmospheric models, using a simplified representation of the soil–plant system. As more sophisticated land surface models (LSMs) that include a phenology component are now available, data assimilation techniques can be implemented to merge vegetation information from the growing variety of satellite-derived products into LSMs.

First, the impact of vegetation on hydrology and how it can be represented in LSMs is presented in this chapter. Second, the chapter describes how key vegetation variables relevant to hydrological processes can now be observed from space. Finally, the chapter discusses the integration of satellite-derived vegetation products in LSMs using sequential data assimilation techniques.

6.1 Impact of vegetation on hydrology

Terrestrial water fluxes are controlled to a large extent by above-ground and below-ground biological processes. Land–atmosphere interactions are constrained by coupled biophysical and biogeochemical processes involving life in all its forms (Bonan, 1995). Vegetation plays a key role in ecosystems by linking biophysical processes—such as absorption

[*] PhD, Habilitation. He joined Centre National de Recherches Météorologiques, Toulouse, in 1994, where he has been the Head of a land surface modeling and remote sensing section since 2003. His research interests include land–atmosphere exchange modeling and the use of remote sensing over land surfaces for meteorology. His most recent works concern the joint analysis of soil moisture and vegetation biomass using data assimilation techniques.

Extreme Hydroclimatic Events and Multivariate Hazards in a Changing Environment.
DOI: https://doi.org/10.1016/B978-0-12-814899-0.00006-7

of solar radiation, rainfall interception, and evapotranspiration—to biogeochemical processes—such as photosynthesis and volatile organic compound emission. Moreover, vegetation links the terrestrial carbon cycle to hydrology through stomatal aperture (Jarvis and McNaughton, 1986), and through other processes such as soil-water extraction by roots (de Jong van Lier et al., 2006). Although these processes are quite complexly related, they can be simulated to a reasonable extent by process-based numerical models.

6.1.1 How is vegetation represented in numerical models?

Numerical LSMs are key components in meteorology, hydrometeorology, and climate applications to monitor and predict environmental conditions and hydroclimatic hazards. LSMs can be coupled to atmospheric models and to river routing models. While LSMs are often used in large-scale hydrology applications, hydro-ecological models, such as the Regional Hydro-Ecologic Simulation System (RHESSys; Tague, 2004), can be used to simulate small catchments and complex terrain regions. For croplands, agro-hydrological models, like the Soil and Water Assessment Tool model (SWAT; Arnold et al., 1998), simulate agricultural practices and soil erosion (Hallouz et al., 2017).

The most advanced LSMs are also used in numerical weather prediction (NWP) and in global water resource assessments to simulate photosynthesis and plant growth. Some examples are the US community Noah LSM with multiparameterization options (Noah-MP; Niu et al., 2011), the joint UK land environment simulator (JULES; Best et al., 2011), and the Interactions between Soil, Biosphere, and Atmosphere (ISBA) model within the French "surface externalisée" (SURFEX) modeling platform (Masson et al., 2013).

Being operated at subhourly time steps, LSMs can represent the diurnal cycle of water and energy fluxes, and surface temperature (Overgaard et al., 2006). LSMs can be used either at the global or regional scale, consistently with the spatial range and spatial resolution of the atmospheric model or atmospheric analysis used to drive them. Typical LSM spatial resolutions range from $0.25° \times 0.25°$ at the global scale to $0.01° \times 0.01°$ at the regional scale. Because of NWP increasing capability in representing subkilometric processed and large eddy simulations of the atmospheric boundary layer, LSMs are used more and more at subkilometric spatial scales (Honnert et al., 2016). The main atmospheric variables needed to drive a LSM are: precipitation, air temperature, air humidity, wind speed, incoming shortwave radiation, incoming longwave radiation, and atmospheric pressure. For LSMs that represent photosynthesis, the CO_2 atmospheric concentration is also needed. Although LSMs solve the energy and water budgets at a subhourly timescale, they can be forced by hourly to 3-hourly atmospheric data.

The main surface water fluxes represented by LSMs are: plant transpiration, soil evaporation, intercepted water evaporation, surface runoff, gravitational drainage from intercepted water to the soil surface and from the top soil layer to deeper soil layers, and capillary rises from deep soil layers to surface soil layers. LSMs also represent phase changes such as condensation, freeze and thaw, snowmelt, and sublimation. LSM simulations are produced on regular grids and subgrid variability can be accounted for.

In contrast to LSMs, hydro-ecological models and agro-hydrological models simulate landscape units or hydrological response units through an object-oriented approach. Land patches are considered rather than grid cells. The simulations are generally produced on a daily basis and the diurnal cycle of water and energy budgets is not represented. LSM and object-oriented approaches could converge in the future.

6.1.2 How are biophysical and biogeochemical processes related?

At the leaf level, transpiration and photosynthesis fluxes are connected because they share a common biological control mechanism called stomatal aperture (Bonan, 1995). Stomatas are small pores at the interface between the leaf tissues and the ambient air. Their aperture responds to bioclimatic variables such as leaf temperature, leaf-to-air saturation deficit, and the solar radiation reaching the leaf. High stomatal conductance triggers a CO_2 flux from the air to the leaf tissues, thus feeding photosynthesis and plant growth. At the same time, water is lost through transpiration, especially at large values of the leaf-to-air saturation deficit. Stomatal aperture is the result of a compromise between photosynthesis and the need for the plant to avoid excessive soil water deficit values that could trigger early senescence. Stomatal aperture can change within minutes. The above-ground and below-ground vegetation architecture is an overarching characteristic of land surfaces. Key variables of the soil–plant system are leaf area index (LAI), canopy height, plant above-ground biomass, rooting depth and root density profiles, and litter mass.

LAI is expressed in units of $m^2 m^{-2}$ and is defined as the one-sided leaf area per unit ground horizontal surface area. LAI is particularly important for water fluxes since it governs transpiration, the energy budget at the soil surface, and soil evaporation. It influences rain interception. While intercepted rain suppresses leaf transpiration to some extent (Tuzet et al., 2017), part of the intercepted rain can be directly evaporated back to the atmosphere. In the case of forests, the biomass of the trunks and branches can store heat and can provide energy to intercepted rain evaporation (e.g., Cisneros Vaca et al., 2018). This process can also impact snow accumulation (Roth and Nolin, 2017). Tree height is key, as tall canopies increase surface roughness, resulting in enhanced heat exchanges with the atmosphere. Rooting depth is also a key variable for water fluxes as it governs soil water uptake and the contribution of groundwater to plant transpiration (Bevan et al., 2014). Finally, litter is often present in forests and grasslands. Litter impacts the energy budget at the soil surface and can intercept rain. Soil organic content is a parameter in state-of-the-art LSMs, related to porosity, soil hydraulic conductivity, and soil water holding capacity.

6.1.3 Do land surface models bring added value in hydrology?

In contrast to empirical hydrological models, LSMs are based on a mechanistic approach able to represent the interactions among various biological and physical processes influencing surface runoff and deep drainage. The explicit representation of vegetation is needed to assess the impact of climate change and of extreme events on surface water fluxes. For example, long-term trends in CO_2 atmospheric concentration can impact LAI and stomatal aperture (Laanaia et al., 2016).

While parameters of hydrological models are generally calibrated using observed river discharge time series, those of LSMs are usually determined a priori from the scientific literature and from maps of land surface properties (Gelati et al., 2018). Since many soil and vegetation parameters are not extensively observed, a parsimonious LSM modeling approach is needed in order to limit the number of parameters. Calibrated hydrological models tend to reproduce river discharge better than uncalibrated LSMs (Beck et al., 2017). However, LSMs may perform better when considering other key hydrologic variables such as evapotranspiration (Haddeland et al., 2012; Schellekens et al., 2017).

Another advantage of LSMs is that they can be used to identify uncertainties and assess the impact of knowledge gaps. Some processes are not completely represented or not represented at all. For example, while soil organic matter is usually hydrophilic, forest soil hydrophobicity favors surface runoff. Hydrophobicity of organic matter can be caused by various factors (Buczko et al., 2006; Mao et al., 2016). In particular the climate and the fire sequence history of the forest stand are key because water-repellent organic molecules can be produced in specific fire intensity conditions (Fisher and Binkley 2000). Forests may also intercept fog and cloud water and increase water availability while moderating the risk of floods (Ellison et al., 2017). Another poorly known process is hydraulic redistribution by roots. Roots may transfer moisture from deep soil layers to surface soil layers (Siqueira et al., 2008) or on the contrary contribute to groundwater recharge (Burgess et al., 2001). Finally, tree density and canopy cover influence groundwater recharge (Ilstedt et al., 2016). Such knowledge gaps can be compensated to some extent by integrating satellite observations of vegetation variables into LSMs.

6.2 Observing vegetation variables from space

Earth observation data enable the implementation of LSMs at the global scale. Land cover maps can be constructed using satellite observations in order to represent the spatial distribution of model parameters. Biophysical variables such as LAI can be produced and further integrated into LSMs.

6.2.1 Land cover maps

Land cover maps are key tools for monitoring the impact of climate change and of human activities on terrestrial surfaces. Since forests can impact surface water yields (Filoso et al., 2017), mapping forest coverage, afforestation, and deforestation in watersheds is of paramount importance. The European Space Agency (ESA) land cover climate change initiative has produced 24 yearly global classifications from 1992 to 2015, at a spatial resolution of 300 m $\times$ 300 m. Such global land cover maps can be used by LSMs to represent the spatial distribution of plant functional types and model parameter values (Duveiller et al., 2018). However, land cover maps have to be interpreted to some extent in terms of land use (e.g., winter crops vs. summer crops) before they can be used in LSMs. Land cover classes have to be converted to plant functional types (Faroux et al., 2013). Using machine-learning techniques, freely available high spatial resolution remotely sensed observations—such as those from the Sentinel-2 and Sentinel-1 satellites of the European Copernicus space program—can be now used to update land cover maps and provide land use information with an increased frequency (Inglada et al., 2015; Clerici et al., 2017).

6.2.2 Vegetation indices

The Normalized Difference Vegetation Index (NDVI) is often used to characterize vegetation activity from space. The fraction of absorbed photosynthetically active radiation is a radiative variable that can also be used provided that illumination conditions are accounted for.

Nevertheless, LAI is more relevant for use in mechanistic models (see Fig. 6.1 for a global map of mean LAI). LAI influences land surface fluxes in various ways and must be accounted for in LSM simulations. LSMs can simulate LAI and this quantity can also be derived from satellite observations. The Copernicus global land service produces satellite-derived global

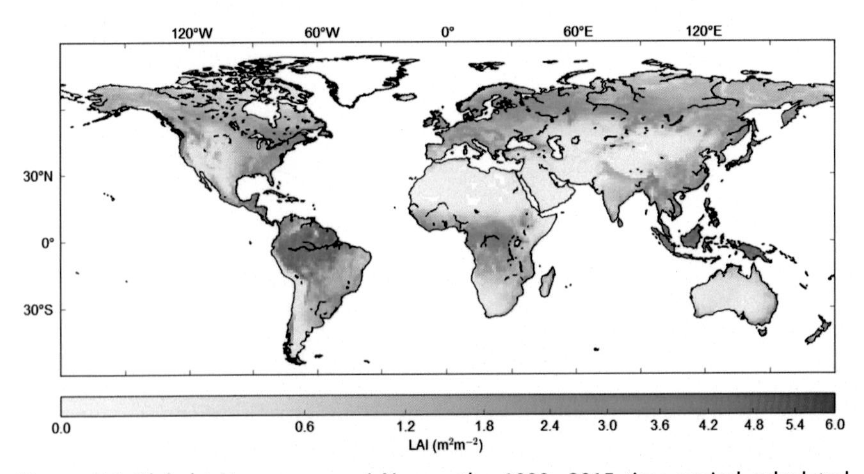

Figure 6.1 Global LAI map: mean LAI over the 1999–2015 time period calculated using the GEOV1 product of the Copernicus global land service (Copernicus, 2015). *LAI*, Leaf area index. *Global LAI map adapted from Munier, S., Carrer, D., Planque, C., Camacho, F., Albergel, C., Calvet J.-C., 2018. Satellite Leaf Area Index: global scale analysis of the tendencies per vegetation type over the last 17 years, Remote Sens. 10 (424), 25. <https://doi.org/10.3390/rs10030424>.*

10-day LAI estimates in near-real-time using a machine learning algorithm at spatial resolutions of 1 km × 1 km and 300 m × 300 m (Baret et al., 2013; Copernicus, 2015). Since this quantity is simulated by LSMs, LAI observations can be integrated in appropriate mechanistic models using data assimilation techniques (Barbu et al., 2014). Because satellite-derived LAI incorporates subpixel heterogeneity, disaggregation techniques can be used to retrieve separate LAI values for each plant functional type (Munier et al., 2018).

A limitation of satellite-derived LAI observations is that data can be missing in cloudy areas. In this case, microwave vegetation optical depth (VOD), only weakly affected by clouds, can be used as a proxy for LAI or NDVI, especially at C-band (Grant et al., 2016).

6.2.3 Forest biomass and tree height

Forests represent the vast majority of the biomass of terrestrial plants. Forest biomass and forest height largely influence the surface water and energy budgets and therefore play a fundamental role in modeling hydroclimatic hazards.

Lidar instruments can observe the Earth surface from space using a laser beam in cloud-free conditions. Forest height can be estimated from such observations. Global forest height maps were derived from the ICESAT satellite, at a spatial resolution of 1 km × 1 km (Simard et al., 2011; ORNL DAAC, 2017). Forest height can be used in conjunction with other information sources to estimate forest biomass. In the microwave wavelengths, VOD can be derived from either active or passive instruments (radars and radiometers, respectively). VOD can be defined as the effective zenith (i.e., nadir) opacity of the vegetation (dimensionless) in the microwave domain. At X-band and C-band, VOD is related to LAI and to photosynthesis activity (Teubner et al., 2018). At larger wavelengths, such as L-band or P-band, the signal tends to be more sensitive to branches and trunks. It was shown that L-band synthetic aperture radars (SAR) such as the phased array type L-band SAR instrument on the Advanced Land Observing Satellite (ALOS) could indirectly give access to the biomass of sparse forests, up to about 100 tons per hectare (Mermoz et al., 2014). The BIOMASS mission of the ESA, to be launched in 2021, is a P-band SAR with a wavelength of 0.68 m, aiming at monitoring the biomass of mature dense forests at a spatial resolution of 200 m × 200 m (Le Toan et al., 2011).

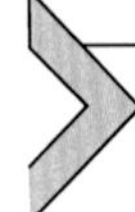

6.3 Integration of satellite-derived vegetation variables into models

The assimilation of satellite-derived vegetation variables into LSMs can be used to estimate key model parameters or to regularly control and correct model simulations.

6.3.1 Data assimilation

A review of the major data assimilation techniques used in terrestrial systems (e.g., ensemble Kalman filter, particle filter, and variational methods) can be found in Montzka et al. (2012). In models considering vegetation variables (e.g., LAI) as parameters derived from look-up tables, such variables cannot be assimilated. Only model state variables can be assimilated, using an observation operator. Therefore, plant growth needs to be explicitly represented by the models in which vegetation variables are integrated through data assimilation techniques. In this case, multivariate assimilation can be used to jointly assimilate vegetation variables and water variables such as surface soil moisture (Barbu et al., 2014). Another difficulty in analyzing the model LAI state is that in many LSMs, LAI is derived from heavily parameterized phenology submodels depending on cumulative empirical indicators such as growing degree days (Yang et al., 2012). In this case, the assimilation can only be used to estimate model parameter values. For example, the carbon cycle data assimilation system was developed by Kaminski et al. (2013) to calibrate an LSM to simulate terrestrial carbon fluxes at a global scale. The main drawback, however, is that not all model errors can be accounted for, especially in extreme conditions rarely or never simulated before. Not all processes can be represented and all LSMs present a simplified representation of reality. Therefore, correcting model states is often more relevant than correcting model parameters in monitoring applications. The land data assimilation system (LDAS) approach permits controlling state variables such as the root-zone soil moisture (RZSM) in order to cope with uncertainties in model parameters, but also in model forcing variables such as precipitation. Barbu et al. (2014) extended the LDAS approach to vegetation variables thanks to the ISBA LSM in the SURFEX modeling platform, using a parsimonious representation of phenology. In the ISBA LSM, phenology is completely driven by photosynthesis or by a deficit in photosynthesis. As a result, very few parameters are needed and the simulated

LAI is sufficiently flexible to be corrected as soon as new LAI observations are available. Albergel et al. (2017) implemented the multivariate sequential assimilation of LAI and surface soil moisture observations at a continental scale using SURFEX. A key advantage of the SURFEX LDAS is that LAI observations are used to analyze RZSM together with LAI and vegetation biomass. At the same time, surface soil moisture observations are used to analyze LAI and vegetation biomass together with RZSM. This multivariate sequential assimilation is a way to combine several observation types and to assess their consistency.

It must be noted that the capability of LAI observations to analyze RZSM in a data assimilation process depends on the model representation of surface hydrology processes and of the effect of soil water deficit on plant transpiration. Albergel et al. (2017) showed that the use of a multilayer discretization of the total soil moisture and soil temperature profile enhances the added value of the assimilation of LAI observations. In particular, RZSM can be analyzed in dry conditions using LAI. In dry conditions, the simulated surface soil moisture is decoupled to a large extent from RSZM and the assimilation of the former to analyze the latter is not very effective (Parrens et al., 2014).

A very important model parameter is the maximum plant rooting depth. In situ observations of this quantity are difficult to obtain. This quantity cannot be directly observed from space. In the case of straw cereals, Dewaele et al. (2017) showed that a LDAS LAI analysis approach could be used to retrieve maximum plant rooting depth thanks to the minimization of LAI analysis increments. In their study, Dewaele et al. (2017) used the LAI product from the European Copernicus global land service (Copernicus, 2015). They showed that annual maximum LAI observations are correlated with the maximum available water content of the soil derived from their LDAS tuning approach.

6.3.2 Vegetation and hydroclimatic hazards

Proper characterization of land use and vegetation properties can affect the modeling and prediction of hydroclimatic hazards. LSMs permit the explicit representation of vegetation and its link to hydroclimatic hazards, such as quick surface runoff and droughts. In particular, LAI is a key driver of evapotranspiration, of rain interception, and can be used to monitor the impact of extreme events such as droughts. LAI observations can be used to diagnose essential vegetation properties such as maximum rooting depth. Climate change can have a marked impact on LAI. Representing this impact in LSMs, as well as the impact of land use change, is crucial.

As hydroclimatic hazards are a global concern, more and more satellite-derived products can be used to monitor vegetation growth and senescence at the global scale with increasingly high resolutions. Such products are now freely available in near-real time and extend to long time periods covering several decades. They can be used to verify and improve LSMs using data assimilation techniques. Sequential assimilation of state variables such as LAI can be used to compensate for modeling and forcing uncertainties. It can also be used to determine key model parameters such as maximum rooting depth. The next generation of satellite-derived biophysical variables will be available at enhanced sampling time and spatial resolution. Land cover mapping and data assimilation methods that were so far applied at a global scale will be used for local and regional applications as well. Space missions such as Sentinel-1, Sentinel-2, and Landsat-8 are particularly promising thanks to large swaths and various spectral bands, allowing monitoring of rapid soil and vegetation changes within individual hydrological response units. The combination of the data from various sensors will facilitate the characterization of the diversity and evolution of vegetation types. For example, the combination of LAI with the microwave VOD in LDAS may improve further the accuracy of LSMs and therefore a better characterization of hydroclimatic hazards.

6.3.3 Impact on river discharge simulations

A large number of case studies have focused on the impact of land cover change on hydrology (e.g., Gyamfi et al., 2016; Welde and Gebremariam, 2017). On the other hand, few studies have investigated the impact of assimilating satellite-derived vegetation variables on hydrological simulations. Fig. 6.2 shows how a LDAS can be connected with hydrological simulations. The LDAS is used to update state variables of an LSM coupled with a hydrological model. In situ river discharge simulations can be used to validate the added value of the assimilation of satellite-derived observations. Fairbairn et al. (2017) used the SURFEX LDAS approach in order to assess the impact of the assimilation of surface soil moisture and LAI on river discharge simulations over France. They showed that correcting wintertime LAI of grasslands had a positive impact on the fit of daily discharge simulations to the observations. This result shows that

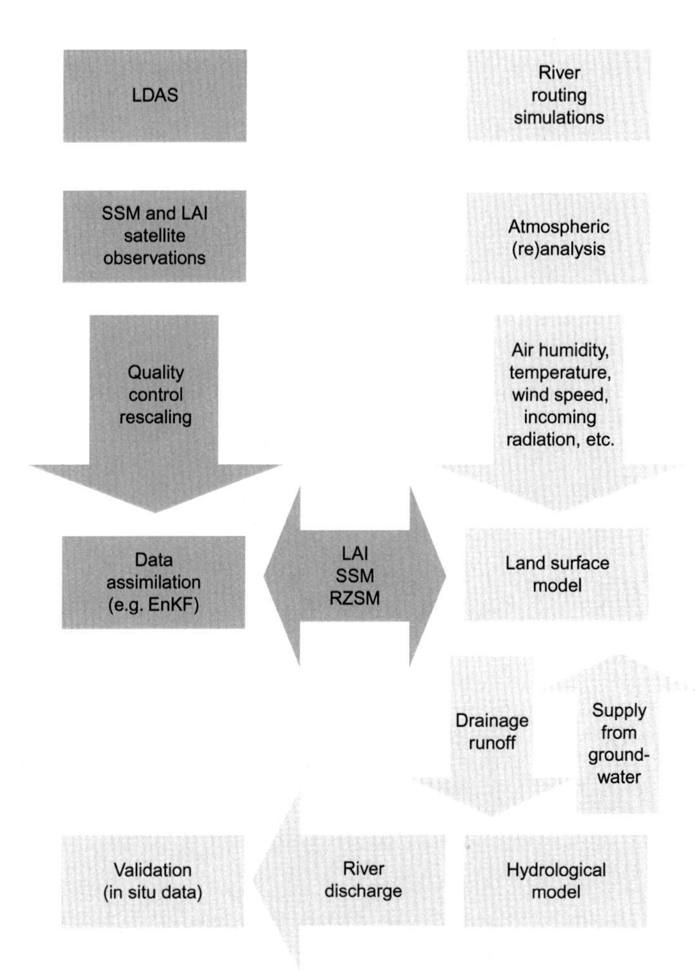

Figure 6.2 Flowchart showing how a LDAS can be connected with hydrological simulations. *LDAS*, Land data assimilation system. *Adapted from Fairbairn, D., Barbu, A.L., Napoly, A., Albergel C., Mahfouf, J.-F., Calvet, J.-C., 2017. The effect of satellite-derived surface soil moisture and leaf area index land data assimilation on stramflow simulations over France. Hydrol. Earth Syst. Sci. 21, 2015–2033. <https://doi.org/10.5194/hess-21-2015-2017>.*

assimilation of LAI can be beneficial to hydrological simulations. On the other hand, the assimilation of surface soil moisture had little or even negative impact on the simulation scores. It must be noticed that they used a simplified surface soil hydrology model. A more positive result was achieved by Albergel et al. (2017) over the Euro-Mediterranean area using a multilayer representation of the soil.

References

Albergel, C., Munier, S., Leroux, D.J., Dewaele, H., Fairbairn, D., Barbu, A.L., et al., 2017. Sequential assimilation of satellite-derived vegetation and soil moisture products using SURFEX_v8.0: LDAS-Monde assessment over the Euro-Mediterranean area. Geosci. Model Dev 10, 3889–3912. Available from: https://doi.org/10.5194/gmd-10-3889-2017.

Arnold, J.G., Srinivasan, R., Muttiah, R.S., Williams, J.R., 1998. Large area hydrologic modelling and assessment, Part 1: Model development. JAWRA 34 (1), 73–90.

Barbu, A.L., Calvet, J.-C., Mahfouf, J.-F., Lafont, S., 2014. Integrating ASCAT surface soil moisture and GEOV1 leaf area index into the SURFEX modelling platform: a land data assimilation application over France. Hydrol. Earth Syst. Sci. 18, 173–192. Available from: https://doi.org/10.5194/hess-18-173-2014. 2014.

Baret, F., Weiss, M., Lacaze, R., Camacho, F., Makhmara, H., Pacholczyk, P., et al., 2013. GEOV1: LAI and FAPAR essential climate variables and FCOVER global time series capitalizing over existing products, Part 1: Principles of development and production. Remote Sens. Environ. 137, 299–309. Available from: https://doi.org/10.1016/j.rse.2012.12.027.

Beck, H.E., van Dijk, A.I.J.M., de Roo, A., Dutra, E., Fink, G., Orth, R., et al., 2017. Global evaluation of runoff from 10 state-of-the-art hydrological models. Hydrol. Earth Syst. Sci. 21, 2881–2903. Available from: https://doi.org/10.5194/hess-21-2881-2017.

Best, M.J., Pryor, M., Clark, D.B., Rooney, G.G., Essery, R.L.H., Ménard, C.B., et al., 2011. The Joint UK Land Environment Simulator (JULES), model description—Part 1: Energy and water fluxes. Geosci. Model Dev. 4, 677–699. Available from: https://doi.org/10.5194/gmd-4-677-2011.

Bevan, S.L., Los, S.O., North, P.R.J., 2014. Response of vegetation to the 2003 European drought was mitigated by height. Biogeosciences 11, 2897–2908. Available from: https://doi.org/10.5194/bg-11-2897-2014.

Bonan, G.B., 1995. Land-atmosphere interactions for climate system models: coupling biophysical, biogeochemical, and ecosystem dynamical processes. Remote Sens. Environ. 51, 58–73.

Buczko, U., Bens, O., Huttl, R.F., 2006. Water infiltration and hydrophobicity in forest soils of a pine–beech transformation chronosequence. J. Hydrol. 331, 383–395. Available from: https://doi.org/10.1016/j.jhydrol.2006.05.023.

Burgess, S.S.O., Adams, M.A., Turner, N.C., White, D.A., Ong, C.K., 2001. Tree roots: conduits for deep recharge of soil water. Oecologia 126, 158–165. Available from: https://doi.org/10.1007/s004420000501.

Cisneros-Vaca, C., van der Tol, C., Ghimire, C.P., 2018. The influence of long-term changes in canopy structure on rainfall interception loss: a case study in Speulderbos, The Netherlands, <https://doi.org/10.5194/hess-2018-54>.

Clerici, N., Valbuena Calderón, C.A., Posada, J.M., 2017. Fusion of Sentinel-1A and Sentinel-2A data for land cover mapping: a case study in the lower Magdalena region, Colombia. J. Maps 13, 718–726. Available from: https://doi.org/10.1080/17445647.2017.1372316.

Copernicus LAI, 2015. GEOV1. Available from: <http://land.copernicus.eu/global/> (last accessed July 2018).

Dewaele, H., Munier, S., Albergel, C., Planque, C., Laanaia, N., Carrer, D., Calvet, J.-C., 2017. Parameter optimisation for a better representation of drought by LSMs: inverse modelling vs. sequential data assimilation. Hydrol. Earth Syst. Sci. 21, 4861–4878. Available from: https://doi.org/10.5194/hess-21-4861-2017.

Duveiller, G., Hooker, J., Cescatti, A., 2018. The mark of vegetation change on Earth's surface energy balance. Nat. Comm. 9 (679). Available from: https://doi.org/10.1038/s41467-017-02810-812 pp.

Ellison, D., Morris, C.E., Locatelli, B., Sheil, D., Cohen, J., Murdiyarso, D., et al., 2017. Trees, forests and water: Cool insights for a hot world. Glob. Environ. Change 43, 51–61.

Fairbairn, D., Barbu, A.L., Napoly, A., Albergel, C., Mahfouf, J.-F., Calvet, J.-C., 2017. The effect of satellite-derived surface soil moisture and leaf area index land data assimilation on stramflow simulations over France. Hydrol. Earth Syst. Sci. 21, 2015–2033. Available from: https://doi.org/10.5194/hess-21-2015-2017.

Faroux, S., Kaptué Tchuenté, A.T., Roujean, J.-L., Masson, V., Martin, E., Le Moigne, P., 2013. ECOCLIMAP-II/Europe: a twofold database of ecosystems and surface parameters at 1 km resolution based on satellite information for use in land surface, meteorological and climate models. Geosci. Model Dev. 6, 563–582. Available from: https://doi.org/10.5194/gmd-6-563-2013.

Filoso, S., Bezerra, M.O., Weiss, K.C.B., Palmer, M.A., 2017. Impacts of forest restoration on water yield: A systematic review. PLoS One 12, e0183210. Available from: https://doi.org/10.1371/journal.pone.0183210. 26 pp.

Fisher, R.F., Binkley, D., 2000. Fire effects, Chapter 10. Ecology and management of forest soils. John Wiley and Sons, New York.

Gelati, E., Decharme, B., Calvet, J.-C., Minvielle, M., Polcher, J., Fairbairn, D., et al., 2018. Hydrological assessment of atmospheric forcing uncertainty in the Euro-Mediterranean area using a land surface model. Hydrol. Earth Syst. Sci. 22, 2091–2115. Available from: https://doi.org/10.5194/hess-22-2091-2018.

Grant, J.P., Wigneron, J.-P., De Jeu, R.A.M., Lawrence, H., Mialon, A., Richaume, P., et al., 2016. Comparison of SMOS and AMSR-E vegetation optical depth to four MODIS-based vegetation indices. Remote Sens. Environ. 172, 87–100. Available from: https://doi.org/10.1016/j.rse.2015.10.021.

Gyamfi, C., Ndambuki, J.M., Salim, R.W., 2016. Hydrological responses to land use/cover changes in the Olifants basin. S. Afr. Water 8, 588. Available from: https://doi.org/10.3390/w8120588.

Haddeland, I., Heinke, J., Voß, F., Eisner, S., Chen, C., Hagemann, S., et al., 2012. Effects of climate model radiation, humidity and wind estimates on hydrological simulations. Hydrol. Earth Syst. Sci. 16, 305–318. Available from: https://doi.org/10.5194/hess-16-305-2012.

Hallouz, F., Meddi, M., Mahé, G., Alirahmani, S., Keddar, A., 2017. Modeling of discharge and sediment transport through the SWAT model in the basin of Harraza (Northwest of Algeria) <https://doi.org/10.1016/j.wsj.2017.12.004> Water Sci. in press.

Honnert, R., Couvreux, F., Masson, V., Lancz, D., 2016. Sampling the structure of convective turbulence and implications for grey-zone parametrizations. Boundary-Layer Meteorol. 160, 133–156. Available from: https://doi.org/10.1007/s10546-016-0130-4.

Ilstedt, U., Bargués Tobella, A., Bazié, H.R., Bayala, J., Verbeeten, E., Nyberg, G., et al., 2016. Intermediate tree cover can maximize groundwater recharge in the seasonally dry tropics. Sci. Rep. 6 (21930). Available from: https://doi.org/10.1038/srep2112 pp.

Inglada, J., Arias, M., Tardy, B., Hagolle, O., Valero, S., Morin, D., et al., 2015. Assessment of an operational system for crop type map production using high temporal and spatial resolution satellite optical imagery. Remote Sens. 7, 12356–12379. Available from: https://doi.org/10.3390/rs70912356.

Jarvis, P.G., McNaughton, K.G., 1986. Stomatal control of transpiration: scaling up from leaf to region. Adv. Ecol. Res. 15, 1–49.

de Jong van Lier, Q., Metselaar, K., van Dam, J.C., 2006. Root water extraction and limiting soil hydraulic conditions estimated by numerical simulation. Vadose Zone J. 5, 1264–1277. Available from: https://doi.org/10.2136/vzj2006.0056.

Kaminski, T., Knorr, W., Schürmann, G., Scholze, M., Rayner, P.J., Zaehle, S., et al., 2013. The BETHY/JSBACH carbon cycle data assimilation system: experiences and challenges. J. Geophys. Res. Biogeosci 118, 1414–1426. Available from: https://doi.org/10.1002/jgrg.20118.

Laanaia, N., Carrer, D., Calvet, J.-C., Pagé, C., 2016. How will climate change affect the vegetation cycle over France? A generic modeling approach. Clim. Risk Manage. 13, 31–42. Available from: https://doi.org/10.1016/j.crm.2016.06.001.

Le Toan, T., Quegan, S., Davidson, M., Balzter, H., Paillou, P., Papathanassiou, K., et al., 2011. The BIOMASS mission: mapping global forest biomass to better understand the terrestrial carbon cycle. Remote Sens. Environ. 115, 2850–2860. Available from: https://doi.org/10.1016/j.rse.2011.03.020.

Mao, J., Nierop, K.G.J., Rietkerk, M., Sinninghe Damsté, J.S., Dekker, S.C., 2016. The influence of vegetation on soil water repellency-markers and soil hydrophobicity. Sci. Total Environ 566–567, 608–620. Available from: https://doi.org/10.1016/j.scitotenv.2016.05.077.

Masson, V., Le Moigne, P., Martin, E., Faroux, S., Alias, A., Alkama, R., et al., 2013. The SURFEXv7.2 land and ocean surface platform for coupled or offline simulation of earth surface variables and fluxes,. Geosci. Model Dev. 6, 929–960. Available from: https://doi.org/10.5194/gmd-6-929-2013.

Mermoz, S., Le Toan, T., Villard, L., Réjou-Méchain, M., Seifert-Granzin, J., 2014. Biomass assessment in the Cameroon savanna using ALOS PALSAR data. Remote Sens. Environ. 155, 109–119. Available from: https://doi.org/10.1016/j.rse.2014.01.029.

Montzka, C., Pauwels, R.N.V., Franssen, H.-J.H., Han, X., Vereecken, H., 2012. Multi variate and multiscale data assimilation in terrestrial systems: a review. Sensors 12, 16291–16333. Available from: https://doi.org/10.3390/s121216291.

Munier, S., Carrer, D., Planque, C., Camacho, F., Albergel, C., Calvet, J.-C., 2018. Satellite Leaf Area Index: global scale analysis of the tendencies per vegetation type over the last 17 years. Remote Sens. 10 (424). Available from: https://doi.org/10.3390/rs1003042425 pp.

Niu, G.-Y., Yang, Z.-L., Mitchell, K.E., Chen, F., Ek, M.B., Barlage, M., et al., 2011. The community Noah land surface model with multiparameterization options (Noah-MP): 1. Model description and evaluation with local-scale measurements. J. Geophys. Res. 116, D12109. Available from: https://doi.org/10.1029/2010JD015139.

ORNL DAAC, 2017. Spatial Data Access Tool (SDAT). ORNL DAAC, Oak Ridge, TN, <https://doi.org/10.3334/ORNLDAAC/1388> (last accessed July 2018).

Overgaard, J., Rosbjerg, D., Butts, M.B., 2006. Land-surface modelling in hydrological perspective—a review. Biogeosciences 3, 229–241. Available from: https://doi.org/10.5194/bg-3-229-2006.

Parrens, M., Mahfouf, J.-F., Barbu, A., Calvet, J.-C., 2014. Assimilation of surface soil moisture into a multilayer soil model: design and evaluation at local scale. Hydrol. Earth Syst. Sci. 18, 673–689. Available from: https://doi.org/10.5194/hess-18-673-2014.

Roth, T.R., Nolin, A.W., 2017. Forest impacts on snow accumulation and ablation across anelevation gradient in a temperate montane environment. Hydrol. Earth Syst. Sci 21, 5427–5442. Available from: https://doi.org/10.5194/hess-21-5427-2017.

Schellekens, J., Dutra, E., Martínez-de la Torre, A., Balsamo, G., van Dijk, A., Sperna Weiland, F., et al., 2017. A global water resources ensemble of hydrological models: the eartH$_2$Observe Tier-1 dataset. Earth Syst. Sci. Data 9, 389–413. Available from: https://doi.org/10.5194/essd-9-389-2017.

Simard, M., Pinto, N., Fisher, J.B., Baccini, A., 2011. Mapping forest canopy height globally with spaceborne lidar. J. Geophys. Res. 116, G04021. Available from: https://doi.org/10.1029/2011JG001708.

Siqueira, M., Katul, G., Porporato, A., 2008. Onset of water stress, hysteresis in plant conductance, and hydraulic lift: scaling soil water dynamics from millimeters to meters. Water Resour. Res. 44, W01432. Available from: https://doi.org/10.1029/2007WR006094.

Tague, C.L., 2004. RHESSys: regional hydro-ecologic simulation system—an object-oriented approach to spatially distributed modeling of carbon, water, and nutrient cyclingEarth Interact. 8 (19), 42 pp . Available from: https://doi.org/10.1175/1087-3562(2004)8 < 1:RRHSSO > 2.0.CO;2.

Teubner, I.E., M. Forkel, M.J., Liu, Y.Y., Miralles, D.G., Parinussa, R., van der Schalie, R., et al., 2018. Assessing the relationship between microwave vegetation optical depth and gross primary production. Int. J. Appl. Earth Obs. Geoinform. 65, 79–91. Available from: https://doi.org/10.1016/j.jag.2017.10.006.

Tuzet, A., Granier, A., Betsch, P., Peiffer, M., Perrier, A., 2017. Modelling hydraulic functioning of an adult beech stand under non-limiting soil water and severe drought condition. Ecol. Model. 348, 56–77. Available from: https://doi.org/10.1016/j.ecolmodel.2017.01.007.

Welde, K., Gebremariam, B., 2017. Effect of land use land cover dynamics on hydrological response of watershed: case study of Tekeze Dam watershed, northern Ethiopia. Int. Soil Water Conserv. Res. 5, 1–16. Available from: https://doi.org/10.1016/j.iswcr.2017.03.002.

Yang, X., Mustard, J.F., Tang, J., Xu, H., 2012. Regional-scale phenology modeling based on meteorological records and remote sensing observations. J. Geophys. Res. 117, G03029. Available from: https://doi.org/10.1029/2012JG001977.

Remote sensing and modeling techniques for monitoring and predicting hydroclimatic hazards: Perspectives and applications

CHAPTER SEVEN

Estimating extreme precipitation using multiple satellite-based precipitation products

Yagmur Derin, Efthymios Nikolopoulos and Emmanouil N. Anagnostou
Department of Civil and Environmental Engineering, University of Connecticut, Mansfield, CT, United States

7.1 Extreme precipitation events

Due to anthropological effects, increased greenhouse gas concentrations have led to rising atmospheric temperatures. The Clausius–Clapeyron relationship suggests that increasing atmospheric temperature leads to an increase in atmospheric water vapor capacity which in turn intensifies the hydrological cycle (Trenberth, 1999). An intensified hydrological cycle is expected to increase the magnitude and intensity of precipitation, leading to enhanced contributions of heavy and extreme events (Trenberth et al., 2003; Allan and Soden, 2008). Such a change has large impacts on water resource management, agriculture, infrastructure planning, etc. Moreover, there is a lack of coherence in what constitutes and defines extreme events across the many disciplines. These disciplines emphasize different aspects of extreme events and have different ultimate motivations (McPhillips et al., 2018). Extreme precipitation events can be characterized by intensity, duration, and frequency. When intensity is used, an extreme event can be extracted by absolute thresholds or quantile thresholds, with quantile thresholds generally ranging from 90th to 99th percentiles. McPhillips et al. (2018) suggest in their study that definitions of extreme events should not be conflated with their impacts or effects.

Accurate quantification of heavy precipitation and its frequency of extremes over remote parts of the world, such as mountainous regions, is crucial for many applications such as the prediction of floods and landslides, which are critical to vulnerability analysis and early warning for

Extreme Hydroclimatic Events and Multivariate Hazards in a Changing Environment.
DOI: https://doi.org/10.1016/B978-0-12-814899-0.00007-9

disaster response (Kirschbaum et al., 2017). Flash flood and debris flows caused by localized heavy precipitation events lead to substantial socioeconomic impacts; hence quantitative information of the spatiotemporal variation of heavy precipitation estimates is of great importance. Due to the sparseness of ground-based in situ networks (e.g., gauges) and the severe limitations of weather radars over complex terrain, a comprehensive measurement of the space−time distribution of global precipitation at high spatial resolution can only be achieved through observations from space.

Satellite-based precipitation products (SPP) have the potential to track and monitor heavy precipitation systems with satellite measurements based on one or more remotely sensed characteristics of clouds, such as reflectivity, cloud-top temperature—visible (VIS), infrared (IR) imagery—or the emission/scattering effects of raindrops, or ice particles—e.g., passive microwave (PMW) radiation (Sapiano and Arkin, 2009; Kidd and Levizzani, 2011). The great success of the Tropical Rainfall Measuring Mission (TRMM) has accelerated the development of rain retrieval algorithms such as the Goddard Profiling Algorithm (GPROF; Kummerow et al., 2015). The PMW algorithms retrieve precipitation based on the scattering effect in the high-frequency channels (equal or higher than 85 GHz) over land. These estimates are normally characterized by uncertainties with more pronounced biases during extreme events, due to the mean observed ratio between ice aloft and the surface rainfall. This underestimation from satellite sensors is more distinct over mountainous regions (Derin et al., 2018; Kwon et al., 2008; Kubota et al., 2009) and is often attributed to the occurrence of shallow but heavy precipitation associated with warm-rain processes (Yamamoto et al., 2017; Bartsotas et al., 2018).

Since VIS, IR, and PMW products are indirect measurements of rainfall, they have strengths and weaknesses that need to be considered (see Chapter 1, Quantitative precipitation estimation from satellite observations, for more detail on SPP). To take advantage of these strengths, the most recent algorithms combine IR and PMW observations to create global-scale multisatellite precipitation products. Among these, well known products are the TRMM Multisatellite Precipitation Analysis (TMPA) gauge adjusted (3B42V07) product from the National Aeronautics and Space Administration Goddard Space Flight Center (Huffman et al., 2007, 2010; Huffman, 2013) which have been substituted by the Integrated MultisatellitE Retrievals for GPM (IMERG) Early, Late, and Final run (gauge adjusted) products. The GPM Core Observatory is an advanced successor of TRMM satellite, with additional sensors and channels like the Dual-frequency Precipitation Radar (DPR) and the GPM Microwave Imager (GMI) with capabilities to sense light rain and falling snow (Hou et al., 2008, 2014). Other currently distributed products are the National Oceanic and Atmospheric Administration (NOAA) Climate Prediction Center Morphing technique (CMORPH) (Joyce et al., 2004) and gauge adjusted product (Xie et al., 2011); the Precipitation Estimation from Remote Sensing Information using Artificial Neural Network (PERSIANN) gauge adjusted product (Sorooshian et al., 2000); and the gauge adjusted Global Satellite Mapping of Precipitation (GSMaP) datasets produced at the Earth Observation Research Center of the Japan Exploration Agency (Kubota et al., 2007; Mega et al., 2014).

SPPs have been evaluated for different continental regimes in the past two decades (Derin et al., 2016; Ebert et al., 2007). Regional studies have been conducted over the continental United States (McCollum et al., 2002; Gottschalck et al., 2005; Hossain and Huffman, 2008; Anagnostou et al., 2010), South America (Su et al., 2008; Dinku et al., 2010; Scheel et al., 2011), Europe (Stampoulis and Anagnostou, 2012; Mei and Anagnostou, 2014; Derin and Yilmaz, 2014), Africa (Dinku et al., 2007, 2008; Hirpa et al., 2010; Habib et al., 2012), and Asia (Chen et al., 2013; Yong et al., 2013). A review of SPPs' accuracy during the TRMM era has been conducted by Maggioni et al. (2016) and showed that in regions characterized by complex orography, SPP detection is weak and the mean error exhibits magnitude dependence. Derin and Yilmaz (2014) found that SPPs have difficulties in representing the rainfall precipitation gradient normal to the orography. They also observed an underestimation along the windward region and an overestimation on the leeward side of the mountains. Overall, there are large uncertainties in SPPs for heavy precipitation events (Anagnostou et al., 2010; Maggioni et al., 2014; Prakash et al., 2016). Stampoulis and Anagnostou (2012) showed that SPPs strongly underestimated heavy precipitation occurring over higher elevations of the Alps during the fall season. Although evaluations of satellite precipitation against gauge observations have been conducted extensively, only a few studies have particularly focused on extreme precipitation (AghaKouchak et al., 2011; Nastos et al., 2013; Lochoff et al., 2014; Prakash et al., 2016; Huang et al., 2017). Petkovic and Kummerow (2015) evaluated the performance of GPM PMW retrievals in an extreme precipitation event and concluded that the satellite underestimated accumulations relative to gauges by 60%. They further concluded that fixing this problem requires better understanding of the ice content in heavy precipitation events.

Heavy precipitation can be produced and intensified by orographic effects. Comprehensive error analysis of extreme precipitation of SPPs across different mountainous regions is needed to evaluate the relative performance characteristics of the available SPP datasets to develop error correction procedures globally. Maggioni et al. (2017) modeled SPP errors over mountainous terrain and showed that their error model was able to correct SPPs over complex terrain successfully. In this chapter, we evaluate six global SPPs in terms of estimating extreme rainfall rates over nine regions in the globe characterized by complex terrain. The evaluated SPPs include the TMPA research product (3B43V07), IMERGV05 final run (gauge adjusted), CMORPH (gauge adjusted), PERSIANN (gauge adjusted), GSMaPV06, and GSMaPV07. The mountainous regions are: Italian and Swiss Alps, French Cévennes, Western Black Sea Region of Turkey, Colombian Andes, Peruvian Andes, US Rocky Mountains, upper Blue Nile, Western Taiwan, and Himalayas over Nepal. High rainfall rate in these regions is represented by the 90th percentile rainfall rate threshold determined from the daily gauge rainfall values at 0.25° SPP grid cells.

The performance of SPPs is assessed using statistical measures and visual comparison methods. The chapter is organized as follows: the details of the study area and datasets are provided in Section 7.2. Evaluation methodology is presented in Section 7.3. The discussion of results is presented in Section 7.4 and the conclusions and recommendations are summarized in Section 7.5.

7.2 Study areas and data

7.2.1 Study domains and rain gauge datasets

The study areas involve nine mountainous ground validation sites exhibiting dense gauge networks (Fig. 7.1).

Every region has a different rain gauge network density and time period of data record. The conditional mean annual precipitation versus elevation of each rain gauge is provided in Fig. 7.2. The extreme precipitation in this study is defined according to the 90th percentile threshold of

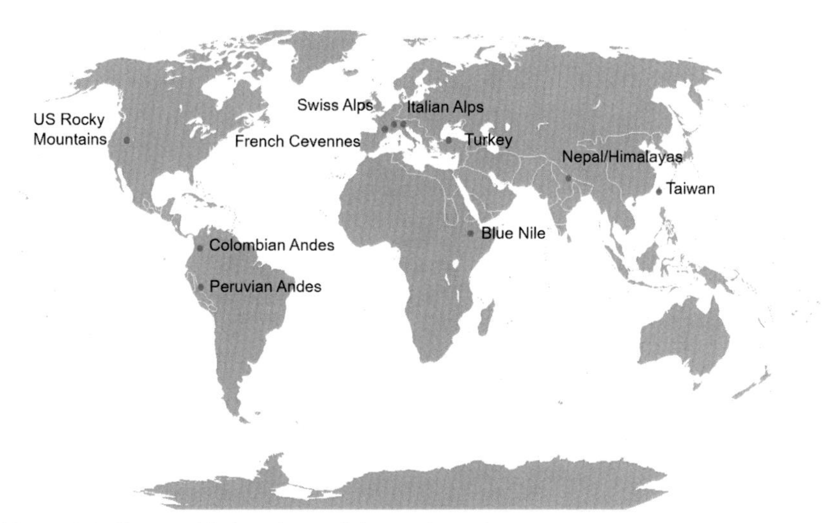

Figure 7.1 Geographic locations of the study regions.

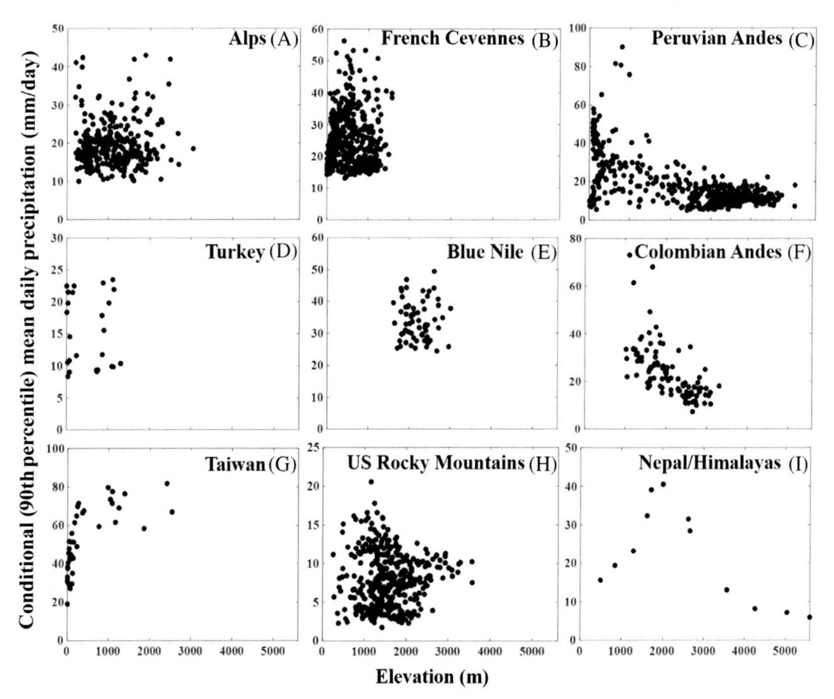

Figure 7.2 Conditional (90th percentile) mean annual precipitation (mm/day) versus gauge elevation (m) for (A) the Alps, (B) the French Cévennes, (C) the Peruvian Andes, (D) Turkey, (E) the Blue Nile, (F) the Colombian Andes, (G) Taiwan, (H) the US Rocky Mountains, and (I) Nepal.

the rainfall magnitude. Conditional (greater than 90th percentile) precipitation will be referred from now on as >Q90.

To ensure consistency across different domains, ground validation datasets are limited to daily time resolution and in situ data records span a minimum of 6 years during 2000–15 (Table 7.1).

Over the Taiwan study area most of the precipitation occurs during the eastern Asia rainy season in April and from typhoons from May to September, while the remaining months represent the dry period exhibiting low precipitation.

The Swiss Alps receive the larger rain amount on the western foothills of the Alps due to westerly winds. Shadow effects result in drier inner alpine valleys, while southern circulation from the Mediterranean generates intense precipitation on the southern side of the Alps in autumn.

Table 7.1 General information of the rain gauge datasets representing the ground validation of the different study regions.

Region	Geographic extent (latitude/longitude)	Number of gauges and periods of observations	Study region-average annual and max daily rainfall	Elevation range (m)
Blue Nile	36.46°–38.5°N 10.00°–12.85°E	70 2000–15	1070 mm/year 312 mm/day	1615–3125
Eastern Italian Alps	10.42°–12.48°N 46.16°–47.20°E	107 2000–15	910 mm/year 214 mm/day	212–2260
Swiss Alps	6.00°–10.50°N 45.80°–47.78°E	83 2002–15	1141 mm/year 243 mm/day	197–3040
French Cévennes	3.00°–4.60°N 43.65°–45.40°E	208[a] 2007–15	1050 mm/year 206 mm/day	140–1567
Turkey	31.48°–34.15°N 40.25°–41.70°E	25 2007–15	640 mm/year 146 mm/day	33–1305
Peruvian Andes	68.65°–79.93°S 4.96°–17.60°W	147 2000–15	960 mm/year 540 mm/day	2007–5020
Colombian Andes	72.38°–78.00°S 0.78°–7.36°E	104 2000–15	1570 mm/year 439 mm/day	286–3666
Taiwan	120.50°–120.90°N 23.20°–23.53°E	40 2002–15	2500 mm/year 1206 mm/day	1040–2540
US Rocky Mountains	103.10°–123.15°S 31.35°–48.80°E	319[a] 2000–15	420 mm/day 83 mm/day	1500–3550
Nepal/ Himalayas	85.15°–88.00°N 26.15°–28.75°E	12 2000–15	530 mm/year 88 mm/day	2660–5600

[a]Area-averaged grid boxes instead of point rain gauge observations.

The Italian Alps region is characterized by steep topographic gradients. Precipitation is primarily attributed to mesoscale convective systems during summer to early fall and frontal systems during fall and early winter (Frei and Schär, 1998; Norbiato et al., 2009). It should be noted that the Italian Alps and Swiss Alps are combined to one region and named as the Alps.

Validation data for the French Cévennes region are available from Météo-France, Électricité de France, and the Service de Prévision des Crues du Grand-Delta in the framework of Cévennes-Vivarais Mediterranean Hydrometeorological Observatory. By using 600 rain gauges over the 2007–15 period, kriging-based area interpolation maps of the region were created (Delrieu et al., 2014). The region is mainly characterized by a Mediterranean climate.

The Turkey study area is located on the south and north side of the mountains and is characterized by a dry/subhumid continental climate (Derin and Yilmaz, 2014). Extension of the mountains effects of the Black Sea determines the region climate types. The mountains, which are parallel to the shoreline, restrict transfer of precipitation to inland zones (south of the mountains), hence decreasing the temperature and precipitation (mean annual value is as low as 400 mm south of the mountains). For this reason, the inland zones observe a more continental climate (Yucel and Onen, 2014).

The Himalayan study area is located in the subtropical zone and is influenced by the monsoon system (Salerno et al., 2015). Orographic effects on precipitation have been studied over this region (Sing and Kumar, 1997; Ichiyanagi et al., 2007; Salerno et al., 2015) and it was found that precipitation increases with altitude below 2000 m a.s.l. and decreases for elevations above 2000 m a.s.l. (Fig. 7.2I).

Validation data for the US Rocky Mountains are based on the Climate Prediction Center (CPC) Unified Gauge-based Analysis of Daily Precipitation over the contiguous United States (CONUS). This dataset is created by combining all information sources available at CPC and taking advantage of the optimal interpolation objective analysis technique at 0.25° daily resolution (Chen et al., 2013). The mountainous regions within CONUS were extracted by removing grid cells with elevation less than 1500 m and without rain gauges. Hence, 319 grid cells were extracted for analysis over the US Rocky Mountains reporting data for the period of 2000–15. The region is characterized by a highland (alpine) climate.

Rain gauges over the Peruvian Andes region are maintained by the national agency Servicio Nacional de Meteorología e Hidrología del Perú. The orographic effect on precipitation can be seen from Fig. 7.2C. Precipitation patterns are controlled by the interaction of synoptic-scale atmospheric currents and the complex Andean topography (Manz et al., 2016). The Pacific coastline is arid and experiences precipitation less than 100 mm/year as a result of the cold von Humboldt current (Manz et al., 2016), which increases towards the Andes mountainous range due to strong topographic gradients resulting in pronounced orographic effects (Espinoza et al., 2015; Espinoza-Villar et al., 2009; Bookhagen and Strecker, 2008).

Rain gauges over the Colombian Andes region are maintained by a national agency, the Instituto de Hidrología, Meteorología Estudios Ambientales. Precipitation in this area is represented by a complex interaction of low-level jet stream with local topography and the Inter Tropical Convergence Zone (ITCZ). Central Colombia experiences a bimodal annual cycle of precipitation with high rain seasons in April–May and September–November months resulting from the double passage of the ITCZ over the region (Villa et al., 2011).

For detailed information related to these regions, readers are recommended to Derin et al. (2016).

7.2.2 Satellite-based precipitation products

We examined six SPPs (Table 7.2) that are produced from different rainfall retrieval algorithms, based on PMW or IR or combined PMW/IR observations.

IMERGV05 final and 3B42V07 use GPROF to compute precipitation estimates for all PMW retrievals as input (Huffman et al., 2015). IMERG introduced a new approach where temporal and spatial gaps in the PMW precipitation estimates are filled by morphing the estimates in between PMW overpasses and incorporating IR estimates with a Kalman filter where the gaps are too long (over about 3 hours) to produce continuous global products. It should be noted that GPM retrieval algorithms use the dual frequency channels of DPR and the high-frequency channels of GMI and hence precipitation products from GPM are different than those from TRMM. The original (version 03) IMERG products (right after the launch of GPM) were highly biased for heavy rain events, while

Table 7.2 Summary of SPPs used in this study.

Abbreviation	Long name	Provider	Spatial resolution	Temporal resolution
IMERGV05 final	Integrated MultisatellitE Retrievals for GPM	NASA GSFC	0.1°	30 min
3B42V07	Post-real-time research	NASA GSFC	0.25°	3 hourly
CMORPH	Gauge adjusted	NOAA	8 km	30 min
GSMaPV06	Reanalysis with gauge-calibration	JAXA	0.1°	1 hourly
GSMaPV07	Standard gauge	JAXA	0.1°	1 hourly
PERSIANN	Gauge adjusted	University of California, Irvine	0.25°	3 hourly

SPP, Satellite-based precipitation products; *NOAA*, National Oceanic and Atmospheric Administration; *CMORPH*, Climate Prediction Center Morphing; *PERSIANN*, Precipitation Estimation from Remote Sensing Information using Artificial Neural Network; *GSMaP*, Global Satellite Mapping of Precipitation; *NASA*, National Aeronautics and Space Administration; *GSFC*, Goddard Space Flight Center; *JAXA*, Japan Exploration Agency.

the upgrades to V04 and V05 progressively reduced this bias (Skofronick-Jackson et al., 2017). Moreover, all GPROF estimates are currently using rain/no-rain thresholds at 0.03 mm/h.

CMORPH derives precipitation estimates from PMW-only satellite estimates, which are propagated by motion vectors derived from IR data (Joyce et al., 2004). GSMaP is very similar to CMORPH where precipitation estimates are derived from PMW estimates and propagated by using IR estimates. The main difference between CMORPH and GSMaP is that GSMaP uses IR estimates at times when PMW estimates are not present (Kubota et al., 2007; Aonashi et al., 2009; Ushio et al., 2009). PERSIANN is IR-based and uses a coincident PMW-calibrated neural network technique to relate IR observations to rainfall estimates (Sorooshian et al., 2000).

All products apply gauge-based corrections to the satellite-only rainfall measurements. 3B42V07 apply monthly gauge adjustments by using Global Precipitation Climatology Project monthly rain gauge analysis developed by the Global Precipitation Climatology Center (Rudolf, 1993) and Climate Assessment and Monitoring System monthly rain gauge analysis developed by the CPC (Xie and Arkin, 1996). CMORPH applies gauge adjustment at a daily timescale using approximately 30,000 stations that are interpolated using an optimal interpolation technique with orographic correction at 0.125° grid resolution, and then averaged to 0.5° latitude/longitude grid cells over the globe.

GSMaP gauge calibrated rainfall algorithm adjusts the GSMaP_MVK estimate with global gauge analysis (CPC Unified Gauge-Based Analysis of Global Daily Precipitation) supplied by NOAA and has spatial and temporal resolution of 0.1° and 1 hour, respectively. The adjustment is applied to the estimation only over land. Version 6 of GSMaP used in this chapter is corrected by reanalysis and gauges (GSMaPV06) which includes orographic rainfall correction method for warm rainfall in coastal area (Yamamoto and Shiege, 2015), update of database, and land surface emission database developed by Japanese DPR/GMI combined team. It should be noted that orographic rainfall correction improves rainfall estimation over the entire Asian region with problems of misdetection and overestimation for GSMaPv06 (Yamamoto et al., 2017). Version 7 of GSMaP used in this chapter is corrected only by rain gauges (GSMaPV07). In addition, GSMaPV07 includes an improvement of the orographic rain correction method by incorporating a variable threshold for orographically-forced upward vertical motion determined based on the mean horizontal wind, using GPM/DPR observations as its database, and improving the gauge-correction method both at near-real time and standard product. Improvements to the orographic rain correction method applied in GSMaPV07 algorithm was able to give a solution to overestimates of orographic rainfall cases over inland regions; however it has been noted in Yamamoto et al.'s (2017) study that the separation from tall orographic rainfall and other types (other than shallow orographic rainfall) of orographic enhancement have not been detected.

It should be noted that all SPPs have different time ranges. 3B42V07 and CMORPH are available for 2000–15, IMERGV05 final and GSMaPV07 are available for 2014–current, GSMaPV06 is available for 2000–14, and PERSIANN is available for 2000–10.

GPM mission science requirements stipulated performance targets for detection, bias, and random error (Hou et al, 2013). For example, rain-rate estimates should exhibit a bias and random error of ≤50%(25%) at rain rates of 1 (10) mm/h for areas of 50 km × 50 km (Skofronick-Jackson et al., 2017). This requirement is checked for all six SPPs by using all regions and rain gauges in Fig. 7.3. Bias in each reference bin is computed as mean relative error (MRE) while for random error centralized root mean square error (CRMSE) is computed.

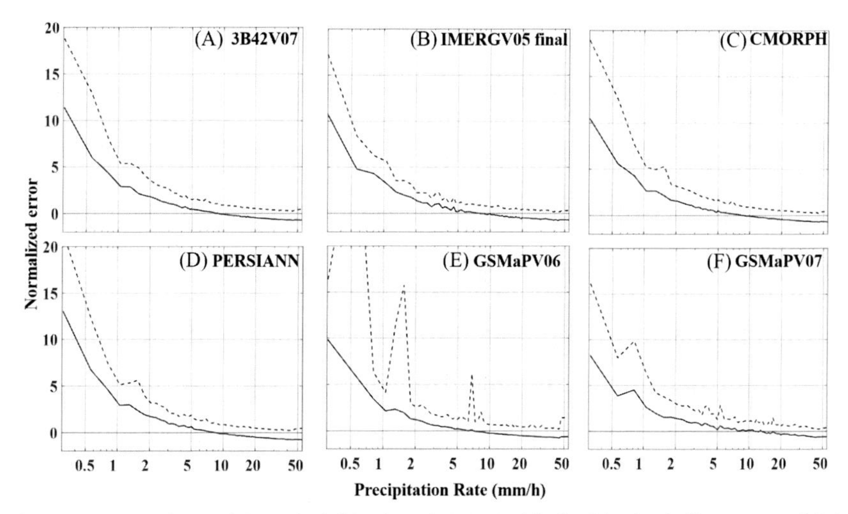

Figure 7.3 Conditional MRE (*solid line*) and CRMSE (*dashed line*) of all regions of (A) 3B42V07, (B) IMERGV05 final, (C) CMORPH, (D) PERSIANN, (E) GSMaPV06, and (F) GSMaPV07 versus the rain gauge rainfall rate (mm/h). *MRE*, Mean relative error; *CRMSE*, centralized root mean square error; *IMERG*, Integrated MultisatellitE Retrievals for GPM; *CMORPH*, Climate Prediction Center Morphing; *PERSIANN*, Precipitation Estimation from Remote Sensing Information using Artificial Neural Network; *GSMaP*, Global Satellite Mapping of Precipitation.

As can be seen from Fig. 7.3, all products' MRE for 1 mm/h is less than 5, however CRMSE values change drastically between products. CRMSE values for IMERGV05 final, CMORPH and GSMaPV07 are considerably lower than the rest of the SPPs for 1 mm/h rain rate. However, in this study we do not focus on the whole range of rainfall values, but rather on extreme precipitation only. As mentioned in the previous section, extreme precipitation is defined based on the 90th percentile of rain gauge rainfall threshold.

7.3 Evaluation methodology

The primary objective of this section is to evaluate the performance of various SPPs for conditional extreme (>Q90) precipitation over complex terrain using the numerous rain gauge networks gathered over the nine different regions.

We assess the performance of SPPs using both statistical and graphical methods. After inspecting some basic time series, probability density functions (PDFs) of daily precipitation extremes from SPPs and rain gauge estimates are compared. The frequency of occurrence of daily precipitation (PDFc) describes the percentages of rain occurrence across the predefined bins. The evaluation is conducted by matching pairs of extreme rain gauge precipitation values with the SPPs at each gauge station for each study region. The evaluations are conducted at daily and annual timescales. The agreement between different products is investigated using quantitative, categorical, and graphical measures. The quantitative statistics include MRE, CRMSE, and correlation coefficient (CORR):

$$\mathrm{MRE} = \frac{\sum_t (S - G)}{\sum_t G} \tag{7.1}$$

$$\mathrm{CRMSE} = \frac{\sqrt{1/M \sum_t \{S - G - 1/M \sum [S - G]\}^2}}{1/M \sum_t G} \tag{7.2}$$

where S and G represent SPPs and rain gauge estimates, respectively, and M represents the total number of days. MRE measures the systematic error component with values greater or smaller than zero indicating over- or underestimation, respectively. CRMSE measures the random component of error, as bias is removed. CORR is an indicator of the temporal similarity between rain gauge and SPP data.

The contingency table-based categorical statistics measure the daily rain-detection capability and include normalized missed rainfall volume (NMRV) and normalized false alarm satellite rainfall volume (NFASRV) error metrics:

$$\mathrm{NMRV} = \text{normalized missed rainfall volume} = \frac{\sum_t (G | G > 0 \& S = 0)}{\sum_t G} \times 100\% \tag{7.3}$$

$$\mathrm{NFASRV} = \text{normalized false alarm satellite rainfall volume} = \frac{\sum_t (S \mid G = 0 \& S > 0)}{\sum_t G} \times 100\% \tag{7.4}$$

NMRV quantifies the missed precipitation volume by SPPs, normalized by the total reference precipitation volume throughout the study period. NFASRV measures the falsely detected precipitation volume by SPPs, normalized by the total reference precipitation volume during the same period of time. A good SPP performance would imply low systematic error (i.e., MRE) accompanied by low magnitude of the random error component (i.e., CRMSE), high covariation (i.e., correlation), and accurate rainfall area detection (i.e., low NMRV and NFASRV).

7.4 Satellite precipitation product performance in complex terrain regions

7.4.1 Regional annual comparisons

Fig. 7.4 shows >Q90 mean annual average precipitation over all regions (3B42V07, IMERGV05 final, CMORPH, PERSIANN, GSMaPV06,

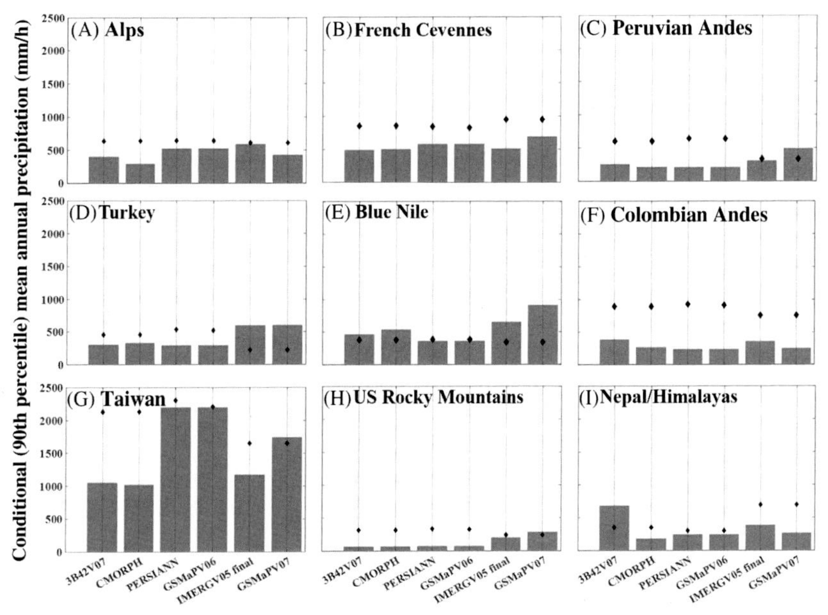

Figure 7.4 Conditional (90th percentile) mean annual precipitation over (A) the Alps, (B) the French Cévennes, (C) the Peruvian Andes, (D) Turkey, (E) the Blue Nile, (F) the Colombian Andes, (G) Taiwan, (H) the US Rocky Mountains, and (I) Nepal/Himalayas for every SPPs (note that *diamond* markers denote mean annual precipitation for rain gauges). Cold season days are included to understand the performances of the total precipitation quantity of the products.

and GSMaPV07) the diamond markers represent the >Q90 mean annual average rain gauge precipitation values. As mentioned earlier, all SPPs have different time records so annual means are calculated for the time record available for each SPPs. Overall (with exceptions) most SPPs underestimate the >Q90 precipitation rain gauge values. In general, IMERGV05 final, GSMaPv06, and GSMaPv07 show the best match to rain gauge in terms of >Q90 mean annual averages with exceptions over Turkey, Blue Nile, and Peruvian Andes, where the products tend to overestimate. IMERGV05 final shows a significant improvement relative to 3B42V07 for >Q90 precipitation values. The performance of GSMaPV06 relative to GSMaPV07 varies considerably with GSMaPV07 showing better results in most cases.

Over the Alps all SPPs underestimate >Q90 mean annual average precipitation (Fig. 7.4A). Generally, IMERGV05 final performance improves relative to 3B42V07. GSMaPV06 and GSMaPV07 performances are very similar and closely follow IMERGV05 final. CMORPH shows the highest underestimation over this region.

Over the French Cévennes, where heavy precipitation from low level warm-rain processes is frequent, all SPPs consistently underestimate >Q90 mean annual average precipitation (Fig. 7.4B). GSMaPV07 performance improves relative to GSMaPV06.

Rain gauges over Turkey are located both along the coast and inland, where the inland region is primarily characterized by rain shadow effects. IMERGV05 final and GSMaPV07 overestimate the rain gauge observations over this region, while the rest of the SPPs slightly underestimate (Fig. 7.4D). Derin et al. (2016) noted that gauge adjustments over Turkey and the Colombian Andes do not improve the performance of satellite products.

The performance evaluation over the Peruvian Andes exhibits significant underestimation of all SPPs except GSMaPV07 (Fig. 7.4C). GSMaPV07 closely follows the rain gauge >Q90 precipitation values. IMERGV05 final shows slight improvement compared to 3B42V07. Over the Colombian Andes, the performance of SPPs is very similar to the Peruvian Andes region (Fig. 7.4F). All SPPs significantly underestimate the >Q90 mean annual rain gauge precipitation.

Over the Blue Nile region, all products overestimate extreme mean annual precipitation with the exception of GSMaPV06 and PERSIANN (Fig. 7.4E). In this region, the performance of GSMaPV07 is the poorest, while GSMaPV06 outperforms all products, which can be explained by the orographic correction that is applied to this product. The low performance of GSMaPV07 should be studied in detail over this region to understand the underlying significant overestimation.

GSMaPV06 and GSMaPV07 performance is significantly better than the rest of the products over Taiwan when compared to conditional mean annual rain gauge precipitation values (Fig. 7.4G). All SPPs significantly underestimate extreme precipitation, except GSMaP products. Orographic enhancement applied by GSMaP works well over this region when compared to other SPPs.

Over the US Rocky Mountains, SPPs underestimate rain gauge observations at the annual timescale (Fig. 7.4H), GSMaPV07 having the most significant underestimation. In general, IMERGV05 final and GSMaPV07 follow the >Q90 mean annual rain gauge precipitation value closely compared to the rest of the SPPs.

Over Nepal all SPPs underestimate the >Q90 mean annual rain gauge precipitation except 3B42V07 (Fig. 7.4I). 3B42V07 has significant overestimation and when IMERGV05 final is compared to this product, we clearly see an improvement. IMERGV05 final performance is very close to PERSIANN over this region.

7.4.2 Daily comparisons

Fig. 7.5 provides the frequency of occurrence of conditional (>Q90) daily precipitation for different SPPs rainfall intensities versus rain gauges. Overall, SPPs are able to capture the distribution of >Q90 daily precipitation accumulations in most regions with an overestimation of the lowest range of rainfall magnitue ($0 < R < 7$ mm/day) and a general underestimation of the middle and highest range rainfall magnitudes. Over all regions the middle range of rainfall magnitude ($10 < R < 40$) is the clear mode of the PDF and the uncertainty is especially large for this range of rainfall values for SPPs. In general over French Cévennes, Turkey, Blue Nile, and Nepal, the performance of the different SPPs is similar. Over the Alps, compared to other SPPs, IMERGV05 final captures the distribution of rainfall well, while CMORPH shows significant overestimation of low ranges and significant underestimation of middle and high ranges of daily precipitation values. Over the Blue Nile, there is a general trend of significant underestimation of middle to high range daily rainfall values and overestimation of the low range daily rainfall values. Over the Peruvian Andes, SPPs significantly overestimate the lowest rainfall range

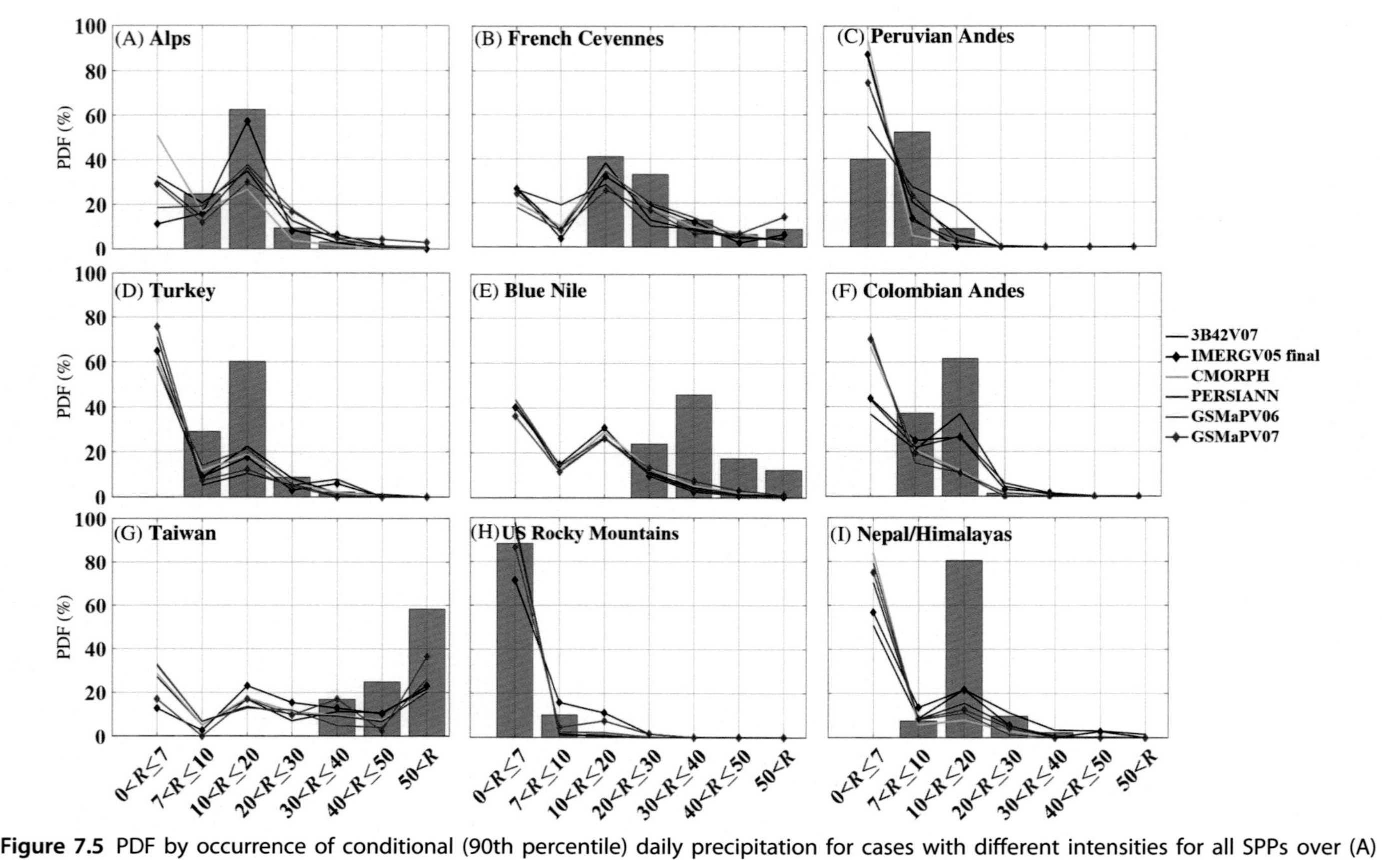

Figure 7.5 PDF by occurrence of conditional (90th percentile) daily precipitation for cases with different intensities for all SPPs over (A) the Alps, (B) the French Cévennes, (C) the Peruvian Andes, (D) Turkey, (E) the Blue Nile, (F) the Colombian Andes, (G) Taiwan, (H) the US Rocky Mountains, and (I) Nepal/Himalayas. *SPP*, Satellite-based precipitation products; *PDF*, probability density function.

($0 < R < 7$ mm/day), while this overestimation turns to significant underestimation in the 7–10 mm/day rainfall range. Over Taiwan uncertainty is large in the 0–30 mm/day rainfall range, whereas relatively better estimates are observed in the 30–40 mm/day range.

7.4.3 Products accuracy and uncertainty

Fig. 7.6 represents the SPPs accuracy in capturing the rainfall occurrence by analyzing NMRV (black colors) and NFASRV (gray colors) values for the conditional (> Q90) precipitation values. SPPs performance in capturing the precipitation occurrence is best over the Blue Nile and French Cévennes regions, followed by Nepal, the Alps, and Turkey. Over almost all regions PERSIANN exhibits the highest NMRV and NFASRV values. Yamamoto et al. (2017) compared GSMaPV06 with GSMaPV07 orographic correction and concluded that the false alarm rate of orographic rainfall conditions was high for GSMaPV06. Our results are in alignment with Yamamoto et al. (2017). GSMaPV06 in general has higher NMFRV values compared to GSMaPV07. GSMaPV07 performance in capturing the precipitation occurrence is better compared to the previous version. Similarly, IMERGV05 final is shown to have better performance in capturing the precipitation occurrence compared to 3B42V07.

To determine the systematic error component of daily SPPs, MRE is presented in Fig. 7.7. Over almost all regions all SPPs underestimate, with the exception of GSMaPV07 and IMERGV05 final over the US Rocky Mountains (Fig. 7.7H). Over the Peruvian Andes, Colombian Andes, Blue Nile, and Turkey all SPPs performances are similar exhibiting underestimation. In general GSMaPV07 and IMERGV05 final performances are better over all regions compared to their previous versions, GSMaPV06 and 3B42V07, respectively.

Over the Alps, the IMERGV05 final underestimation is the lowest compared to other SPPs. The highest MRE values are shown for CMORPH and PERSIANN over this region. This result is in agreement with the findings of Mei and Anagnostou (2014) who showed that CMORPH and PERSIANN strongly underestimated gauge rainfall values in all quantile ranges, especially during winter periods. Over the French Cévennes the lowest MRE values correspond to GSMaPV07 and GSMaPV06 followed by IMERGV05 final. 3B42V07, CMORPH, and PERSIANN products have similar performances exhibiting nearly 50% underestimation. GSMAP products over the Taiwan study area exhibit

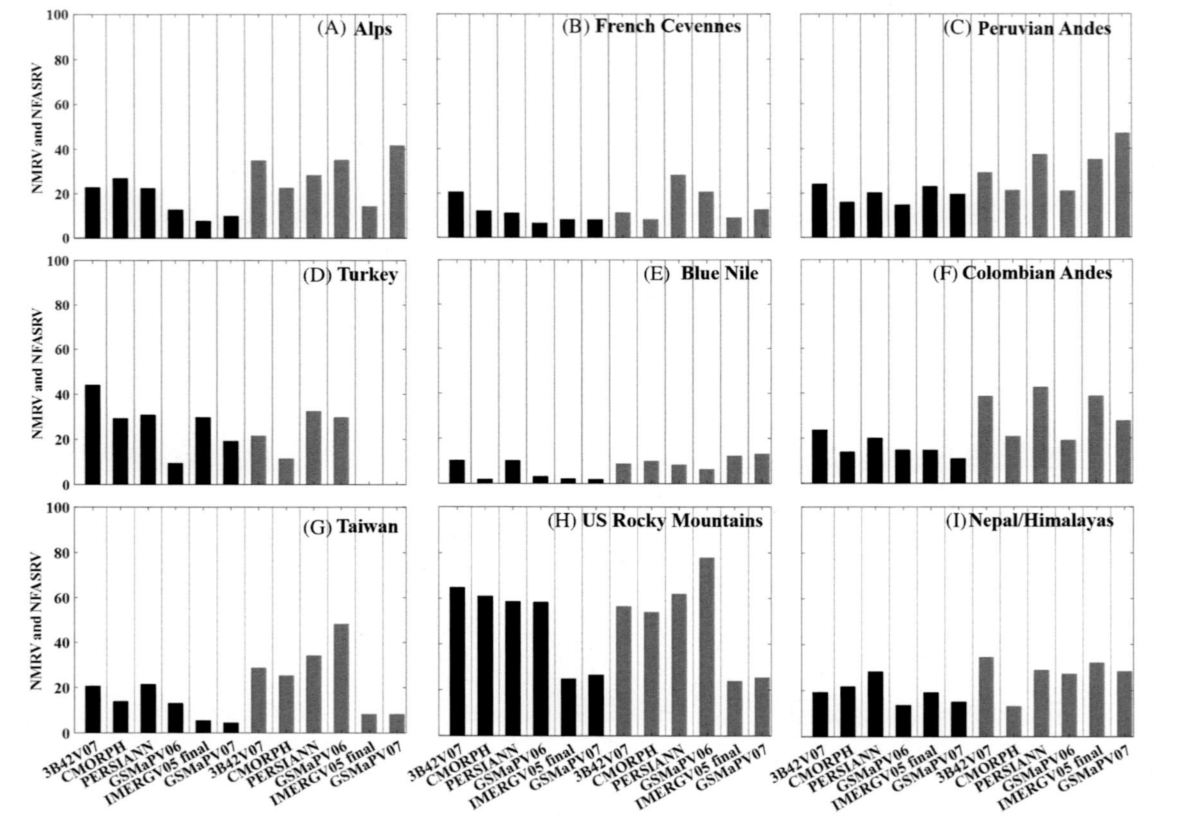

Figure 7.6 NMRV and NFASRV values over (A) the Alps, (B) the French Cévennes, (C) the Peruvian Andes, (D) Turkey, (E) the Blue Nile, (F) the Colombian Andes, (G) Taiwan, (H) the US Rocky Mountains, and (I) Nepal/Himalayas and every SPP (black colors represent NMRV and gray colors represent NFASRV). *SPP*, Satellite-based precipitation products; *NMRV*, normalized missed rainfall volume; *NFASRV*, normalized false alarm satellite rainfall volume.

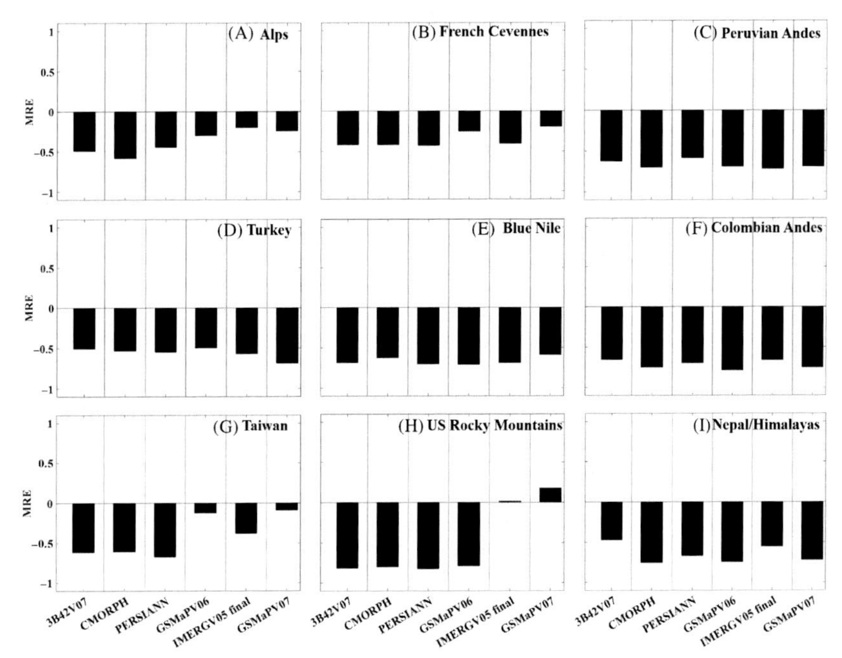

Figure 7.7 MRE of the conditional (90th quantile) rain gauge over (A) the Alps, (B) the French Cévennes, (C) the Peruvian Andes, (D) Turkey, (E) the Blue Nile, (F) the Colombian Andes, (G) Taiwan, (H) the U.S. Rocky Mountains, and (I) Nepal/Himalayas for every SPP.

the lowest underestimation compared to other SPPs. Our results are in alignment with Yamamoto et al. (2017) who concluded that orographic/nonorographic rainfall classification schemes improve estimates of rainfall over the entire Asian region. However, over Nepal IMERV05 and 3B42V07 MRE results are lower compared to GSMaP. Over Taiwan 3B42V07, CMORPH, and PERSIANN performances are the lowest with underestimation exeeding 50%. In Nepal 3B42V07 has the lowest underestimation, followed by IMERGV05 final. These results are in agreement with Derin et al. (2016) that compared similar SPPs over same regions for 99th quantiles.

Fig. 7.8 captures the CORR of SPPs for the >Q90 daily rainfall accumulations. Correlation values are generally low. IMERGV05 final and GSMaPV07 CORR is higher compared to other SPPs. Over the Alps, Taiwan, US Rocky Mountains, and Nepal, CORR values of GSMaPV07 are higher compared to GSMaPV06. Over the Alps, the highest CORR value corrosponds to IMERGV05 final (0.5) followed by

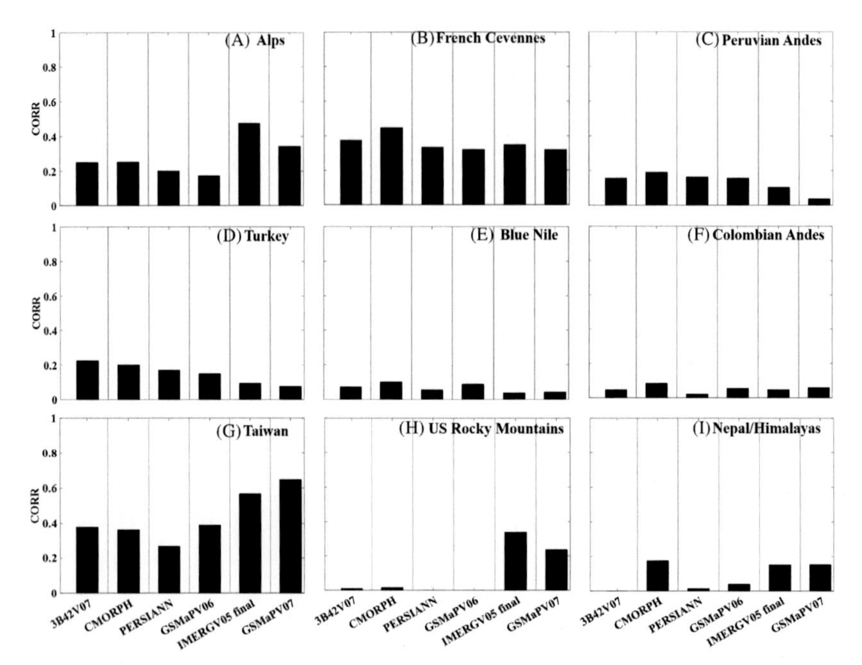

Figure 7.8 As in Fig. 7.7, but for correlation coefficient.

GSMaPV07 (0.4). Both these product performances are improved compared to their previous versions. The lowest CORR over this region is reported by GSMaPV06 (0.2). Over Turkey we observe a shift in the SPPs performance; 3B42V07 has the best CORR value (0.2), whereas the lowest CORR values are reported for IMERGV05 final and GSMaPV07 (less than 0.1). The lowest CORR values are reported over the BlueNile, US Rocky Mountains, and Colombian Andes regions, where the highest CORR is below 0.1.

Lastly Fig. 7.9 shows the random component of the error (CRMSE) for daily conditional (>Q90) rainfall values. In general, over all regions CRMSE values are low with exceptions for Taiwan (specifically, GSMaPV06 in Fig. 7.9G) and the US Rocky Mountains (specifically, GSMaPV07 product in Fig. 7.9H). Beside these cases, the performance of all SPPs is very similar to each other over almost all regions. Over the Peruvian Andes, Turkey, Blue Nile, and Colombian Andes, all CRMSE values are similar to each other and relatively low. In the Alps, the largest random error corresponds to GSMaPV06, followed by PERSIANN and GSMaPV07. For the rest of the SPPs, CRMSEs are very similar and less than 1. Over the French Cévennes region, the highest CRMSE

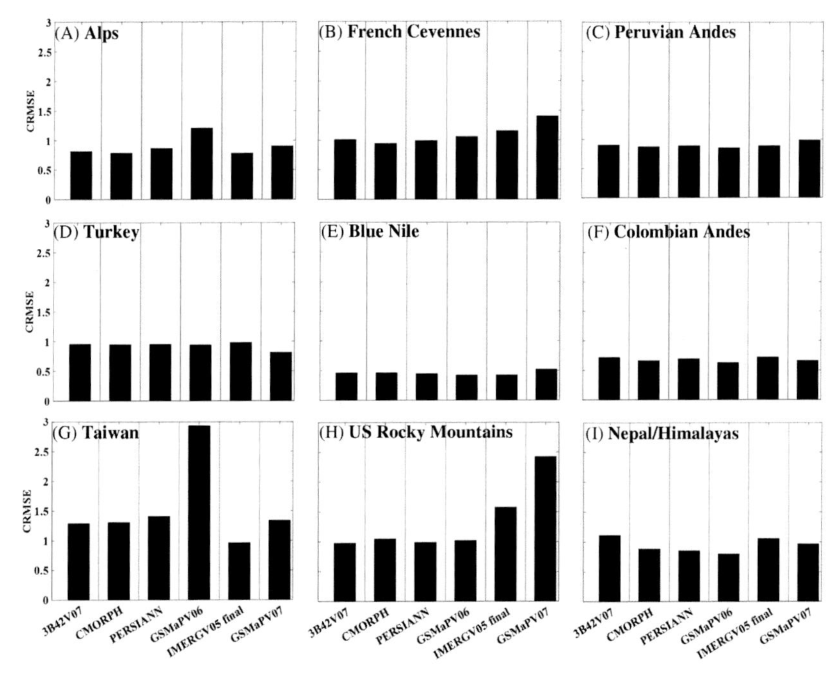

Figure 7.9 As in Fig. 7.7, but for CRMSE. *CRMSE*, Centralized root mean square error.

corresponds to GSMaPV07, followed by IMERGV05 final and the rest of the SPPs report CRMSE values of 1. Over Taiwan, the GSMaPV06 CRMSE is 3, which could be due to an overestimation of orographic rainfall whereby estimates of rainfall reach the specified limit (Yamamoto et al., 2017). High random errors could be related to high NFSARV as discussed earlier. GSMaPV07 has the highest CRMSE over the US Rocky Mountains, followed by IMERGV05 final, whereas the rest of the SPPs performances are similar to each other with CRMSE of 1.

7.5 Conclusions

This chapter evaluated six satellite-based rainfall products over nine mountainous regions around the globe. The error analysis focused on high daily rainfall accumulations (values exceeding the 90th quantiles) and was based on daily rain gauge rainfall data from more than 6 years (within the period 2000–15). The evaluation was conducted by matching pairs of extreme average rain gauge precipitation values and SPPs (available at 0.25° regular Cartesian grid) for each study region. The evaluated SPPs are 3B42V07 (gauge-adjusted), gauge-adjusted CMORPH, gauge-adjusted PERSIANN, gauge adjusted GSMaPV06, gauge-adjusted GSMaPV07 and gauge adjusted IMERGV05 final run. The study sites were the Alps, French Cévennes, Western Black Sea Region of Turkey, Colombian Andes, Peruvian Andes, US Rocky Mountains, Blue Nile, Western Taiwan, and Himalayas over Nepal. Various evaluation metrics were used to assess the performance of the satellite-based rainfall products in terms of systematic and random errors.

The study regions are characterized by different hydroclimatic characteristics. The study showed that over all regions SPPs exhibit consistent underestimation of values greater than the 90th percentiles of reference rainfall, which is in agreement with the findings in past studies, e.g., Derin et al (2018), Kwon et al. (2008), Kubota et al. (2009), and Maggioni et al. (2017). This underestimation is attributed to the occurrence of shallow, but high accumulation precipitation, related to warm-rain processes (Yamamoto et al., 2017). The study also showed that the enhancement to the SPP algorithms for GSMaPV07 and IMERGV05 final provided statistical improvements (approximately 40% for CRMSE, 35% for NFARV, 30% for CORR, and 30% for MRE). The new orographic rainfall classification in the GSMaPv07 algorithm was able to improve the detection of orographic rainfall, the rainfall amounts, and error metrics (10% decrease in CRMSE and 20% increase in CORR) relative to GSMaPV06. It should also be noted that the improvement of GSMaPV07 was demonstrated beyond the Asian land region.

All SPPs examined herein are gauge adjusted and Derin et al. (2016) has concluded that the performance of gauge adjusted SPPs are highly dependent on the representativeness of the local rain gauges. Hence, over data sparse regions and complex terrain, gauge adjustment could have limited improvements compared to satellite-alone rainfall products. Taking into account all presented results in this study, we can conclude the following:

1. SPPs are able to capture the distribution of conditional (90th percentile) rainfall rates for most regions. However, there is a general trend of underestimating the occurrence of high rain rates.
2. SPP performance for precipitation detection is best observed over the Blue Nile region, followed by the Alps, French Cévennes, Taiwan, and Peruvian Andes. In general, GSMaPV06 had the highest NFASRV, which was improved with GSMaPV07.

3. All products underestimated the reference >Q90 precipitation values (MRE values of −0.1 to −0.8) with low CORR (0.1−0.7) and generally low CRMSE (0.4−1.5). There were exceptions: for example GSMaPV06 had the highest random component over Taiwan, which has been significantly improved by GSMaPV07.

In summary, we have shown that the performance of integrated satellite rainfall products for precipitation values higher than 90th quantile (considered extreme rainfall) highly depends on the satellite precipitation algorithms. Derin et al. (2018) showed that the correlation and biases of the new version of GPROF (GPROFV05) are significantly improved over its previous version (GPROFV04). Improvements to the current algorithms advance the detection of orographic rainfall and the precipitation amount, which can result in a major advancement in the utilization of satellite data in monitoring hydroclimatic hazards. Petkovic et al. (2018) tested the use of large-scale environmental variables that are associated with the potential for the atmosphere to produce heavy precipitation. Analysis suggested that system morphology and retrieval biases could indeed be linked through the use of environmental predictors in the atmospheric states which is favorable for convection. By complementing the a priori information by collocated environment properties, the overall pixel-level bias for heavy precipitation is reduced by 20%−30%. These improvements are accompanied by a noticeable reduction in the random error as well.

This chapter showed that GSMaPV07 and IMERGV05 final statistically outperformed CMORPH, PERSIANN, 3B42V07, and GSMaPV06 for precipitation values higher than the 90th quantile. However, it should be noted that in situ observations, while extremely important, provide limited representation of heavy precipitation over complex terrain, and spatial scale mismatch between satellite and gauges should be considered. Therefore, future work should consider utilizing more advanced sensors, e.g., X-band polarimetric radar observations, which provide better characterization of spatial variability of precipitation over complex terrain and thus can enhance investigation of satellite-rainfall error characteristics. Moreover, future research work in this area should also include the evaluation of instantaneous PMW-based precipitation datasets, which constitute the basis of most combined satellite precipitation algorithms. Such analysis can give further insight about the error characteristics of SPPs in estimating extreme precipitation. Finally, investigating possible links between the errors and physiographic features within each region can provide more detailed error characteristics, which can provide useful information to algorithm developers and data users.

References

AghaKouchak, A., Behrangi, A., Sorooshian, S., Hsu, K., Amitai, E., 2011. Evaluation of satellite-retrieved extreme precipitation rates across the central United States. J. Geophys. Res. Atmos. 116 (D2), 1984–2012.

Allan, R.P., Soden, B.J., 2008. Atmospheric warming and the amplification of precipitation extremes. Science 32, 1481–1494.

Anagnostou, E.N., Maggioni, V., Nikolopoulos, E., Meskele, T., Hossain, F., Papadopoulos, A., 2010. Benchmarking high-resolution global satellite raingall products to radar and rain-gauge rainfall estimates. IEEE Trans. Geosci. Remote Sens. 48 (4), 1667–1683.

Aonashi, K., et al., 2009. GSMaP passive microwave precipitation retrieval algorithm: algorithm description and validation. J. Meteorol. Soc. Jpn. 87A, 119–136.

Bartsotas, N.S., Anagnostou, E.N., Nikolopoulos, E.I., Kallos, G., 2018. Investigating satellite precipitation uncertainty over complex terrain. J. Geophys. Res. Atmos.

Bookhagen, B., Strecker, M.R., 2008. Orographic barriers, high-resolution TRMM rainfall, and relief variations along the eastern Andes. Geophys. Res. Lett. 35, L06403.

Chen, S., Hong, Y., Cao, Q., Kirstetter, P.E., Gourley, J.J., Qi, Y., et al., 2013. Performance of evaluation of radar and satellite rainfalls for typhoon morakot over Taiwan: are remote-sensing products ready for gauge denial scenario of extreme events? J. Hydrol. 506, 4–13.

Delrieu, G., Wijbrans, A., Boudevillain, B., Faure, D., Bonnifait, L., Kirstetter, P.E., 2014. Geostatistical radar–raingauge merging: a novel method for the quantification of rain estimation accuracy. Adv. Water Resour. 71, 110–124.

Derin, Y., Yilmaz, K.K., 2014. Evaluation of multiple satellite-based precipitation products over complex topography. J. Hydrometeorol. 15, 1498–1516.

Derin, Y., Anagnostou, E., Berne, A., Borga, M., Boudevillain, B., Buytaert, W., et al., 2016. Multi-regional satellite precipitation products evaluation over complex terrain. J. Hydrometeorol. 17, 1817–1836.

Derin, Y., Anagnostou, E., Anagnostou, M.N., Kalogiros, J., Casella, D., Marra, A.C., et al., 2018. Passive microwave rainfall error analysis using high-resolution X-band dual-polarization radar observations in complex terrain. IEEE Trans. Geosci. Remote Sens. 56 (5), 2565–2586.

Dinku, T., Ceccato, P., Grover-Kopec, E., Lemma, M., Connor, S.J., Ropelewski, C.F., 2007. Validation of satellite rainfall products over East Africa's complex topography. Int. J. Remote Sens. 28, 1503–1526.

Dinku, T., Chidzambwa, S., Ceccato, P., Connor, S.J., Ropelewski, C.F., 2008. Validation of high-resolution satellite rainfall products over complex terrain. Int. J. Remote Sens. 29, 4097–4110.

Dinku, T., Connor, S.J., Ceccato, P., 2010. Comparison of CMORPH and TRMM-3B42 over mountainous regions of Africa and South America. In: Gebremichael, M., Hossain, F. (Eds.), Satellite Rainfall Applications for Surface Hydrology. Springer, pp. 193–204.

Ebert, E.E., Janowiak, J.E., Kidd, C., 2007. Comparison of near-real-time precipitation estimates from satellite observations and numerical models. Bull. Am. Meteorol. Soc. 88, 47–64.

Espinoza, J.C., Chavez, S., Ronchail, J., Junquas, C., Takahashi, K., Lavado, W., 2015. Rainfall hotspots over the southern tropical Andes: spatial distribution, rainfall

intensity, and relations with large-scale atmospheric circulation. Water Resour. Res. 51, 3459–3475.
Espinoza-Villar, J.C., Ronchail, J., Guyot, J.L., Cochonneau, G., Naziano, F., Lavado, W., et al., 2009. Spatio-temporal rainfall variability in the Amazon basin countries (Brazil, Peru, Bolivia, Colombia, and Ecuador). Int. J. Climatol. 29 (11), 1574–1594.
Frei, C., Schär, C., 1998. A precipitation climatology of the Alps from high-resolution rain-gauge observations. Int. J. Climatol. 18, 873–900.
Gottschalck, J., Meng, J., Rodell, M., Houser, P., 2005. Analysis of multiple precipitation products and preliminary assessment of their impact on global land data assimilation system land surface states. J. Hydrometeorol. 6, 573–598.
Habib, E., Elsaadani, M., Haile, A.T., 2012. Climatology-focused evaluation of CMORPH and TMPA satellite rainfall products over the Nile basin. J. Appl. Meteorol. Climatol. 51, 2105–2121.
Hirpa, F., Gebremichael, M., Hopson, T., 2010. Evaluation of high-resolution satellite precipitation products over very complex terrain in Ethiopia. J. Appl. Meteorol. Climatol 49, 1044–1051.
Hossain, F., Huffman, G.J., 2008. Investigating error metrics for satellite rainfall data at hydrologically relevant scales. J. Hydrometeorol. 9, 563–575.
Hou, A.Y., Skofronick-Jackson, G., Kummerow, C., Shepherd, J.M., 2008. Global precipitation measurement. In: Michaelides, S. (Ed.), Precipitation: Advances in Measurement and Prediction. Springer-Verlag, pp. 131–169.
Hou, A.Y., et al., 2013. NASA GPM science implementation plan. NASA Goddard Space Flight Center, 162 pp. Available online from: <http://pmm.nasa.gov/reserouces/documents/gpm-science-implmenetation-plan>.
Hou, A.Y., et al., 2014. The global precipitation measurements mission. Bull. Am. Meteorol. Soc. 95, 701–722.
Huang, X., Wang, D., Liu, Y., Feng, Z., Wang, D., 2017. Evaluation of extreme precipitation based on satellite retrievals over China. Front. Earth Sci. Available from: https://doi.org/10.1007/S11707-017-0643-2.
Huffman, G.J., 2013. README for accessing experimental realtime TRMM Multi-Satellite Precipitation Analysis (TMPA-RT) data sets. NASA Doc., 11 pp. Available online from: <ftp://mesoa.gsfc.nasa.gov/pub/trmmdocs/rt/3B4XRT_README.pdf>.
Huffman, G.J., Adler, R.F., Bolvin, D.T., Gu, G., Nelkin, E.J., Bowman, K.P., et al., 2007. The TRMM Multisatellite Precipitation Analysis (TMPA): quasi-global, multiyear, combined-sensor precipitation estimates at fine scales. J. Hydrometeorol 8, 38–55. Available from: https://doi.org/10.1175/JHM560.1.
Huffman, G.J., Adler, R.F., Bolvin, D.T., Gu, G., Nelkin, E.J., 2010. The TRMM Multi-satellite Precipitation Analysis (TMPA). In: Gebremichael, M., Hossain, F. (Eds.), Satellite Rainfall Applications for Surface Hydrology. Springer, pp. 3–22.
Huffman, G.J., Bolvin, D.T., Braithwaite, D., Hsu, K., Joyce, R., Xie, P., 2015. NASA Global Precipitation Measurement (GPM) Integrated Multi-Satellite Retrievals for GPM (I-MERG). Algorithm Theoretical Basis Doc. (ATBD), Version 4.5, Greenbelt, MD, 26 pp. Available online from: <http://pmm.nasa.gov/sites/default/failes/document_files/IMERG_ATBD_V4.5.pdf>.
Ichiyanagi, K., Yamanaka, M.D., Muraji, Y., Vaidya, B.K., 2007. Precipitation in Nepal between 1987 and 1996. Int. J. Climatol. 27, 1753–1762.
Joyce, J.R., Janowiak, E.J., Arkin, P.A., Xie, P., 2004. CMORPH: a method that produces global precipitation estimates from passive microwave and infrared data at high spatial and temporal resolution. J. Hydrometeorol. 5, 487–503.
Kidd, C., Levizzani, V., 2011. Status of satellite precipitation retrievals. Hydrol. Earth Syst. Sci. 15, 1109–1116.

Kirschbaum, D.B., et al., 2017. NASA's remotely sensed precipitation: a reservoir for applications users. Bull. Am. Meteorol. Soc. 98, 1169–1184.
Kubota, T., Shige, S., Hashizume, H., Aonashi, K., Takahashi, N., Seto, S., et al., 2007. Global precipitation map using satellite-borne microwave radiometers by the GSMaP project: production and validation. IEEE Trans. Geosci. Remote Sens. 45 (7), 2259–2275.
Kubota, T., Ushio, T., Shige, S., Kida, S., Kachi, M., Okamoto, K., 2009. Verification of high-resolution satellite-based rainfall estimates around Japan using a gauge-calibrated ground-radar dataset. J. Meteorol. Soc. Jpn. 87A, 203–222.
Kummerow, C.D., Randel, D.L., Kulie, M., Wang, N.-Y., Ferraro, R., Munchak, S.J., et al., 2015. The evaluation of Goddard Profiling Algorithm to a fully parametric scheme. J. Atmos. Oceanic Technol. 32, 2265–2280.
Kwon, E.-H., Sohn, B.-J., Chang, D.-E., Ahn, M.-H., Yang, S., 2008. Use of numerical forecasts for improving TMI rain retrievals over the mountainous area in Korea. J. Appl. Meteor. Climatol. 47, 1995–2007.
Lochoff, M., Zolina, O., Simmer, C., Schulz, J., 2014. Evaluation of satellite-retrieved extreme precipitation over Europe using Gauge observations. J. Clim. 27 (2), 607–623.
Maggioni, V., Sapiano, M.R., Adler, R.F., Tian, Y., Huffman, G.J., 2014. An error model for uncertainty quantification in high-time-resolution precipitation products. J. Hydrometeorol. 15 (3), 1274–1292.
Maggioni, V., Meyers, P.C., Robinson, M.D., 2016. A review of merged high-resolution satellite precipitation product accuracy during the Tropical Rainfall Measuring Mission (TRMM) era. J. Hydrometeorol. 17 (4), 1101–1117.
Maggioni, V., Nikolopoulos, E.I., Anagnostou, E.N., Borga, M., 2017. Modeling satellite precipitation errors over mountainous terrain: the influence of gauge density, seasonality, and temporal resolution. IEEE Trans. Geosci. Remote Sens. 55 (7), 4130–4140.
Manz, B., Buytaert, W., Zulkafli, Z., Lavado, W., Willems, B., Robles, L.A., et al., 2016. High-resolution satellite-gauge merged precipitation climatologies of the Tropical Andes. J. Geophys. Res. Atmos. 121. Available from: https://doi.org/10.1002/2015JDO23788.
McCollum, J.R., Krajewski, W.F., Ferraro, R.R., Ba, M.B., 2002. Evaluation of biases of satellite rainfall estimation algorithms over the continental united states. J. Appl. Meteorol. 41, 1065–1080.
McPhillips, L.E., Chang, H., Chester, M.V., Depietri, Y., Friedman, E., Grimm, N.B., et al., 2018. Defining extreme events: a cross-disciplinary review. Earth Future 6, 1002/2017EF000686.
Mega, T., Ushio, T., Kubota, T., Kachi, M., Aonashi, K., Shige, S., 2014. Gauge adjusted global satellite mapping of precipitation (GSMaP_Gauge). Proc. Gen. Assem. Sci. Symp.
Mei, Y., Anagnostou, E.N., 2014. Error analysis of satellite precipitation products in mountainous basins. J. Hydrometeorol. 15, 1778–1793.
Nastos, P.T., Kapsomenakis, J., Douvis, K.C., 2013. Analysis of precipitation extremes based on satellite and high-resolution gridded data set over Mediterranean basin. Atmos. Res. 131, 46–59.
Norbiato, D., Borga, M., Merz, R., Blöschl, G., Carton, A., 2009. Controls on event runoff coefficients in the eastern Italian Alps. J. Hydrol. 375, 312–325.
Petkovic, V., Kummerow, C.D., 2015. Performance of the GPM passive microwave retrieval in the Balkan flood event of 2014. J. Hydrometeorol 16, 2501–2518.
Petkovic, V., Kummerow, C.D., Randel, D.L., Pierce, J.R., Kodros, J.K., 2018. Improving the quality of heavy precipitation estimates from satellite microwave rainfall retrievals. J. Hydrometeorol. 19, 69–85.

Prakash, P., Mitra, A.K., Mitra, D.S., AghaKouchak, A., 2016. From TRMM to GPM: how well can heavy rainfall be detected from space? Adv. Water Resour. 88, 1–7.
Rudolf, B., 1993. Management and analysis of precipitation data on a routine basis. In: Sevruk, B., Lapin, M. (Eds.), Proceedings of International Symposium on Precipitation and Evaporation, vol. 1, Slovak Hydrometeorology Institution, pp. 69–76.
Salerno, F., Guyennon, N., Thakuri, S., Viviano, G., Romano, E., Vuillermoz, E., et al., 2015. Weak precipitation, warm winters and springs impact glaciers of south slopes of Mt. Everest (central Himalaya) in the last 2 decades (1994–2013). Cryosphere 9, 1229–1247.
Sapiano, M.R.P., Arkin, P., 2009. An intercomparison and validation of high-resolution satellite precipitation estimates with 3-hourly gauge data. J. Hydrometeorol. 10, 149–166.
Scheel, M.L.M., Rohrer, M., Huggel, C., Santos Villar, D., Silvestre, E., Huffman, G.J., 2011. Evaluation of TRMM multi-satellite precipitation analysis (TMPA) performance in the central Andes region and its dependency on spatial and temporal resolution. Hydrol. Earth Syst. Sci. 15, 2649–2663.
Sing, P., Kumar, N., 1997. Effect of orography on precipitation in the western Himalayan region. J. Hydrol. 199, 183–206.
Skofronick-Jackson, G., et al., 2017. The global precipitation measurement (GPM) mission for science and society. Bull. Am. Meteorol. Soc. 98, 1679–1695.
Sorooshian, S., Hsu, K., Gao, X., Gupta, H.V., Imam, B., Braithwaite, D., 2000. Evaluation of PERSIANN system satellite-based estimates of tropical rainfall. Bull. Am. Meteorol. Soc. 81, 2035–2046.
Stampoulis, D., Anagnostou, E.N., 2012. Evaluation of global satellite rainfall products over continental Europe. J. Hydrometeorol 13, 588–603.
Su, F., Hong, Y., Lettenmaier, D.P., 2008. Evaluation of TRMM Multisatellite Precipitation Analysis (TMPA) and its utility in hydrologic prediction in the La Plata Basin. J. Hydrometeorol 9, 622–640.
Trenberth, K.E., 1999. Conceptual framework for changes of extremes of the hydrological cycle with climate change. Clim. Change 42, 327–339.
Trenberth, K.E., Dai, A., Rasmussen, R.M., Parsons, D.B., 2003. The changing character of precipitation. Bull. Am. Meteorol. Soc. 84, 1205–1217.
Ushio, T., et al., 2009. A Kalman filter approach to the Global Satellite Mapping of Precipitation (GSMaP) from combined passive microwave and infrared radiometric data. J. Meteorol. Soc. Jpn. 87A, 137–151.
Villa, O.D.A., Velez, J.I., Poveda, G., 2011. Improved long-term mean annual rainfall fields for Colombia. Int. J. Climatol. 31, 2194–2212.
Xie, P., Arkin, P.A., 1996. Gauge-based monthly analysis of global land precipitation from 1971 to 1994. J. Geophys. Res. 101, 19023–19034.
Xie, P., Yoo, S.H., Joyce, R.J., 2011. Bias-corrected CMORPH: a 13-year analysis of high-resolution global precipitation. <http://ftp.cpc.ncep.noaa.gov/precip/CMORPHV1.0/REF/EGU1104Xie bias-CMORPH.pdf>.
Yamamoto, M.K., Shiege, S., 2015. Implementation of an orographic/nonorographic rainfall classification scheme in the GSMaP algorithm for microwave radiometers. Atmos. Res. 163, 36–47.
Yamamoto, M.K., Shiege, S., Yu, C.-K., Cheng, L.-W., 2017. Further improvement of the heavy orographic rainfall retrievals in the GSMaP Algorithm for Microwave Radiometers. J. Appl. Meteor. Climatol. 56, 2607–2619.
Yong, B., Ren, L., Hong, Y., Gourley, J.J., Tian, Y., Huffman, G.J., et al., 2013. First evaluation of climatological calibration algorithm in the real-time TMPA precipitation

estimates over two basins at high and low latitudes. Water Resour. Res. 49, 2461–2472.

Yucel, I., Onen, A., 2014. Evaluating a mesoscale atmosphere model and a satellite-based algorithm in estimating extreme rainfall events in northwestern Turkey. Nat. Hazard. Earth Syst. Sci. 14, 611–624.

Further reading

Xie, P., Yatagai, A., Chen, M., Hayasaka, T., Fukushima, Y., Liu, C., et al., 2007. A gauge-based analysis of daily precipitation over East Asia. J. Hydrometeorol. 8, 607–626.

CHAPTER EIGHT

Evaluating the sp
pattern of concentr
aggressiveness and s
of precipitation over B
with time–series Tropica all
Measuring Mission data

Ashraf Dewan[1], Kexiang Hu[1], Mohammad Kamruzzaman[2] and Md. Rafi Uddin[3]

[1]School of Earth and Planetary Sciences, Curtin University, Perth, WA, Australia
[2]Institute of Water Modelling, Dhaka, Bangladesh
[3]Bangladesh University of Engineering and Technology, Dhaka, Bangladesh

8.1 Introduction

Surface air temperatures around the world continue to rise in response to increased anthropogenic activities, with the rate of warming highly pronounced between 1951 and 2012 (0.12°C per decade), and this has major implications for the global precipitation distribution [Intergovernmental Panel on Climate Change (IPCC), 2014]. Held and Soden (2006), for instance, estimated a 1.5%–2% increase in global mean precipitation per Kelvin surface warming. Since climate change is expected to enhance evaporation and atmospheric moisture content, worldwide changes in mean and extreme precipitation have been projected (Wentz et al., 2007; Emori and Brown, 2005). Using satellite observations and model simulations, Allan and Soden (2008) demonstrated that heavy rain events are increasing during warm periods and decreasing in cold periods. Specifically, there is a strong tendency for wet areas to get wetter and dry areas to become drier in response to global warming (Trenberth, 2011; Seager et al., 2010). However, observed changes in precipitation vary according to latitudinal bands (Zhang et al., 2007),

Extreme Hydroclimatic Events and Multivariate Hazards in a Changing Environment.
DOI: https://doi.org/10.1016/B978-0-12-814899-0.00008-0

mainly because of dynamic (e.g., due to atmospheric motion) and thermodynamic mechanisms (e.g., due to atmospheric moisture content) (Emori and Brown, 2005).

A rise in the number of intense precipitation events in many locations worldwide is already evident (Coumou and Rahmstorf, 2012; Alexander et al., 2006) and has been linked to anthropogenic forcing (Min et al., 2011). Since precipitation is a vital component of the hydrological cycle, further changes in its pattern and amount could increase the risk of floods and droughts at many locations worldwide (Hajani et al., 2017). Andersen and Shepherd (2013) noticed that flooding in relation to extreme precipitation events is becoming more frequent and intense. The Louisiana flood of August 2016 is one such example (Wiel et al., 2017). Likewise, Hirsch and Archfield (2015) confirmed that floods have become more frequent because of a significant increase in heavy rain events. However, the effects of atmospheric warming on precipitation distribution are not geographically uniform (Rummukainen, 2013; Zhang et al., 2007) and appear to be scale dependent due to the differences in local factors such as geomorphology, elevation, and climatic patterns (Tan et al., 2017). Therefore, regional assessment of the statistical patterns of precipitation has been receiving increasing attention (Fenta et al., 2017; Valdés-Pineda et al., 2016).

To estimate the effect of atmospheric warming on precipitation, several indices have been developed. These indices can generally be divided into precipitation extreme indices (http://etccdi.pacificclimate.org/index.shtml), concentration indices (Wang and Li, 2016), and aggressiveness (García-Barrón et al., 2018). While extreme indices are useful to determine the trends of intense events (Jones et al., 1999), concentration indices help identify the potential for droughts and floods over a geographical region (Li et al., 2011). These methods have been used extensively in different regions to understand the spatiotemporal variability of precipitation regimes (García-Barrón et al., 2018; Fenta et al., 2017; Razeiei, 2017; Chatterjee et al., 2016; Wang and Li, 2016; Jiang et al., 2016; Deshpande et al., 2016; Abolverdi et al., 2016; Li et al., 2011; Dash et al., 2009; Alexander et al., 2006). Generally, these studies are unequivocal in confirming that changes in the behavior of precipitation, both at global and regional levels, are due to anthropogenic climate change (Min et al., 2011). In addition to concentration and extreme indices, rainfall erosivity [e.g., Fournier index (FI) or its variants] and seasonality indices [e.g., seasonality index (SI)] are also important measures for estimating precipitation aggressiveness and seasonality (see García-Barrón et al., 2018). These indices contribute to the understanding of periodic variation of rainfall at any location. Assessment of both concentration and aggressiveness/seasonality indices is therefore not only valuable for identifying possible modifications in the water cycle, but also for evaluating flooding/drought potential (Valdés-Pineda et al., 2016; García-Barrón et al., 2015).

With an estimated population of more than 168 million, Bangladesh is one of the countries most vulnerable to anthropogenic climate change (Maplecroft, 2016). The degree of vulnerability appears to be intensifying with continued changes in climatic variables such as temperature and precipitation (Rahman and Lateh, 2017; Siddik and Rahman, 2014). Using both regional and global climate change models, Chowdhury and Ndiaye (2017) demonstrated a rise in both temperature and annual rainfall in Bangladesh. Any changes in the intensity, frequency, and amount of rainfall could be a significant threat for Bangladesh. On one hand, extreme events may enhance the risk of localized floods, and on the other hand, they have the potential to affect millions of people across various sectors (e.g., socioeconomic) due to their low adaptation capacity (Maplecroft, 2016). Given the number of people (nearly 75% of the total population) engaged in agricultural activities, quantifying the statistical patterns of precipitation may be extremely beneficial for water resources planning, management of floods and droughts, as well as providing a deeper understanding of climate change (Li et al., 2018; García-Barrón et al., 2015; Huang et al., 2013).

Because of its importance to agriculture and the economy, the investigation of spatial and temporal patterns of precipitation has gained increased attention. Although rain gauge data have been a major source of information for assessing annual and seasonal characteristics of precipitation in specific regions (particularly the northwest) and across the country (Bari and Hussain, 2017; Nowreen et al., 2015; Kumar et al., 2014; Endo et al., 2015; Shahid, 2011; Kripalani et al., 1996), a few works have utilized satellite-based data (such as those from the Tropical Rainfall Measuring Mission—TRMM) to investigate rainfall climatology (Tarek et al., 2017; Takahashi, 2016; Islam and Uyeda, 2007), and to examine the potential for flood assessment (Prasanna et al., 2014; Rahman et al., 2012). Three key observations may be made from works that have used in situ precipitation records: (1) annual rainfall shows an upward trend (Rahman and Lateh, 2017; Nowreen et al., 2015; Shahid, 2011); (2) frequency of extreme rainfall events is increasing with a concurrent rise in the number of wet days and wet months, though Basher et al. (2017) showed a decreasing trend in all extreme indices over the northeast region; and (3) there is an increase in annual and seasonal precipitation under climate warming conditions. However, estimates of future rainfall appear to be contradictory at both annual and seasonal scales (Fahad et al., 2018; Rahman and Lateh, 2017; Chowdhury and Ndiaye, 2017; Nowreen et al., 2015; Shahid, 2011). Compared to these studies, knowledge of precipitation concentration, seasonality, and aggressiveness is extremely poor for Bangladesh. Two popular measures, the precipitation concentration index (PCI) and the SI, have been used to assess spatiotemporal variations of precipitation concentration and seasonality using station-based records. The findings from these studies are: (1) there has been no significant change in the precipitation seasonality in northern Bangladesh (Bari and Hussain, 2017); and (2) PCI across the country is declining (Shahid and Khairulmaini, 2009). Regardless, station-based results seem to vary, which may be due to variation in data quality, inconsistent records, and the sparse distribution of weather stations (Alexander et al., 2006, Cash et al., 2008). In addition, spatial interpolation of point data may not provide accurate results, especially in areas where the distribution of rainfall is driven by convection and orography (Fenta et al., 2017). Although quite a few satellite-based works exist, Tarek et al. (2017) and Islam and Uyeda (2007) demonstrated that TRMM-based precipitation estimates in particular closely match with rain gauge records. Further, TRMM provides superior results over point-based observations in the identification of heavy rainfall leading to floods in Bangladesh (Prasanna et al., 2014). Hence, utilizing TRMM products provides significant advantages over point-based records, since they offer spatially consistent time—series precipitation data.

Based on elevation and hydroclimatic homogeneity, Bangladesh is divided into seven unique hydrological regions (Rahman et al., 2012) (Fig. 8.1). Among these regions, the northwest is the largest, with an area of 31,604 km^2, and the southeast is the smallest (15,440 km^2). Average elevation, derived from Shuttle Radar Topographic Mission data (Farr et al., 2007), ranges from 6 m (in the southcentral region) to 110 m (in the eastern hills). Though analysis of precipitation over the entire country has been thoroughly conducted, results of these works may not reveal true regional characteristics (Dash et al., 2009). This is because there is considerable spatiotemporal variation in the amount of precipitation across the different hydrological regions in Bangladesh. Consequently, the occurrence of floods/droughts differs substantially in space and time (Rahman et al., 2018). Therefore, assessment of the statistical behaviors of

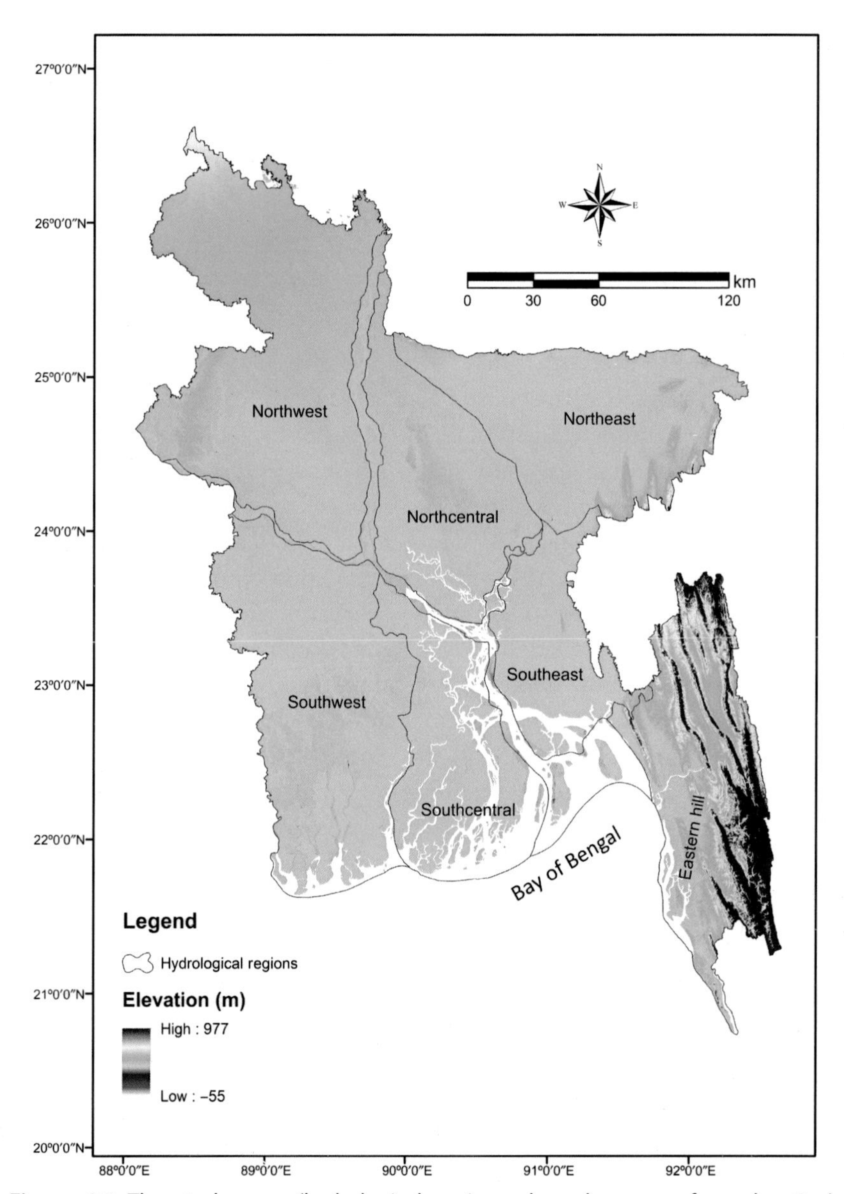

Figure 8.1 The study area (hydrological regions draped over surface elevation). *Adapted from Shuttle Radar Topographic Mission (SRTM).*

precipitation may provide crucial information for water resources management, for designing water infrastructures at the catchment scale (García-Barrón et al., 2018), and to support policy formulation, especially in the context of global warming (Dash et al., 2009). In summary, the primary aim of this chapter is to examine spatiotemporal patterns and trends of annual precipitation, and its concentration, aggressiveness, and seasonality in different hydrological regions across Bangladesh.

8.2 Methods and materials

8.2.1 Study area

Bangladesh is located between 88.05° and 92.74°E longitude and 20.45° and 26.63°N latitude (Fig. 8.1). Its topography is characterized by lowlands seldom rising to 10 m above sea level (Ohsawa et al., 2000). The climate of the country is generally characterized by cool dry winters, hot humid summers, and a rainy monsoon season. Pre-monsoon (March–May), monsoon (June–September), post-monsoon (October–November), and winter (December–February) are the four distinct seasons. However, Bangladesh's climate is dominated by the summer (June–September) and winter (December–February) monsoons with marked seasonality in rainfall and temperature due to reversal of wind circulation—a dominant feature of the South Asian summer monsoon (Salahuddin et al., 2006). Mean temperatures range from 27.8°C to 29°C in summer and from 18.5°C to 21°C in winter.

The distribution of precipitation varies spatially to a significant degree, with the heaviest rainfall occurring in the northeast under the influence of the bordering Meghalaya Plateau (Khatun et al., 2016). About 20%, 62.5%, 15.5%, and 2% of annual rainfall occurs during pre-monsoon, monsoon, post-monsoon, and winter, respectively, signifying high seasonality in the precipitation regime (Islam and Uyeda, 2007). Because of low topography and heterogeneity in yearly precipitation, floods are common and their extent and severity is often determined by the duration and intensity of precipitation (Cash et al., 2008). Past floods, particularly those that occurred in 1987, 1988, 1998, 2004, and 2007, were closely associated with heavy precipitation (Prasanna et al., 2014; Mirza, 2011). The projected increase in heavy precipitation due to climatic warming is therefore a matter of serious concern for Bangladesh, as more frequent floods of large magnitude are likely (see Mirza, 2011). Studying the statistical patterns of precipitation, in terms of its concentration, aggressiveness, and seasonality, is of particular interest in improving the monitoring and prediction of extreme events.

8.2.2 Data

The TRMM, launched in 1997, is equipped with a microwave imager, a precipitation radar, visible and infrared (IR) scanners, a lightning imaging sensor, and a clouds and earth radiant energy system (https://pmm.nasa.gov/TRMM/mission-overview). It is a joint program of USA and Japan, dedicated to providing rainfall measurements over significantly under-sampled tropical and subtropical oceans and continents (Liu et al., 2012). Since its launch, TRMM has provided a wealth of information, particularly global rainfall estimates that have supported a range of hydrometeorological studies and other applications. As measurements from a single satellite offer inadequate spatiotemporal coverage, a blended approach has been utilized to overcome this issue, combining multisatellite and multi-sensors datasets (Liu, 2015). This approach has been used to produce the multisatellite precipitation analysis (TMPA) products of TRMM (Huffman et al., 2007). TMPA includes two products, the 3-hourly near-real-time data set (3B42RT) with a spatial coverage of 60°N–60°S and the research-grade product (3B42), extending from 50°N–50°S (Huffman et al., 2015). Although both products are widely used for a variety of applications, the 3B42 product provides more accurate estimates, and is thus highly suitable for research (Liu, 2015).

The daily TMPA 3B42 product has been utilized in this work. TMPA 3B42 is generated by merging TRMM high quality, microwave/IR precipitation, and root-mean-square precipitation-error estimates, with availability from January 1, 1998 on a 0.25° grid (Liu, 2015; Huffman and Bolvin, 2013). The daily data were combined into monthly and annual time–series for the years 1998–2016. A shapefile, containing hydrological regions, was provided by the Bangladesh Water Resources Planning Organization and was used to extract precipitation for 19 years over the Bangladesh landmass and for each of its hydrological regions at the two timescales, monthly and annual.

8.2.3 Precipitation seasonality

Walsh and Lawler (1981) proposed the following rainfall SI to determine the degree of monthly precipitation variation:

$$\mathrm{SI} = \frac{1}{R}\sum_{n=1}^{12}\left|x_n - \frac{R}{12}\right| \tag{8.1}$$

where x_n = monthly total and R = annual total precipitation. SI is a useful measure for assessing seasonal contrasts in monthly precipitation amounts, instead of defining wet and dry seasons in absolute terms (Walsh and Lawler, 1981). The SI values vary from 0, if all months have equal precipitation, to 1.83, if all the rainfall occurs in a single month. To identify precipitation regimes over a region, a classification scheme has been suggested as follows: very equable ($SI \leq 0.19$), equable but with a definite wetter season ($SI = 0.20-0.39$), rather seasonal with a short duration ($SI = 0.40-0.59$), seasonal ($SI = 0.60-0.79$), markedly seasonal with a long drier season ($SI = 0.80-0.99$), most rain in three months or less ($SI = 1.00-1.19$), and extreme seasonality (almost all rain in 1–2 months) where $SI \geq 1.20$. We computed SI for individual years and also averaged SI for the entire time series (1998–2016) to produce long-term means. The index was subsequently used to examine interannual variability.

8.2.4 Precipitation aggressiveness

The FI (Fournier, 1960) was originally developed for estimating rainfall erosivity, or R-factor, and later modified by Arnoldus (1980), known as modified FI (MFI). It takes annual precipitation accumulation into account to determine the aggressiveness over a region (Valdés-Pineda et al., 2016; García-Barrón et al., 2015). MFI is defined as:

$$\mathrm{MFI} = \frac{\sum_{n=1}^{12} (P_i)}{p_t} \tag{8.2}$$

where P_i = monthly precipitation and p_t corresponds to mean annual precipitation. Higher values refer to a greater aggressivity, whilst lower values represent lower aggressivity. Because of its usefulness in determining the aggressive nature of precipitation, many previous studies have used it to examine the changing behavior of precipitation (García-Barrón et al., 2015, 2018; Valdés-Pineda et al., 2016; Nunes et al., 2016; Elagib, 2011). Using monthly time–series data, MFI was computed for each year to examine the temporal pattern of precipitation aggressiveness, and was also averaged over the study period. Interannual variability was studied using annual MFI.

8.2.5 Precipitation concentration period and degree

To examine temporal inhomogeneity of annual intense precipitation and theoretical evaluation of floods and droughts in a region, Zhang and Qian (2004) proposed two statistical measures: (1) precipitation concentration degree (PCD); and (2) precipitation concentration period (PCP). These indices consider monthly precipitation as a synthetic vector quantity, comprised of magnitude and direction, in which a year is considered to be a circle (360°) and a month has a value of 30° (360/12). PCD corresponds to the vector magnitude and reflects the degree to which annual total precipitation is concentrated across the 12 months. The PCD and PCP can be defined as:

$$R_{xi} = \sum r_{ij} \cdot \cos\theta_j \tag{8.3}$$

$$R_{yi} = \sum r_{ij} \cdot \sin\theta_j \tag{8.4}$$

$$\mathrm{PCD}_{i,j} = \frac{\sqrt{R^2{}_{xi} + R^2{}_{yi}}}{R_i} \tag{8.5}$$

$$\mathrm{PCP}_{i,j} = \arctan\left(\frac{R_{xi}}{R_{yi}}\right) \tag{8.6}$$

where i equates to the year, j represents the month of a year, and θ_j refers to the azimuth of month j. Annual PCD ranges from 0 to 1, where 0 reflects a homogenous distribution of precipitation over a year, and a value of 1 indicates a heterogeneous distribution, e.g., if annual precipitation is concentrated within a specific month in a year. Annual PCP indicates the month with the maximum precipitation in a year. As for the other two indices, PCD and PCP values were computed for each year and averaged to derive a long-term mean for Bangladesh and for each of its hydrological regions. Annual PCD and PCP data were used to derive interannual information.

8.2.6 Trend and correlation analyses

Several statistical techniques exist to study the influence of climate change on the spatiotemporal pattern of precipitation indices. They may be categorized as parametric (e.g., linear regression) and nonparametric tests (e.g., Mann–Kendall, MK). Although both methods are useful in detecting trends in hydrometeorological time–series (Sonali and Kumar, 2013; Kundzewicz and Robson, 2004), parametric tests such as linear regression appear to be less reliable than the MK test (Zhang et al., 2004). Interested readers should refer to Sonali and Kumar (2013) and Kundzewicz and Robson (2004) for more information about these methods. The MK test does not require any particular distribution of data, is robust in the presence of outliers, and has the same statistical power as parametric tests (Yue and Pilon, 2004). Therefore, we have used the nonparametric MK method (Kendall, 1975; Mann, 1945) to detect nonlinear trends of precipitation concentration, seasonality, and aggressiveness. Trends were analyzed for all of the annual SI, MFI, PCD, and PCP indices as well as monthly precipitation statistics. To determine the relationship between annual precipitation and individual indices, the Pearson Correlation Coefficient was adopted.

8.3 Results and discussion

8.3.1 Statistical patterns of precipitation

To understand precipitation characteristics, mean annual rainfall, average number of rainy days, and coefficient of variation (CV) were computed (Fig. 8.2A–C). A clear northeast to southwest gradient in the degree of annual precipitation is evident between 1998 and 2016. Most of the northeast region had the highest precipitation (>3300 mm) among all hydrological regions, whilst a portion of the northwest and southwest regions had the lowest precipitation (<2200 mm), with distinct spatial variation (Fig. 8.2A). However, parts of northeast, southeast, and eastern hill regions had >3000 mm of average precipitation between 1998 and 2016. The number of rainy days, on average, varied from 90 to 211 days across Bangladesh (Fig. 8.2B) with the highest number occurring in the northeast region (>143 days), followed by the eastern hill region and parts of the coastal areas. A considerable portion of the southwest and northwest regions experienced a mean number of rainy days of 90–103 days. The analysis further indicated that both mean annual precipitation and rainy days were highest in the northeast region, whereas the lowest values occurred in the southeast region with a mean number of rainy days of 106. The northwest region had the lowest number of rainy days (Table 8.1). In contrast, the CV of the precipitation distribution is relatively inhomogeneous. The high values of CV (>17%) are largely concentrated in the northern regions and low values (<12%) can be found in southcentral, southeast, and parts of the eastern hill regions (Fig. 8.2C). This may be attributed to a range of factors, including orography, prevailing winds, and local precipitation systems (Rafiuddin et al., 2013; Islam and Uyeda, 2007; Ohsawa et al., 2000).

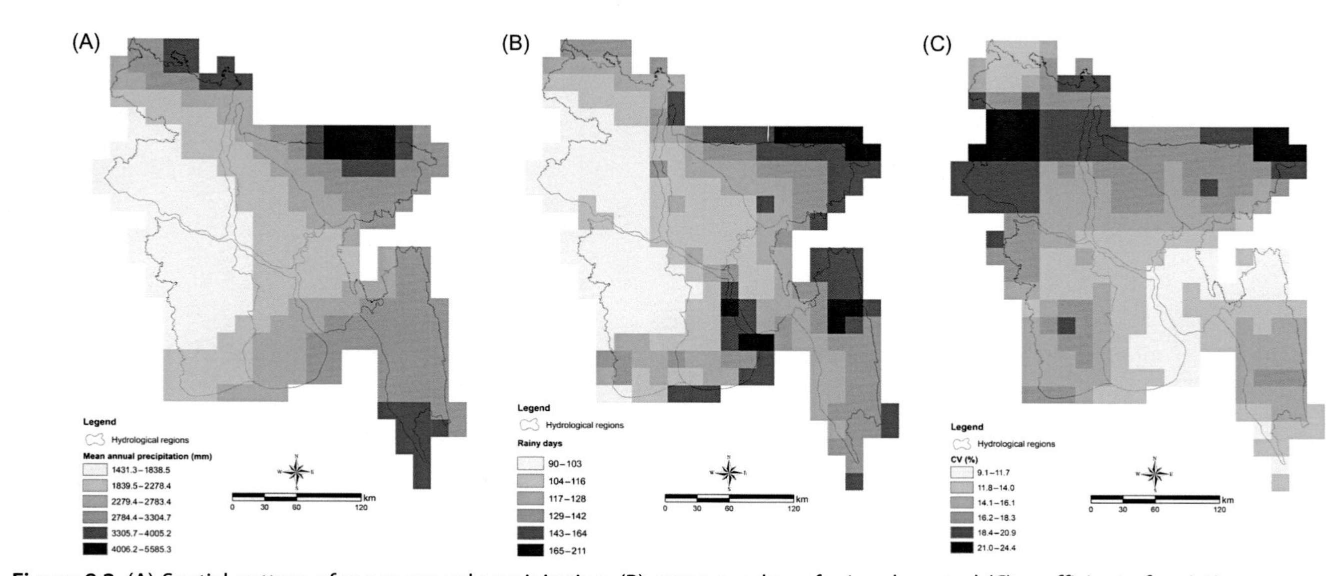

Figure 8.2 (A) Spatial pattern of mean annual precipitation; (B) mean number of rainy days; and (C) coefficient of variation.

Table 8.1 Mean annual precipitation and number of rainy days according to hydrological region.

Hydrological region	Mean annual precipitation (mm)	Mean number of rainy days
Northwest	1972.2	104
Northcentral	2062.1	114
Northeast	3196.0	136
Southeast	1765.9	106
Southwest	2344.5	125
Eastern Hill	2378.7	121
Southcentral	3073.0	141

The three statistical properties noted above clearly indicated inconsistencies in the spatial distribution of precipitation over different hydrological regions. This necessitates evaluating the statistical pattern of precipitation in more detail, if its direct influence on severe flood and drought events is to be better determined, especially for those that have occurred recently (Rahman et al., 2018; Rahman and Lateh, 2017; Rahman et al., 2012; Mirza, 2011).

8.3.2 Trends for monthly precipitation

The trends for monthly precipitation in different hydrological regions and for the whole of Bangladesh is shown in Table 8.2. Except for February and December, negative trends were consistent across Bangladesh between 1998 and 2016; note that none of these monthly figures is statistically significant. However, monthly precipitation trends vary according to hydrological region (Table 8.2). No statistically significant trend, either increasing or decreasing, was observed in the southwest, eastern hill, and southcentral regions. Statistically significant increasing (decreasing) trends, however, may be seen in the other four regions, viz. the northwest, northcentral, northeast, and southeast. For example, September and November rainfall data showed negative trends, whilst December data had a positive trend in the northwest region. In the northeast region, October data has a significant negative trend, and in the northcentral region, May and October data showed significant negative trends, but December data had a positive trend (at the $\geq$90% Confidence Interval (CI)). The southeast region had a negative trend during September–November (at the 90% CI). Generally, the findings are consistent in the sense that precipitation is likely to decrease at subtropical

Table 8.2 Trends of monthly precipitation in hydrological regions and over Bangladesh, 1998–2016.

Month	BD	NW	NC	NE	SE	SW	EH	SC
January	−0.559	0.839	−0.279	−0.279	0.419	0.070	−0.4898	−0.2099
February	0.134	0.489	0.699	0.349	0.629	0.629	0.2799	0.6297
March	−0.489	−0.629	−0.629	−0.979	0.279	0.000	−0.6997	−0.070
April	−0.349	−1.399	−1.049	−0.209	−0.769	−0.909	−0.4198	−0.6297
May	−0.979	−0.979	−1.679[a]	−0.419	−0.349	−0.489	−0.7697	−1.5394
June	−0.629	−1.329	−0.139	−0.839	0.070	0.070	−0.1399	0.070
July	−1.329	−1.469	−0.909	−1.189	0.070	−0.139	−1.1195	−1.0496
August	−0.141	−0.279	0.349	−0.279	0.699	0.419	0.000	−0.5598
September	−0.419	−1.959[b]	−0.279	0.699	−1.749[a]	−0.979	0.4198	0.5598
October	−1.318	−1.819	−1.959[b]	−1.819[a]	−1.399[a]	−1.399	−0.8397	0.2099
November	−0.710	−1.679[a]	−0.489	−0.349	−0.839[a]	−0.489	−0.6997	−1.1895
December	0.560	1.609[a]	1.889[b]	1.329	0.699	0.769	0.8397	−0.2099

BD, Bangladesh; *NW*, Northwest; *NC*, Northcentral; *NE*, Northeast; *SE*, Southeast; *SW*, Southwest; *EH*, Eastern Hill; *SC*, Southcentral.

[a]90% significant level.

[b]95% significant level.

latitudes (Zhang et al., 2007; Held and Soden, 2006; Emori and Brown, 2005). Also, results are in line with regional studies (Basher et al., 2017; Rahman and Lateh, 2017; Roxy et al., 2015), but differ with others (Fahad et al., 2018; Endo et al., 2015; Shahid, 2011), supporting the observation of Cash et al. (2008), who stated that the absence of rain gauge data influences both qualitative and quantitative conclusions drawn regarding precipitation in Bangladesh. Our findings of decreasing monthly trends may be attributed to the weakening of monsoon circulation (Dash et al., 2009), a possible reduction in the land–sea thermal contrast due to enhanced Indian Ocean warming (Roxy et al., 2015), and an increase in anthropogenic aerosol emissions (Bollasina et al., 2011). Nevertheless, observed variations in monthly precipitation trends among hydrological regions may have stemmed from local geographical features, land surface conditions, as well as individual precipitation systems (Ono and Takahashi, 2016; Rafiuddin et al., 2013; Islam and Uyeda, 2007).

8.3.3 Spatial pattern and trends of precipitation concentration, seasonality, and aggressiveness

The mean SI over the study period (1998–2016) ranges from 0.79 to 1.01 (Fig. 8.3A), suggesting that the precipitation regime in Bangladesh is generally characterized by long drier periods with marked seasonality (Walsh and Lawler, 1981). Because of the marked spatial variability, the hydrological regions exhibited greater heterogeneity in terms of SI. The northwest region showed long drier seasons, as indicated by the SI, with mean values ranging from 0.75 to 1.10. This finding is in agreement with Bari and Hussain (2017). Since the precipitation regime is highly influenced by monsoons, most of the hydrological regions (e.g., northcentral, northeast, southwest, and southcentral regions) are dominated by a seasonal SI classification scheme. In contrast, the southeast region showed a short drier season (SI = 0.595–1.057), whilst the eastern hill region has a mostly seasonal distribution, with SI ranging from 0.77 to 1.025 (Fig. 8.3). An interesting feature of the SI distributions during the period 1998–2016 is that none of the hydrological regions fell into the very equable precipitation category, nor that defined by all rainfall concentrated into 1–2 months, according to Walsh and Lawler (1981).

The MK test of SI across the country showed an increasing trend, but statistically significant cells are location-specific. For instance, Fig. 8.3B shows that only a few locations in the southwest, southcentral, northwest, and northeast regions, have statistically significant trends (≥90% CI), and

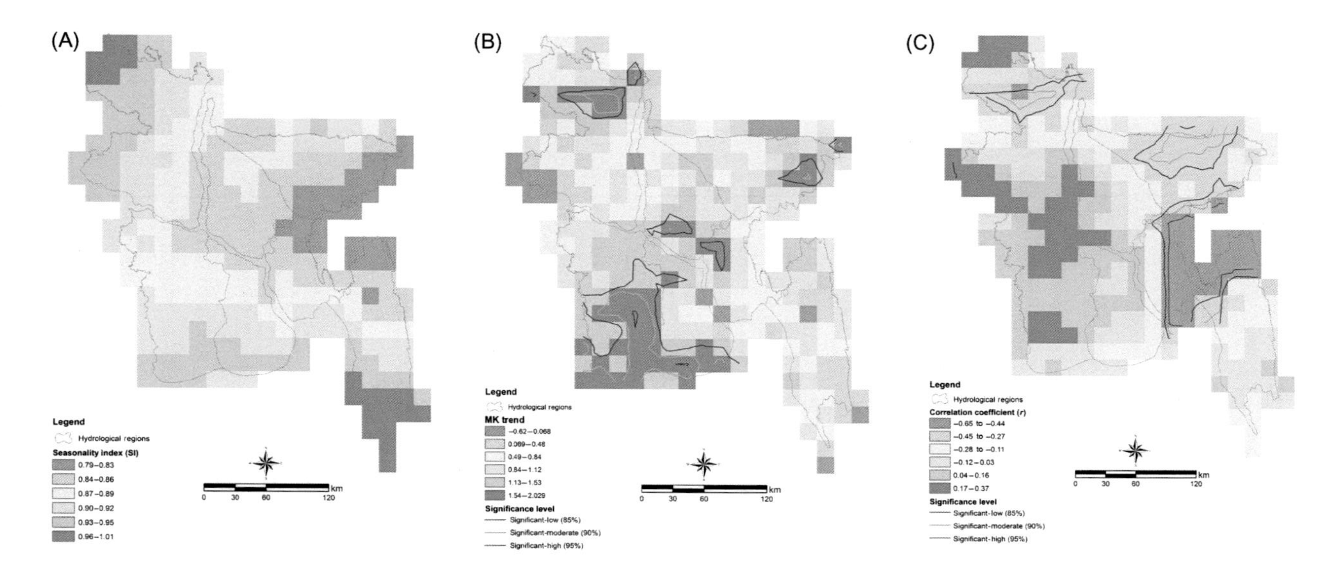

Figure 8.3 (A) Spatial distribution of SI. (B) Spatial patterns of Mann–Kendall trends of SI. (C) Spatial correlation map of annual precipitation and SI. *SI*, Seasonality index.

these demonstrate an increase in seasonality. The spatial correlation between annual rainfall and SI has been examined in order to understand the changing behavior of precipitation (Fig. 8.3C). This indicates a mixed association across the country, wherein negative correlations are dominant in the southeast, northeast, and upper portion of northwest and eastern hilly regions. On the other hand, the southwest, parts of northcentral, and lower parts of northwest regions exhibit a positive correlation. However, few grids in the northeast, northwest, eastern hill, and southeast regions exhibited statistically significant relationships at the 90% confidence level (Fig. 8.3C). Year-to-year variability and irregular distribution of precipitation are possibly factors affecting such associations (García-Barrón et al., 2015; Elagib, 2011).

Precipitation aggressiveness distributions, as estimated by the MFI, indicate that higher MFI values are distributed over locations where annual rainfall and number of rainy days were high (Fig. 8.4A). Nunes et al. (2016) observed a similar phenomenon in southern Portugal and identified a strong link between annual precipitation depth and the distribution of MFI. Because of orography and the route of monsoon progression, northeast and eastern hill regions receive significantly higher annual precipitation (Islam and Uyeda, 2007), implying that higher accumulation of annual totals leads to greater aggressiveness (Valdés-Pineda et al., 2016). On the other hand, a portion of the northwest and southwest regions (22.0°–25.0°N latitude and 88.0°–89°E longitude) exhibited low aggressiveness, possibly attributed to low precipitation and high annual variability (Shahid, 2011).

The trend of MFI and its statistical significance at different confidence levels (CI) are shown in Fig. 8.4B, which indicates a decrease in MFI between 1998 and 2016 in the northwest region. This is particularly prominent between 24.0°–26.0°N latitude and 88.0°–90.5°E longitude, but the result is statistically insignificant. Portions of the northwest, eastern hill, southeast hydrological regions, and most of the northcentral region showed statistically significant trends (≥90% CI). The MK test also exhibited an increasing (but statistically insignificant) MFI trend in the southwest and northeast regions (Fig. 8.4B). Spatial correlation between annual rainfall and MFI showed positive associations in Bangladesh at the 99% CI (Fig. 8.4C). Because all grids within the country indicated a positive correlation with a high significance, we did not use contouring.

The analysis of PCP showed that its spatial pattern usually corresponds to the annual distribution of precipitation, with values ranging from 163°

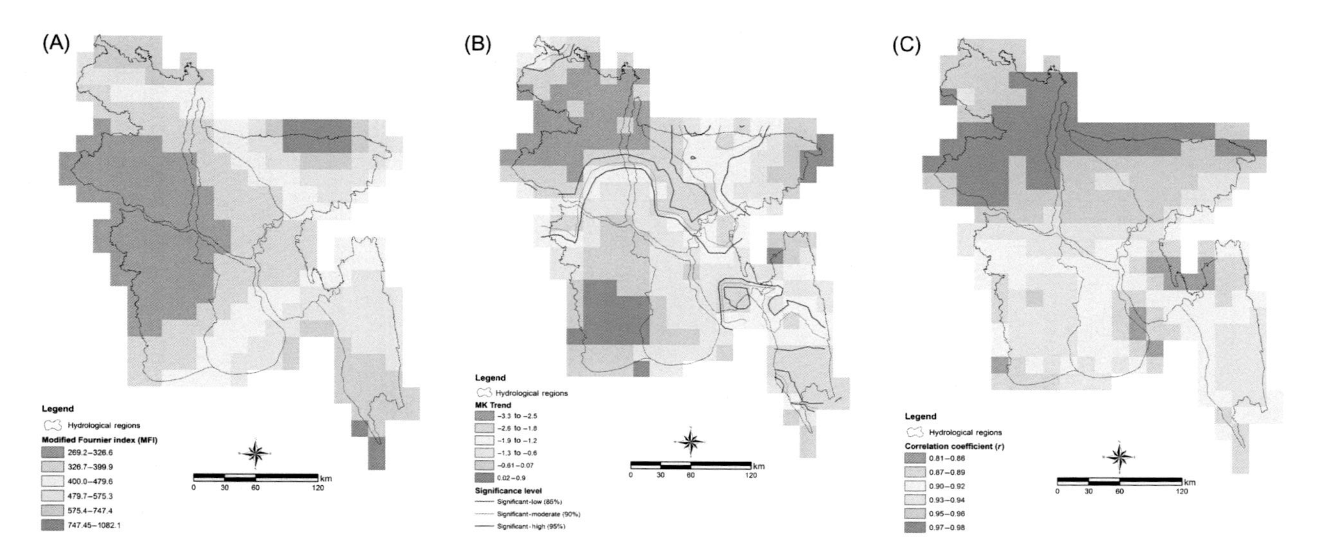

Figure 8.4 (A) Spatial distribution of MFI. (B) Spatial patterns of Mann–Kendall trends of MFI. (C) Spatial correlation map of annual precipitation and MFI. *MFI*, Modified Fournier index.

to 198°. The results accord with the months of the summer monsoon, which brings ubiquitous rainfall over the country. As a result, regions that receive high rainfall during the monsoon have low PCP (163°–175°) and areas that experience low precipitation have high PCP values (187°–198°) (Fig. 8.5A). The mean annual PCP is higher in the western hydrological regions than in the eastern ones, suggesting an uneven distribution of rainfall. A similar gradient in PCP was observed in Xinjiang (Li et al., 2011) and in the Yangtze basin of China (Wang et al., 2013). The PCP in Bangladesh is possibly influenced by a seasonal transition of precipitation characteristics (Ono and Takahashi, 2016), as well as active and break monsoon periods (Ohsawa et al., 2000), thus giving a distinct spatial pattern.

The MK test showed a declining trend for PCP over Bangladesh despite the fact that hydrological regions have shown varied trends (i.e., both increasing and decreasing) (Fig. 8.5B). Generally, PCP is decreasing in many locations of the northern, southwest, and southeast regions whilst an increasing trend is observed in the eastern hills, southeast regions, and along the areas bordering northern Bangladesh, where orography has marked effects on precipitation occurrence (Islam and Uyeda, 2007). Only a few cells, located at the edges of the southwest and northwest regions, were significant at the ≥90% confidence level (Fig. 8.5B). The spatial correlation map shows that negative correlations exist between mean annual rainfall and PCP in most eastern regions, whereas a positive correlation may be seen in many locations of the northcentral, northwest, and southeast regions (Fig. 8.5C), implying that the annual distribution of precipitation plays an important role in determining the degree of precipitation concentration in different hydrological regions.

PCD results are shown in Fig. 8.6A, which indicates that precipitation is highly concentrated in specific months and the mean PCD values, which varied from 0.52 to 0.73 across Bangladesh, indicate spatial inhomogeneity of precipitation. Higher PCD values are distributed over the extreme northern and southeastern parts of Bangladesh with a mean value of 0.71 between 1998 and 2016. In contrast, low PCD (0.55–0.61) grids are located in the northeast, northcentral, and southeast regions, suggesting that PCD is influenced by summer monsoon months. Roughly, both PCP and PCD have a similar spatial pattern that may be related to moisture content in the atmosphere. Since moisture is invariably higher during the monsoon than in other seasons due to the maritime nature of clouds (Choudhury et al., 2016), this may explain the observed pattern of PCD

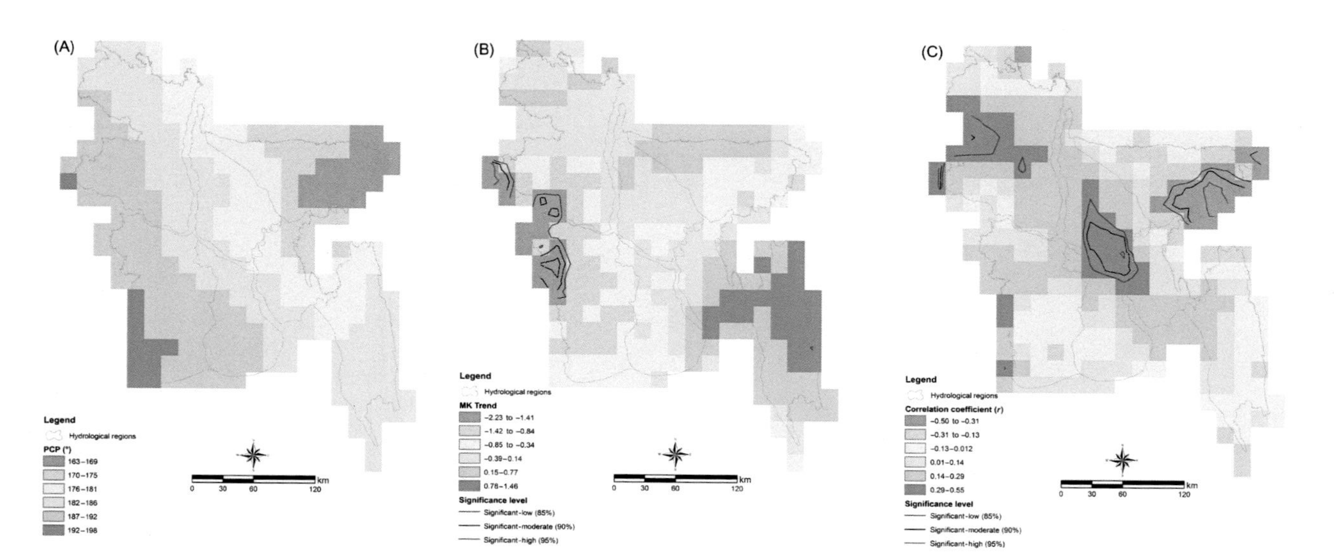

Figure 8.5 (A) Spatial distribution of PCP. (B) Spatial patterns of Mann–Kendall trends of PCP. (C) Spatial correlation map of annual precipitation and PCP. *PCP*, Precipitation concentration period.

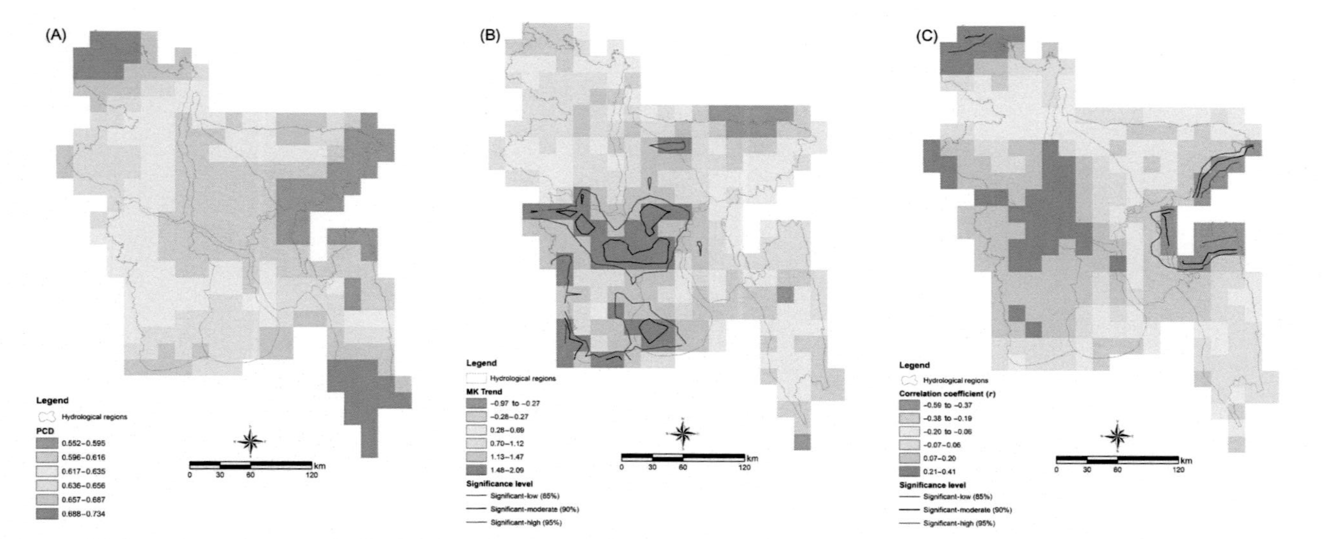

Figure 8.6 (A) Spatial distribution of PCD. (B) Spatial patterns of Mann–Kendall trends of PCD. (C) Spatial correlation map of annual precipitation and PCD. *PCD*, Precipitation concentration degree.

over Bangladesh. An important point to note is that no cells across Bangladesh had a zero (0) PCD value, which means that the precipitation distribution over a year varies substantially in time and space.

The pattern of the MK trend for PCD is shown in Fig. 8.6B. This indicates that PCD has an increasing trend over most of the hydrological regions. The increasing trend is particularly pronounced in the northwest, northcentral, southwest and southcentral regions and in many locations it is statistically significant (at the $\geq 90\%$ CI). On the other hand, northeast, eastern region, and northwest extreme locations have a declining, nonsignificant trend (Fig. 8.6B). Because most northcentral, southwest, and southeast areas are relatively low-lying lands, an increasing trend for PCD could potentially lead to substantial flooding (Zhang and Qian, 2004). The correlation map (Fig. 8.6C) shows that many grids in the southwest, southeast, northcentral, and northwest have positive correlations between annual precipitation and PCD, although the relationship is not statistically significant. Given that these locations coincide with increasing MK trends, there may be a high probability of large floods under atmospheric warming conditions (Wang et al., 2013). The northeast and southeast hydrological regions are mainly dominated by significant ($\geq 90\%$ CI) negative correlations as inferred from Fig. 8.6C.

A summary of the trends, either increasing or decreasing, of the four precipitation indices is shown in Table 8.3, which was obtained by using interannual values. An interesting outcome may be observed for the entirety of Bangladesh. Overall, SI and PCD are on the rise, though the trend statistics vary depending on the hydrological regions. Despite PCP

Table 8.3 Summary of precipitation seasonality, aggressiveness, and concentration in hydrological regions and over Bangladesh.

Hydrological regions	SI	MFI	PCP	PCD
Northwest	+	−	−	+
Northcentral	+	−	+	+
Northeast	+	−	−	+
Southeast	+	−	−	+
Southwest	+	−	+	+
Eastern Hill	+	−	+	+
Southcentral	+	−	−	+
Bangladesh	**+**	−	**+**	**+**

Plus (+) sign denotes increasing while negative (−) sign represents decreasing trend. *SI*, Seasonality index; *MFI*, modified Fournier index; *PCP*, precipitation concentration period; *PCD*, precipitation concentration degree.

having a varying trend over the study period, PCD and SI showed consistently rising trends (Table 8.3). This implies that: (1) precipitation is becoming more seasonal (i.e., specific to several months), which may lead to frequent droughts in areas experiencing low rainfall; and (2) a rise in PCD may have severe impacts, such as recurrent flooding.

8.4 Implications for multi-hazard management

Since Bangladesh is affected by a range of natural hazards, including geophysical (e.g., landslide, riverbank erosion) and hydroclimatological (e.g., floods, cyclone, droughts) hazards every year, spatiotemporal quantification of precipitation indices generated in this chapter are likely to contribute significantly to the area of multihazard risk management due to the fact that many of them have direct or indirect associations with precipitation patterns. Any irregularity in the distribution and amount of precipitation is, therefore, a potential threat to Bangladesh's growing population and economy. As climate change is intensifying extreme weather events [World Meteorological Organisation (WMO), 2018], our results have scientific and practical relevance to the management of various natural hazards that affect Bangladesh.

Precipitation variability is shown to affect ecosystem productivity (Qadir et al., 2013). Besides being a freshwater and groundwater source, homogeneous distribution of precipitation is essential for maintaining soil moisture and vegetation growth. Decreasing trends in monthly precipitation may be of particular concern for Bangladesh as further drying could lead to a myriad of climate-related issues. For example, its climate is characterized by a long dry spell of 8 months due to a strong seasonality of rainfall (Shahid, 2011). A decrease in monthly precipitation totals and an increase in seasonality could lengthen existing dry spells, which could in turn lead to more frequent droughts (Rahman and Lateh, 2017), a decline in groundwater table (Dey et al., 2017), and scarcity of freshwater resources (Grafton et al., 2013)—all of these detrimental to environmental sustainability. In addition, projected heat waves in South Asia could result in higher energy demands and may therefore elevate the degree of vulnerability of populations to climate extremes (Im et al., 2017). This information could provide valuable insights into the development of region-specific mitigation measures for adaptation with climate-induced extreme events, such as drought.

The most dramatic effect of precipitation extremes may be more frequent and intense flooding events. As mentioned above, the precipitation regime in different hydrological regions varies both spatially and temporally, which has a significant impact on the varying depths and duration of floods (Mirza, 2011). Even though water flow in the major rivers, in combination with the degree of precipitation upstream and downstream, determines flood severity, increasing precipitation concentration trends have been shown to be associated with large flood events (Wiel et al., 2017; Rahman et al., 2012; Zhang and Qian, 2004). In 2017, Bangladesh experienced three large floods (i.e., March–April, July, and August 2017) that affected at least eight million people and devastated the agricultural sector—all these flood events were related to intense precipitation of a shorter duration than average (FAO, 2017). The northwest region experienced 360 mm of rainfall between August 11 and 12, 2017 that critically affected the life and livelihoods of millions (Reliefweb, 2017). Similarly, Dhaka megacity was heavily inundated following the 135 mm of rainfall that occurred during September 10–11, 2017, which caused disruptions to economic activities and destroyed infrastructure. Therefore, increases in precipitation concentration under a warmer climate, as identified in this chapter, could lead to intense flooding, which destroys crops, infrastructure, and ultimately causes human suffering. The results of this study could help monitor the progress of floods and promote effective management of recurrent hydroclimatic disasters.

Seasonality information may also be highly useful for a country like Bangladesh, since it identifies uncertainty in the intensity, arrival, and duration of seasonal rainfall (Feng et al., 2013), and hence water availability for agriculture (Fenta et al., 2017). As the temporal trend of SI appears to be increasing (Fig. 8.2B), the traditional crop calendar may be difficult to maintain due to the fact that crops, such as rice, are sensitive to climate (Sarker et al., 2017). Consequently, agricultural production, especially rain-fed agriculture, may be hampered (Kattelus et al., 2016). Further, increasing seasonality could enhance drought conditions, meaning that increased stresses on the groundwater-based irrigation system may be anticipated. As indicated by Li et al. (2011), PCP has a substantial influence on crop irrigation, since heterogeneous precipitation concentration could lead to difficulties in timely sowing and harvesting.

Energetic input of rainfall is one of the key factors controlling water erosion, and land degradation (Panagos et al., 2017). Because of the alluvial nature of soil, land loss by riverbank and coastal erosion are persistent issues in Bangladesh [Bangladesh Bureau of Statistics (BBS), 2016] that may be related to precipitation aggressiveness. Even though aggressiveness shows a decreasing trend in this study, erosion along major rivers or at the coast may still be significant as the erosive power of soil depends on other factors, such as vegetation cover and soil type. Therefore, the MFI results should be used with caution, since they are largely influenced by total annual rainfall and do not take extreme events into account (García-Barrón et al., 2015). Additionally, extreme events are also occasional (Martin-Vide, 2004). It is, however, important to note that a decreasing trend in precipitation (Table 8.2) may be detrimental for vegetation growth and enhance the risk of topsoil erosion, and thus create a higher potential for riverbank and coastal erosion. Since dry areas become drier under warming air conditions, the MFI results may also be useful for agencies working with irrigation-based agriculture, particularly in the northwest and southwest hydrological regions.

Geophysical hazards, such as landslide occurrences, are a relatively recent phenomenon, and are directly related to intense precipitation events. Between 1999 and 2016, the eastern hill region experienced a number of devastating landslides, triggered by heavy precipitation, that killed 472 people and injured 411 [Ahmed, 2017; Bangladesh Bureau of Statistics (BBS), 2016]. A record precipitation of 365 mm on June 11, 2017 also initiated severe landslide events in the eastern hill region that killed 162 and injured 88, and 15 people were killed (including those missing) and 5 were injured, following further heavy rainfall events (Ahmed, 2017). Our finding of increasing trends in PCD and PCP in this region may provide important insights in this regard. Thus, the concentration maps may be useful to support the prediction and management of landslides in Bangladesh, particularly in the eastern hill region.

8.5 Conclusions

The chapter analyzed statistical patterns of precipitation concentration, aggressiveness, and seasonality in Bangladesh. Time–series TMPA data between 1998 and 2016 were used to quantify spatiotemporal precipitation distributions and a nonparametric MK test was utilized to examine any trends. The analysis indicated that most months, with a few exceptions, have declining trends in the total amount of monthly rainfall over Bangladesh and in many smaller scale hydrological regions. Seasonality and PCD indices were found to be increasing over time, whilst aggressiveness index showed a declining trend across various hydrological regions, although statistically significant trends were location-specific. The decrease in monthly precipitation amount and aggressiveness is partly offset by the increase in seasonality and precipitation concentration indices, with regard to their hazardous effects. The results are expected to have major implications for management of floods, droughts, and landslide hazards, and could assist in the development of appropriate countermeasures for adapting to climatic change.

References

Abolverdi, J., Ferdosifar, G., Khalili, D., Kamgar-Haghighi, A.A., 2016. Spatial and temporal changes of precipitation concentration in Fars province, southwestern Iran. Meteorol. Atmos. Phys. 128 (2), 181–196.

Ahmed, B., 2017. Community Vulnerability to Landslides in Bangladesh (Ph.D. dissertation). University College London, UK.

Alexander, L.V., Zhang, X., Peterson, T.C., Caesar, J., Gleason, B., Klein Tank, A.M.G., et al., 2006. Global observed changes in daily climate extremes of temperature and precipitation. J. Geophys. Res. Atmos. 111 (D5).

Allan, R.P., Soden, B.J., 2008. Atmospheric warming and the amplification of precipitation extremes. Science 321 (5895), 1481–1484.

Andersen, T.K., Shepherd, M., 2013. Floods in a changing climate. Geogr. Compass 7 (2), 95–115.

Arnoldus, H.M.J., 1980. An approximation of the rainfall factor in the universal soil loss equation. In: De Boodt, M., Gabriels, D. (Eds.), Assessment of Erosion. John Wiley, Chichester, pp. 127–132.

Bangladesh Bureau of Statistics (BBS), 2016. Bangladesh Disaster–related Statistics 2015: Climate Change and Natural Disaster Perspectives. Ministry of Planning, Dhaka.

Bari, S.H., Hussain, M.M., 2017. Rainfall variability and seasonality in northern Bangladesh. Theor. Appl. Climatol. 129 (3-4), 995–1001.

Basher, M.A., Stiller-Reeve, M.A., Islam, A.S., Bremer, S., 2017. Assessing climatic trends of extreme rainfall indices over northeast Bangladesh. Theor. Appl. Climatol. 1–12.

Bollasina, M.A., Ming, Y., Ramaswamy, V., 2011. Anthropogenic aerosols and the weakening of the South Asian summer monsoon. Science 334 (6055), 502–505.

Cash, B.A., Rodó, X., Kinter III, J.L., Fennessy, M.J., Doty, B., 2008. Differing estimates of observed Bangladesh summer rainfall. J. Hydrometeorol. 9 (5), 1106–1114.

Chatterjee, S., Khan, A., Akbari, H., Wang, Y., 2016. Monotonic trends in spatio-temporal distribution and concentration of monsoon precipitation (1901–2002), West Bengal, India. Atmos. Res. 182, 54–75.

Choudhury, H., Roy, P., Kalita, S., Sharma, S., 2016. Spatio-temporal variability of the properties of mesoscale convective systems over a complex terrain as observed by TRMM sensors. Int. J. Climatol. 36 (6), 2615–2632.

Chowdhury, M., Ndiaye, O., 2017. Climate change and variability impacts on the forests of Bangladesh—a diagnostic discussion based on CMIP5 GCMs and ENSO. Int. J. Climatol. 37 (14), 4768–4782.

Coumou, D., Rahmstorf, S., 2012. A decade of weather extremes. Nat. Clim. Change 2 (7), 491.
Dash, S.K., Kulkarni, M.A., Mohanty, U.C., Prasad, K., 2009. Changes in the characteristics of rain events in India. J. Geophys. Res. Atmos. 114 (D10).
Deshpande, N.R., Kothawale, D.R., Kulkarni, A., 2016. Changes in climate extremes over major river basins of India. Int. J. Climatol. 36 (14), 4548–4559.
Dey, N.C., Saha, R., Parvez, M., Bala, S.K., Islam, A.S., Paul, J.K., et al., 2017. Sustainability of groundwater use for irrigation of dry-season crops in northwest Bangladesh. Groundwater Sustainable Dev. 4, 66–77.
Elagib, N.A., 2011. Changing rainfall, seasonality and erosivity in the hyper-arid zone of Sudan. Land Degrad. Dev. 22 (6), 505–512.
Emori, S., Brown, S.J., 2005. Dynamic and thermodynamic changes in mean and extreme precipitation under changed climate. Geophys. Res. Lett. 32 (17), L17706.
Endo, N., Matsumoto, J., Hayashi, T., Terao, T., Murata, F., Kiguchi, M., et al., 2015. Trends in precipitation characteristics in Bangladesh from 1950 to 2008. Sci. Online Lett. Atmos. 11, 113–117.
Fahad, M.G.R., Islam, A.S., Nazari, R., Hasan, M.A., Islam, G.T., Bala, S.K., 2018. Regional changes of precipitation and temperature over Bangladesh using bias-corrected multi-model ensemble projections considering high-emission pathways. Int. J. Climatol. 38 (4), 1634–1648.
Farr, T.G., Rosen, P.A., Caro, E., Crippen, R., Duren, R., Hensley, S., et al., 2007. The shuttle radar topography mission. Rev. Geophys. 45 (2), 2005RG000183.
Feng, X., Porporato, A., Rodriguez-Iturbe, I., 2013. Changes in rainfall seasonality in the tropics. Nat. Clim. Change 3 (9), 811.
Fenta, A.A., Yasuda, H., Shimizu, K., Haregeweyn, N., Kawai, T., Sultan, D., et al., 2017. Spatial distribution and temporal trends of rainfall and erosivity in the Eastern Africa region. Hydrol. Processes 31 (25), 4555–4567.
Food and Agricultural Organisation (FAO), 2017. GIEWS: Global Information and Early Warning System on Food and Agriculture. FAO, Rome.
Fournier, F., 1960. Climat et Erosion. Presse Universitaire de France, Paris.
García-Barrón, L., Camarillo, J.M., Morales, J., Sousa, A., 2015. Temporal analysis (1940–2010) of rainfall aggressiveness in the Iberian Peninsula basins. J. Hydrol. 525, 747–759.
García-Barrón, L., Morales, J., Sousa, A., 2018. A new methodology for estimating rainfall aggressiveness risk based on daily rainfall records for multi-decennial periods. Sci. Total Environ. 615, 564–571.
Grafton, R.Q., Pittock, J., Davis, R., Williams, J., Fu, G., Warburton, M., et al., 2013. Global insights into water resources, climate change and governance. Nat. Clim. Change 3 (4), 315.
Hajani, E., Rahman, A., Ishak, E., 2017. Trends in extreme rainfall in the state of New South Wales, Australia. Hydrol. Sci. J. 62 (13), 2160–2174.
Held, I.M., Soden, B.J., 2006. Robust responses of the hydrological cycle to global warming. J. Clim. 19 (21), 5686–5699.
Hirsch, R.M., Archfield, S.A., 2015. Flood trends: not higher but more often. Nat. Clim. Change 5 (3), 198.
Huang, J., Sun, S., Zhang, J., 2013. Detection of trends in precipitation during 1960–2008 in Jiangxi province, southeast China. Theor. Appl. Climatol. 114 (1-2), 237–251.
Huffman, G.J., Bolvin, D.T., 2013. TRMM and Other Data Precipitation Data Set Documentation. NASA, Greenbelt, MD, p. 28.
Huffman, G.J., Bolvin, D.T., Nelkin, E.J., Wolff, D.B., 2007. The TRMM multisatellite precipitation analysis (TMPA): quasi-global, multiyear, combined-sensor precipitation estimates at fine scales. J. Hydrometeorol. 8 (1), 38–55.

Huffman, G.J., Bolvin, D.T., Nelkin, E.J., 2015. Integrated multi-satellitE retrievals for GPM (IMERG) technical documentation. NASA/GSFC Code, 612. p. 47.

Im, E.S., Pal, J.S., Eltahir, E.A., 2017. Deadly heat waves projected in the densely populated agricultural regions of South Asia. Sci. Adv. 3 (8), e1603322.

Intergovernmental Panel on Climate Change (IPCC), 2014. Climate change 2014: assessment synthesis report. Available at: <http://www.ipcc.ch/report/ar5/syr> (accessed 01.02.18.).

Islam, M.N., Uyeda, H., 2007. Use of TRMM in determining the climatic characteristics of rainfall over Bangladesh. Remote Sens. Environ. 108 (3), 264–276.

Jiang, P., Yu, Z., Gautam, M.R., Acharya, K., 2016. The spatiotemporal characteristics of extreme precipitation events in the western United States. Water Resour. Manage. 30 (13), 4807–4821.

Jones, P.D., Horton, E.B., Folland, C.K., Hulme, M., Parker, D.E., Basnett, T.A., 1999. The use of indices to identify changes in climatic extremes. Clim. Change 42 (1), 131–149.

Kattelus, M., Salmivaara, A., Mellin, I., Varis, O., Kummu, M., 2016. An evaluation of the Standardized Precipitation Index for assessing inter-annual rice yield variability in the Ganges–Brahmaputra–Meghna region. Int. J. Climatol. 36 (5), 2210–2222.

Kendall, M.G., 1975. Rank Correlation Methods. Charless Griffin, London.

Khatun, M.A., Rashid, M.B., Hygen, H.O., 2016. Climate of Bangladesh. Norwegian Meteorological Institute and Bangladesh Meteorological Department, Dhaka.

Kripalani, R.H., Inamdar, S., Sontakke, N.A., 1996. Rainfall variability over Bangladesh and Nepal: comparison and connections with features over India. Int. J. Climatol. 16 (6), 689–703.

Kumar, D., Arya, D.S., Murumkar, A.R., Rahman, M.M., 2014. Impact of climate change on rainfall in Northwestern Bangladesh using multi-GCM ensembles. Int. J. Climatol. 34 (5), 1395–1404.

Kundzewicz, Z.W., Robson, A.J., 2004. Change detection in hydrological records—a review of the methodology. Hydrol. Sci. J. 49 (1), 7–19.

Li, X., Jiang, F., Li, L., Wang, G., 2011. Spatial and temporal variability of precipitation concentration index, concentration degree and concentration period in Xinjiang, China. Int. J. Climatol. 31 (11), 1679–1693.

Li, X., Wang, X., Babovic, V., 2018. Analysis of variability and trends of precipitation extremes in Singapore during 1980–2013. Int. J. Climatol. 38 (1), 125–141.

Liu, Z., 2015. Comparison of precipitation estimates between Version 7 3-hourly TRMM Multi-Satellite Precipitation Analysis (TMPA) near-real-time and research products. Atmos. Res. 153, 119–133.

Liu, Z., Ostrenga, D., Teng, W., Kempler, S., 2012. Tropical Rainfall Measuring Mission (TRMM) precipitation data and services for research and applications. Bull. Am. Meteorol. Soc. 93 (9), 1317–1325.

Mann, H.B., 1945. Nonparametric tests against trend. Econometrica J. Econometric Soc. 245–259.

Maplecroft, 2016. Climate change vulnerability index. Available at: <https://maplecroft.com/about/news/ccvi.html> (accessed 07.02.18.).

Martin-Vide, J., 2004. Spatial distribution of a daily precipitation concentration index in peninsular Spain. Int. J. Climatol. 24 (8), 959–971.

Min, S.K., Zhang, X., Zwiers, F.W., Hegerl, G.C., 2011. Human contribution to more-intense precipitation extremes. Nature 470 (7334), 378.

Mirza, M.M.Q., 2011. Climate change, flooding in South Asia and implications. Reg. Environ. Change 11 (1), 95–107.

Nowreen, S., Murshed, S.B., Islam, A.S., Bhaskaran, B., Hasan, M.A., 2015. Changes of rainfall extremes around the haor basin areas of Bangladesh using multi-member ensemble RCM. Theor. Appl. Climatol. 119 (1-2), 363–377.

Nunes, A.N., Lourenço, L., Vieira, A., Bento-Gonçalves, A., 2016. Precipitation and erosivity in southern Portugal: seasonal variability and trends (1950–2008). Land Degrad. Dev. 27 (2), 211–222.

Ohsawa, T., Hayashi, T., Mitsuta, Y., Matsumoto, J., 2000. Intraseasonal variation of monsoon activities associated with the rainfall over Bangladesh during the 1995 summer monsoon season. J. Geophys. Res. Atmos. 105 (D24), 29445–29459.

Ono, M., Takahashi, H.G., 2016. Seasonal transition of precipitation characteristics associated with land surface conditions in and around Bangladesh. J. Geophys. Res. Atmos. 121 (19).

Panagos, P., Borrelli, P., Meusburger, K., Yu, B., Klik, A., Lim, K.J., et al., 2017. Global rainfall erosivity assessment based on high-temporal resolution rainfall records. Sci. Rep. 7 (1), 4175.

Prasanna, V., Subere, J., Das, D.K., Govindarajan, S., Yasunari, T., 2014. Development of daily gridded rainfall dataset over the Ganga, Brahmaputra and Meghna river basins. Meteorol. Appl. 21 (2), 278–293.

Qadir, M., Noble, A.D., Chartres, C., 2013. Adapting to climate change by improving water productivity of soils in dry areas. Land Degrad. Dev. 24 (1), 12–21.

Rafiuddin, M., Uyeda, H., Kato, M., 2013. Development of an arc-shaped precipitation system during the pre-monsoon period in Bangladesh. Meteorol. Atmos. Phys. 120 (3-4), 165–176.

Rahman, H.T., Hickey, G.M., Ford, J.D., Egan, M.A., 2018. Climate change research in Bangladesh: research gaps and implications for adaptation-related decision-making. Reg. Environ. Change . Available from: https://doi.org/10.1007/s10113-017-1271-9.

Rahman, M.M., Sarkar, S., Najafi, M.R., Rai, R.K., 2012. Regional extreme rainfall mapping for Bangladesh using L-moment technique. J. Hydrol. Eng. 18 (5), 603–615.

Rahman, M.R., Lateh, H., 2017. Climate change in Bangladesh: a spatio-temporal analysis and simulation of recent temperature and rainfall data using GIS and time series analysis model. Theor. Appl. Climatol. 128 (1-2), 27–41.

Raziei, T., 2017. An analysis of daily and monthly precipitation seasonality and regimes in Iran and the associated changes in 1951–2014. Theor. Appl. Climatol. Available from: https://doi.org/10.1007/s00704-017-2317-0.

ReliefWeb, 2017. <https://reliefweb.int/sites/reliefweb.int/files/resources/Situation_Report%20of%20Flood%20_Updated%20on%20August_22%2C%202017.pdf> (accessed 02.04.18.).

Roxy, M.K., Ritika, K., Terray, P., Murtugudde, R., Ashok, K., Goswami, B.N., 2015. Drying of Indian subcontinent by rapid Indian Ocean warming and a weakening land–sea thermal gradient. Nat. Commun. 6, 7423.

Rummukainen, M., 2013. Climate change: changing means and changing extremes. Clim. Change 121 (1), 3–13.

Salahuddin, A., Isaac, R.H., Curtis, S., Matsumoto, J., 2006. Teleconnections between the sea surface temperature in the Bay of Bengal and monsoon rainfall in Bangladesh. Global Planet. Change 53 (3), 188–197.

Sarker, M.A.R., Alam, K., Gow, J., 2017. Performance of rain-fed Aman rice yield in Bangladesh in the presence of climate change. Renewable Agric. Food Syst. Available from: https://doi.org/10.1017/S1742170517000473.

Seager, R., Naik, N., Vecchi, G.A., 2010. Thermodynamic and dynamic mechanisms for large-scale changes in the hydrological cycle in response to global warming. J. Clim. 23 (17), 4651–4668.

Shahid, S., 2011. Trends in extreme rainfall events of Bangladesh. Theor. Appl. Climatol. 104 (3-4), 489–499.

Shahid, S., Khairulmaini, O.S., 2009. Spatio-temporal variability of rainfall over Bangladesh during the time period 1969–2003. Asia-Pac. J. Atmos. Sci. 45 (3), 375–389.
Siddik, M.A.Z., Rahman, M., 2014. Trend analysis of maximum, minimum, and average temperatures in Bangladesh: 1961–2008. Theor. Appl. Climatol. 116 (3-4), 721–730.
Sonali, P., Kumar, D.N., 2013. Review of trend detection methods and their application to detect temperature changes in India. J. Hydrol. 476, 212–227.
Takahashi, H.G., 2016. Seasonal and diurnal variations in rainfall characteristics over the tropical Asian monsoon region using TRMM-PR data. SOLA 12, 22–27.
Tan, M.L., Ibrahim, A.L., Cracknell, A.P., Yusop, Z., 2017. Changes in precipitation extremes over the Kelantan River Basin, Malaysia. Int. J. Climatol. 37 (10), 3780–3797.
Tarek, M.H., Hassan, A., Bhattacharjee, J., Choudhury, S.H., Badruzzaman, A.B.M., 2017. Assessment of TRMM data for precipitation measurement in Bangladesh. Meteorol. Appl. 24 (3), 349–359.
Trenberth, K.E., 2011. Changes in precipitation with climate change. Clim. Res. 47 (1/2), 123–138.
Valdés-Pineda, R., Pizarro, R., Valdés, J.B., Carrasco, J.F., García-Chevesich, P., Olivares, C., 2016. Spatio-temporal trends of precipitation, its aggressiveness and concentration, along the Pacific coast of South America (36–49 S). Hydrol. Sci. J. 61 (11), 2110–2132.
Walsh, R.P.D., Lawler, D.M., 1981. Rainfall seasonality: description, spatial patterns and change through time. Weather 36 (7), 201–208.
Wang, R., Li, C., 2016. Spatiotemporal analysis of precipitation trends during 1961–2010 in Hubei province, central China. Theor. Appl. Climatol. 124 (1-2), 385–399.
Wang, W., Xing, W., Yang, T., Shao, Q., Peng, S., Yu, Z., et al., 2013. Characterizing the changing behaviours of precipitation concentration in the Yangtze River Basin, China. Hydrol. Processes 27 (24), 3375–3393.
Wentz, F.J., Ricciardulli, L., Hilburn, K., Mears, C., 2007. How much more rain will global warming bring? Science 317 (5835), 233–235.
Wiel, K., Kapnick, S.B., van Oldenborgh, G.J., Whan, K., Philip, S., Vecchi, G.A., et al., 2017. Rapid attribution of the August 2016 flood-inducing extreme precipitation in south Louisiana to climate change. Hydrol. Earth Syst. Sci. 21 (2), 897–921.
World Meteorological Organisation (WMO), 2018. <https://public.wmo.int/en/media/press-release/2017-set-be-top-three-hottest-years-record-breaking-extreme-weather> (accessed 12.04.18.).
Yue, S., Pilon, P., 2004. A comparison of the power of the *t* test, Mann–Kendall and bootstrap tests for trend detection. Hydrol. Sci. J. 49 (1), 21–37.
Zhang, L.J., Qian, Y.F., 2004. A study on the feature of precipitation concentration and its relation to flood-producing in the Yangtze River Valley of China. Chin. J. Geophys. 47 (4), 709–718.
Zhang, X., Zwiers, F.W., Li, G., 2004. Monte Carlo experiments on the detection of trends in extreme values. J. Clim. 17 (10), 1945–1952.
Zhang, X., Zwiers, F.W., Hegerl, G.C., Lambert, F.H., Gillett, N.P., Solomon, S., et al., 2007. Detection of human influence on twentieth-century precipitation trends. Nature 448 (7152), 461.

Further reading

World Meteorological Organisation (WMO), 2013. WMO Statement on the Status of the Global Climate in 2012. WMO, Geneva.

CHAPTER NINE

Characterizing meteorological droughts in data scare regions using remote sensing estimates of precipitation

Mauricio Zambrano-Bigiarini[1,2] and Oscar Manuel Baez-Villaneuva[3,4]
[1]Department of Civil Engineering, Universidad de la Frontera, Temuco, Chile
[2]Center for Climate and Resilience Research, Universidad de Chile, Santiago, Chile
[3]Institute for Technology and Resources Management in the Tropics and Subtropics (ITT), TH Köln, Cologne, Germany
[4]Faculty of Spatial Planning, TU Dortmund University, Dortmund, Germany

9.1 Defining droughts

Drought is a *creeping phenomenon* (Tannehill, 1947; Gillette, 1950), which slowly appears over a region and intensifies its effects with time. There is no a unique definition of drought (Yevjevich, 1967; Glantz and Katz, 1977; Dracup et al., 1980; Wilhite, 2000; Tsakiris et al., 2007; Mishra and Singh, 2010; Lloyd-Hughes, 2014; Van Loon et al., 2016a,b), which hampers the identification of its onset, duration, severity, and spatial extension (Hayes et al., 2012). In general terms, a drought is *an exceptional lack of water compared to normal conditions* (Van Loon et al., 2016a), and different disciplines study droughts with varied methodological approaches (Yevjevich, 1967), depending on the environmental process that is affected. Wilhite and Glantz (1985) completed an exhaustive review of drought definitions, and finally grouped them into five types:

Meteorological drought refers to a reduction in precipitation that might occur along with an increase in potential evapotranspiration. It extends over a large area and usually spans a long period of time, reducing water availability and leading to soil moisture depletion and agricultural drought at short timescales (Mishra et al., 2015). Definitions of meteorological drought are region specific, because of the highly variable atmospheric conditions that lead to a reduction in precipitation. *Agricultural drought*

Extreme Hydroclimatic Events and Multivariate Hazards in a Changing Environment.
DOI: https://doi.org/10.1016/B978-0-12-814899-0.00009-2

refers to a deficit of soil moisture in the root zone which reduces the supply of moisture to the vegetation and as a consequence, the plant water demand is not met. When an agricultural drought persists and aggravates over time, it might lead to crop failure and livestock deaths, triggering severe food shortages. A comprehensive definition of agricultural drought should include the variability in crop characteristics during different stages of crop development. *Hydrological drought* refers to dry spells with impacts on surface and groundwater availability. The impacts of hydrological droughts are usually linked to below-normal water levels in rivers, lakes, reservoirs, aquifers, and to declining wetland areas (Van Loon, 2015). The *socioeconomic drought* is related to the impacts caused by the aforementioned types of drought. It refers to either a failure of the water systems to meet the demand or to health or ecological impacts related to drought scenarios. Recently, Crausbay et al. (2017) defined *ecological drought* as an episodic deficit in water availability that drives ecosystems beyond thresholds of vulnerability, impacts ecosystem services, and triggers feedbacks in natural and/or human systems. Fig. 9.1, adapted from Wilhite (2000) and

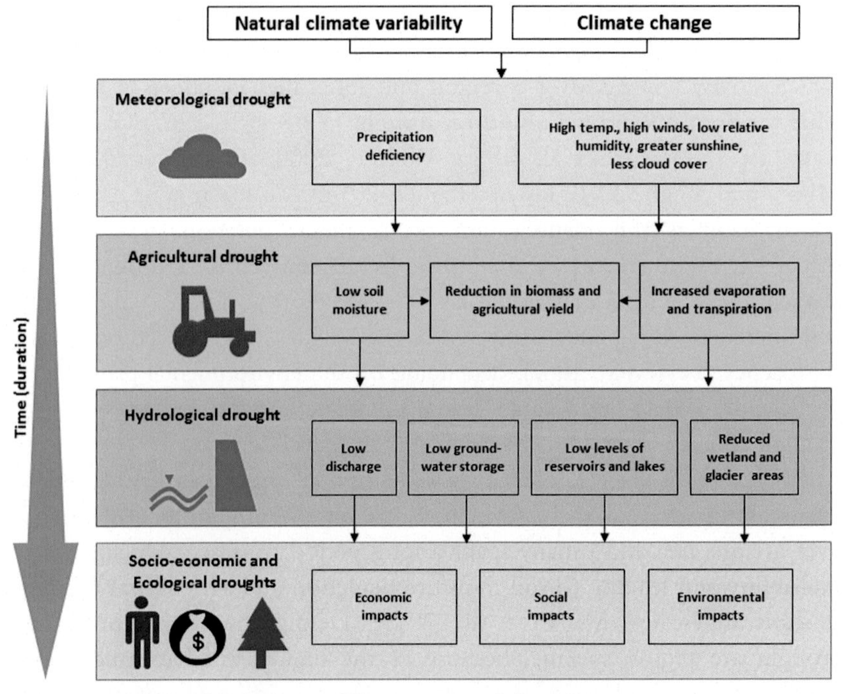

Figure 9.1 Relationship between different types of drought.

Van Loon (2015), shows the relationship between the aforementioned types of droughts.

Recently, Van Loon and Van Lanen (2013) highlighted the importance of making the distinction between *drought* and *water scarcity* in the context of river basin management; the first is referred as an episodic phenomenon and the second results from a long-term imbalance between water demand and water availability.

9.2 Impact of climate change

Droughts have economic and ecological impacts which affect millions of people each year (Wilhite, 2000). During the 1980s, over half a million people in Africa lost their lives due to drought-related disasters, while during the 1990s, around 17.6 million people were affected by them in Latin America and the Caribbean (LAC) (Kallis, 2008). The recent unprecedented drought in California (2012−16) impacted the surface water supplies and increased the groundwater extraction resulting in land subsidence (Flint et al., 2018). Also, the so-called megadrought in Chile (2010−15) presented ~30% deficit in precipitation, which caused a decrease in runoff, a substantial loss in agricultural productivity, and a reduction in snow water equivalent (Garreaud et al., 2017).

Moreover, according to the Intergovernmental Panel on Climate Change, the duration and/or extension of meteorological and agricultural droughts are projected to increase by the end of the 21st century in some regions (Jiménez Cisneros et al., 2014). This due to the reduction in precipitation which might be combined with an increase of the evaporative demand. Therefore, more frequent dry spells related to higher temperatures combined with an increase in the evaporative demand suggest that drought conditions could worsen for many regions in the world (Naumann et al., 2018). In dry regions, the increase of drought intensities will put additional stress on water supply systems, whereas in wetter regions that increase could be managed by taking adaptive measures within an integrated water resources management approach.

Dai (2011) suggests that precipitation may become more intense but less frequent under global warming. This would increase the occurrence of flash floods but reduce soil moisture, and therefore increase the risk of agricultural drought. Severe and more widespread droughts over many land areas are expected as a consequence from decreased precipitation and/or increased evapotranspiration (Dai, 2013), which is in agreement with Wang (2005).

By the end of the 21st century approximately two-thirds of the total population may face a progressive increase in drought-related hazards with climate change (Naumann et al., 2018). The limited capacity of developing countries to cope with natural hazards is becoming an important issue, and therefore efforts must be taken to tackle the expected negative impacts of climate change. Moreover, the increment in global population by the end of this century would endanger food security, due to the projected increase in agricultural drought as a consequence of soil moisture reduction. Therefore, actions must be taken to build anticipatory adaptation and mitigation strategies to increase societal resilience to projected impacts of climate change.

9.3 Drought indices

Droughts have diverse spatiotemporal dynamics, making it difficult to use a single index to characterize them. In the last decades, there has been much debate on which drought index should be used in a particular climate and for which application, resulting in a plethora of indices to quantitatively describe droughts. Excellent reviews are provided by Byun and Wilhite (1999), Heim Jr (2002), Keyantash and Dracup (2002), Zargar et al. (2011), and WMO-GWP (2016). Existing indices process data from one or more hydrological variable(s) (*indicator*), such as precipitation or streamflow, to produce a single numerical value representative of drought intensity. These indices are commonly categorized based on the type of impacts that they are related to (meteorological, agricultural, and hydrological droughts). The development of such indices heavily depends on data availability.

Some indices widely used for analyzing *meteorological droughts* are the percent of normal, which describes the drought as the precipitation deviation from the average; the deciles method (Gibbs and Maher, 1967), which divides the distribution of monthly precipitation into 10% parts, and the Standardized Precipitation Index (SPI; McKee et al., 1993).

However, in order to establish a better correlation with drought impacts, additional meteorological variables have been considered. This is the case of the Standardized Precipitation Evapotranspiration Index (SPEI; Vicente-Serrano et al., 2010) which is sensitive to long-term changes in temperature, and the Palmer Drought Severity Index (SPDI; Palmer, 1965), which is based on a supply–demand concept rather than analyzing the precipitation anomalies.

For *agricultural droughts* it is crucial to monitor a subsequent deficit in the soil water balance. The Crop Moisture Index (CMI; Palmer, 1968) considers moisture deficit in the top (5 ft) of the soil column. The Soil Moisture Deficit Index (SMDI; Narasimhan and Srinivasan, 2005) characterizes soil moisture deficit at different depths, while the Evapotranspiration Deficit Index (EDTI; Narasimhan and Srinivasan, 2005) considers the deficit in evapotranspiration. Also, the Normalized Difference Vegetation Index (NDVI; Tucker, 1979), the Enhanced Vegetation Index (EVI; Liu and Huete, 1995), the Temperature Condition Index (TCI; Kogan, 1990), and the Normalized Difference Water Index (NDWI; Gao, 1996) are remote-sensing based vegetation indices used to monitor the vegetation health (Zargar et al., 2011).

Finally, for *hydrological droughts* it is important to provide a characterization of delayed hydrological impacts. For this purpose, the Palmer Hydrological Drought Index (PHDI; Palmer, 1965) considers precipitation, evapotranspiration, recharge, runoff, and soil moisture; the Surface Water Supply Index (SWSI; Shafer and Dezman, 1982) calculates the weighted average of the Standardized anomalies of precipitation, reservoir storage, snowpack, and runoff, while the Remote Sensing Drought Index (RSDI; Stahl, 2001) calculates the deficiency in streamflow and then uses cluster analysis to delineate the drought affected areas. However, these indices are data demanding, which is an important problem in data scarce regions. On the other hand, the Streamflow Drought Index (SDI; Nalbantis and Tsakiris, 2009) is an SPI-based index which only uses high quality streamflow data and therefore is a good proxy for hydrological drought analysis if streamflow data are available.

Among the existing indices, the World Meteorological Organization (WMO) recommended in 2009 the use of the SPI as the standard index to characterize and monitor meteorological droughts at different timescales (Hayes et al., 2011; WMO, 2012), and since then, it has been used worldwide for drought monitoring (e.g., Meroni et al., 2017; Zhang et al., 2013; Zambrano et al., 2016, 2017).

9.4 Case study

Drought impacts are less obvious than other natural disasters such as floods, earthquakes, and wildfires (Guha-Sapir et al., 2004). However, droughts are one of the most important natural triggers of food shortages, malnutrition, and famine. Between 1970 and 1999, droughts affected 58 million people in the LAC region (Charvériat, 2000), where drought frequency and severity are expected to increase by the end of the 21st century, with Chile as a possible hotspot of future water security issues (Prudhomme et al., 2014).

Traditionally, droughts have been characterized using ground-based data (Hayes et al., 1999; Hu and Willson, 2000; Vicente-Serrano and López-Moreno, 2005; Chang and Kleopa, 1991; Barker et al., 2016). However, in the last decades remote sensing techniques have become a promising alternative to provide a spatial characterization of drought-related variables and to quantify drought impacts (AghaKouchak et al., 2015).

Therefore, this chapter evaluates the suitability of state-of-the-art remote sensing estimates of precipitation to characterize droughts. Specifically, we present a case study that uses satellite data to identify the spatial extension, duration, and severity of the so-called *megadrought*, which has affected Chile during the last decade (Boisier et al., 2016; Garreaud et al., 2017; CR2, 2015). In particular, three regions with different climate, topography, and land cover are used as study areas. Duration and severity are used to characterize meteorological drought events, based on the SPI drought index. Results of this work aim at providing a cost-effective drought monitoring and early warning framework, which might be used to characterize ongoing drought events and to foster the elaboration of improved drought adaptation and mitigation policies at the regional scale.

9.4.1 Study areas

Fig. 9.2 shows the three study areas selected for this work: Valparaíso, Maule, and Araucanía in Chile. These regions present differences in climate, topography, land cover, and economic activities. According to the world map of the Köppen–Geiger climate classification (Rubel et al., 2017), the Valparaíso region is mainly characterized by temperate weather with hot and dry summer (Csb), cold and arid steppe (BSk), cold

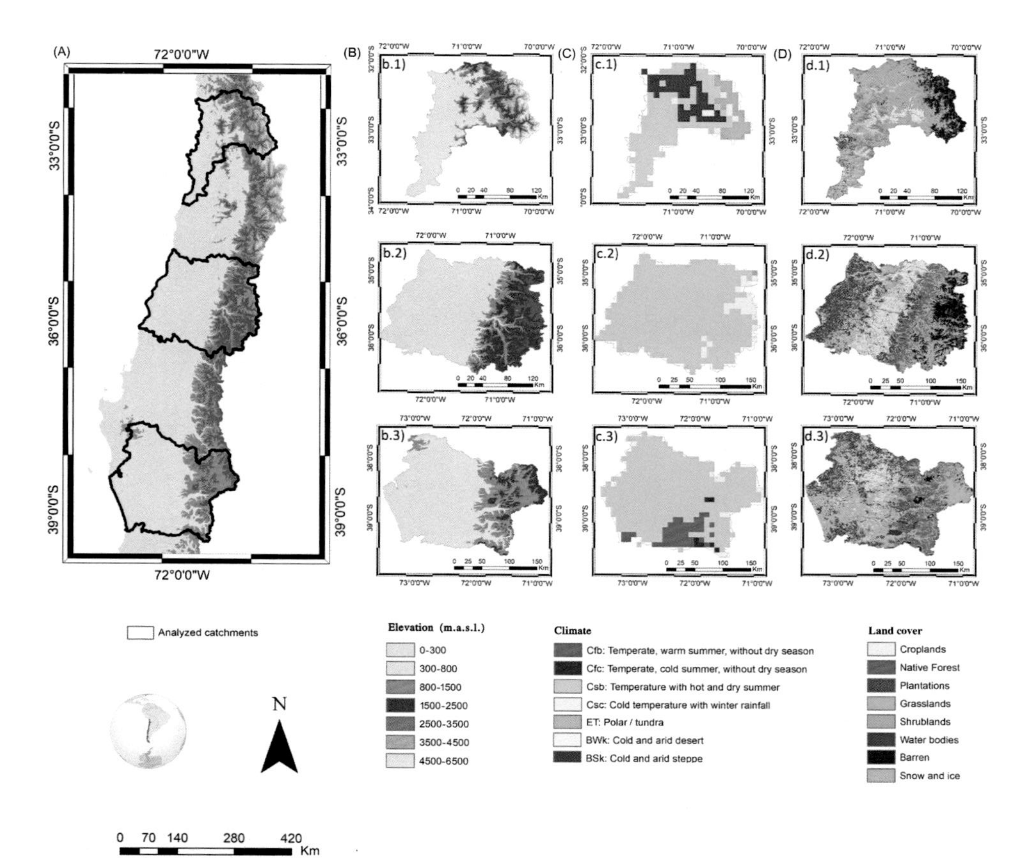

Figure 9.2 Study areas. From left to right: (A) Localization of the study areas across Chile; (B) digital elevation model (Jarvis et al., 2008); (C) climate zones based on Köppen−Geiger classification (Rubel et al., 2017); and (D) land cover (Zhao et al., 2016).

temperature with winter rainfall (Csc), and polar/tundra (ET). Maule is also characterized by the temperate weather with hot and dry summer (Csb), cold temperature with winter rainfall (Csc), and polar/tundra (ET). Finally, the temperate weather with hot and dry summer (Csb) dominates the Araucanía region, with temperate, warm summer, without dry season (Cbf) in the southern part and minor presence of temperate, cold summer, without dry season (Cfc) and cold temperature with winter rainfall (Csc).

According to the most updated land cover map of Chile (Zhao et al., 2016) (see Fig. 9.2D), the native forest and grasslands dominate the Valparaíso Region, with a smaller area covered by plantations, croplands, and barren. In the case of Maule and Araucanía regions, the plantations and croplands are more important in extension, with native forest, grasslands, and barren lands also observed in the upper part of both regions. For all the study areas the wet season starts in May and ends in August, while the dry season starts in September and ends in April (see Fig. 9.3).

The Valparaíso region is currently facing an incipient water crisis. The avocado industry is using large amounts of water, reducing water

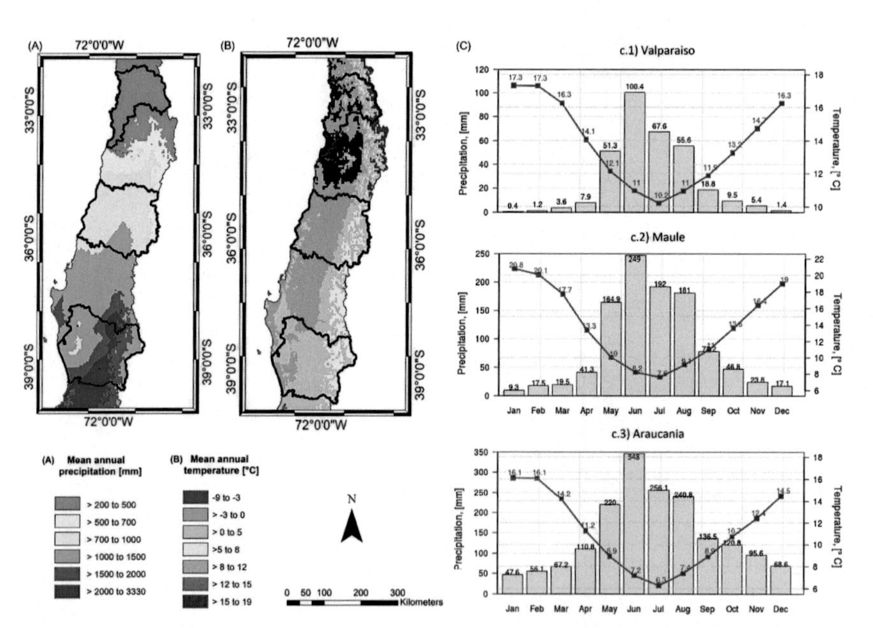

Figure 9.3 Study areas. From left to right: (A) mean annual precipitation (Fick and Hijmans, 2017); (B) mean annual temperature (Fick and Hijmans, 2017); and (C) climographs of each study area. Valparaíso, Maule, and Araucanía regions are denoted by the numbers 1, 2, and 3, respectively.

availability for small farmers and communities (El Mercurio, 2018; CIPER, 2018). In particular, Cabildo, a commune located in the northern part of Valparaíso, might be the first area in Chile to reach a condition of total water scarcity (El Desconcierto, 2018). On the other hand, Maule is the most important agricultural region of Chile, primarily with citrus, fruits, and wine grapes. In 2005 the agrifood sector represented 11% of Chilean GDP and the total employment related to the agricultural production is approximately 22% (CONICYT, 2007), therefore, the agricultural sector is extremely important for this region. Finally, the Araucanía region is facing an accelerated agriculture expansion and a recent increment in regional temperature. During the last decade the Araucanía, historically a region with a large amount of precipitation, has also been affected by the Chilean megadrought, and nowadays is the region with the highest usage of trucks for water supply in Chile. Because the aforementioned reasons, a detailed analysis regarding the spatial extent, duration, and severity of the Chilean megadrought is required for these regions, in order to improve our understanding of the different types of droughts and to cope with this relevant hazard.

9.4.1.1 Valparaíso region

The Valparaíso region is located between the latitudes 33°57′S and 32°01′S, and between the longitudes 71°50′W and 69°50′W, with elevations going from sea level to 6049 m a.s.l. This region has a total area of 16,396 km^2, with a population of 1,790,219 inhabitants, ranking third among all Chilean regions. The mean annual temperature is 11.8°C, while the temperature in summer increases to 17°C on average. June presents the highest amount of rain (about 100 mm) and January is the driest month (see Fig. 9.3). The city of Valparaíso also has the main port of Chile, with 10 million tons annually shipped, and is also important to the tourism sector, with 50 cruises and 150,000 passengers on average annually.

9.4.1.2 Maule region

The Maule region is located between the latitudes 36°32′S and 34°41′S, and between the longitudes 72°47’W and 70°18’W. Elevations range from sea level to 4774 m a.s.l. June and January are the wettest (~250 millimeters) and driest (~9.3 millimeters) months, respectively. This region has a total area of 30,269 km^2, with a population of 1,044,950 inhabitants, ranking fourth among the Chilean regions. The annual precipitation is ~708 mm, distributed mainly during winter (June–July–August, see Fig. 9.3). Mean monthly temperatures range from 8.5 to 10°C (Rubilar et al., 2008).

9.4.1.3 *Araucanía region*

The Araucanía Region is located between the latitudes 39°38'S and 37°34'S, and between the longitudes 73°31'W and 70°49'W. Elevations range from 0 to 3723 m a.s.l. This region has an area of about 31,842 km^2, with a population of 938,626 inhabitants, ranking fifth among the Chilean regions. As in the aforementioned regions, June and January are the wettest (~348 mm) and driest (~47.6 mm) months, respectively. The average annual precipitation is about 1200–1600 mm. The most important economic activities for this region are farming (especially wheat), cattle raising, and lumbering.

9.4.2 Data

Developing countries usually do not have a dense meteorological network to allow a reliable characterization of the spatiotemporal variability of key meteorological variables related to droughts. Therefore, the interpolation using point-based information of these parameters is subject to large uncertainties (Woldemeskel et al., 2013).

Recent advances in remote sensing techniques have allowed quantitative estimates of drought-related variables (AghaKouchak et al., 2015). Based on an exhaustive assessment of several state-of-the-art precipitation products available for Chile (Zambrano-Bigiarini et al., 2017), the Climate Hazards Group InfraRed Precipitation with Station data version 2.0 (hereafter, CHIRPSv2) was selected as the most representative precipitation data source for the three study areas. CHIRPSv2 is a new quasi-global (50°S–50°N, 180°E–180°W), high resolution (0.05°) and long-term (1981 to near-present) precipitation product (Funk et al., 2015). CHIRPSv2 combines infrared (IR) and passive microwave (PMW) measurements with ground station records to calibrate global Cold Cloud Duration (CCD) rainfall estimates (Funk et al., 2015).

9.4.3 Methodology

9.4.3.1 *Selected drought index*

Several indices are currently proposed for monitoring meteorological, hydrological, agricultural, and socioeconomical droughts (e.g., Heim Jr, 2002; Keyantash and Dracup, 2002; Zargar et al., 2011; Dai, 2011; WMO-GWP, 2016). In general, the selection of an index for a particular application is based on the availability of required input data, the ease of computation and communication of the obtained results (Masud et al., 2017). In this work, we follow the Lincoln Declaration on Drought Indices (Hayes et al., 2011), which allowed the WMO to recommend in 2009 the SPI (McKee et al., 1993) as the standard index to characterize and monitor meteorological droughts at different timescales WMO (2012). In order to allow the statistical comparison of wetter and drier climates, the SPI transforms the accumulated precipitation into a standard normal variable, that is, a random variable with zero mean and variance equal to one.

SPI compares the accumulated precipitation during a period of time with the long-term rainfall average for the same period at that location, and therefore it can be used to compare deficit or excess across regions with markedly different climates. To compute the numeric value of SPI, the first step is to select a probability density function to represent the long-term time-series of precipitation accumulated over the desired accumulation period, as well as a method to estimate the parameters of the chosen distribution (e.g., Hosking and Wallis, 1995). Then, the cumulative distribution function (CDF) is obtained from the fitted distribution. Finally, an equipercentile inverse transformation is applied to the CDF to obtain the standard normal scores (i.e., the SPI values) corresponding to the scores of the fitted distribution. The resulting numeric values of SPI represent the number of standard deviations that cumulative precipitation deviates from the long-term average (JRC, 2011), that is, an SPI-1 equal to −2 in July 2015 means that the accumulated precipitation in that particular month is 2 standard deviations smaller than the long-term average (e.g., 1981–2010) of monthly precipitation recorded in July. A given SPI value simultaneously represents drought intensity and frequency, because using the standard normal distribution is possible to associate probabilities to each SPI value, that is , users can expect the SPI to be within one standard deviation of the mean about 68% of the time, and so on. Notwithstanding Guttman (1999) and Vicente-Serrano (2006) suggesting the use of Pearson Type III distribution, the SPI values are computed in this work using the two-parameter Gamma distribution, as recently recommended by Stagge et al. (2015), in combination with the unbiased Probability Weighted Moments (ub-pwm; Hosking and Wallis, 1995) to estimate the parameters of the Gamma distribution.

Satellite-based precipitation data (Section 4.2) are used to compute the SPI using the *SPEI* (Beguería and Vicente-Serrano, 2017), *raster* (Hijmans, 2017), and *hydroTSM* (Zambrano-Bigiarini) R packages in R 3.4.4 (R Core Team, 2018).

9.4.3.2 Timescales

The SPI is computed on a monthly basis using a moving window of n months, where n indicates the accumulation period (i.e., the *timescale*), typically 1, 3, 6, 9, 12, 24, or 48 months, and the corresponding SPIs are denoted using the same month numbers (e.g., SPI-1 and SPI-48 for the SPI values computed using 1 and 48 months, respectively). The aforementioned accumulation periods are arbitrary but represent typical timescales where precipitation deficits have an impact on water resources such as streamflow, reservoirs storage, soils moisture, groundwater, and snowpack (McKee et al., 1993). SPI at timescale less than 9 months has been linked to agricultural drought (Quiring and Ganesh, 2010). In this work, we computed SPI at 3-, 6-, 9-, and 12-month timescales. Strictly speaking SPI-12 is a meteorological drought index, however, it is often used as a proxy for hydrological drought.

9.4.3.3 Drought characteristics

In this work drought events are described (Mishra and Singh, 2010) in terms of their duration and severity, using the *theory of runs* (Yevjevich, 1967). A run is defined as a subset of a time series in which all values are either below or above a selected truncation level X_o (Yevjevich, 1967).

Following the drought classes proposed by McKee et al. (1993), a threshold of -1.0 is used to identify (moderate, severe, and extreme) drought events using the SPI index; that is, the *onset* of a drought event occurs when the SPI is less than -1.0, its *duration* D is the number of continuous months where the SPI values satisfy the aforementioned criterion, while its *severity* S is defined as the cumulative sum of all the SPI values over the duration D.

$$S_j = \sum_{i=1}^{D} \mathrm{SPI}_i < -1.0 \tag{9.1}$$

For each grid cell within the study area, the total severity (S_y) and duration (D_y) for each year are computed as follows:

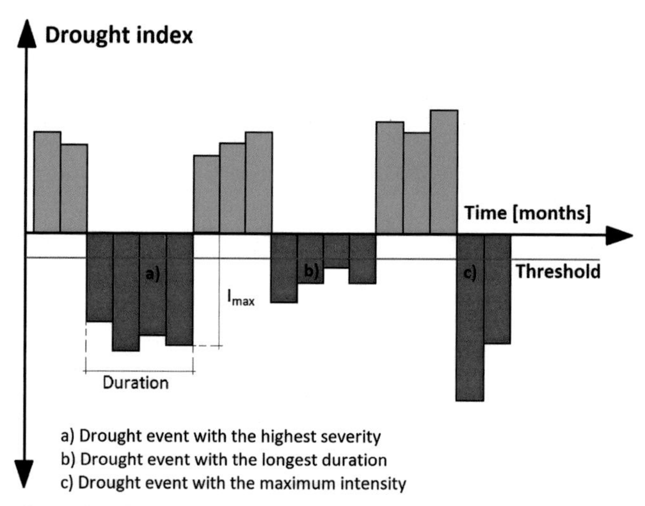

Figure 9.4 Drought characteristics using the run theory. X_o is the threshold level selected to define the onset and ending time of a drought event.

$$S_y = \sum_{j=1}^{N} S_j \tag{9.2}$$

$$D_y = \sum_{j=1}^{N} D_j \tag{9.3}$$

where S_j and D_j represent the severity and duration of the *j-th* drought event, respectively, and N is the total number of drought events identified during the year y. A schematic illustration of the drought characteristics (such as severity, duration, and intensity) previously described is shown in Fig. 9.4.

9.4.4 Results

The SPI was computed at 3-, 6-, 9-, and 12-month timescales to identify the spatial extension, duration, and severity of the so-called Chilean *mega-drought* in the selected study areas.

9.4.4.1 Megadrought duration

The total duration of drought events identified with SPI-3, -6, -9, and -12 was computed for Valparaíso, Maule, and Araucanía regions. All the regions show an increment in drought duration for all timescales since 2008 onwards. However, only some representative figures are shown in this section.

Fig. 9.5 shows the duration of drought events in the Valparaíso region, based on SPI-3 (as proxy of agricultural drought) for the 2001–16 period. A 3-month SPI reflects short- and medium-term soil moisture conditions and provides an estimation of seasonal precipitation. The highest duration values are found during the 2008–12 period. However, drought duration obtained with SPI-6, SPI-9, and SPI-12 indicate 2011 as the year with highest duration, reaching between 9 and 12 months for most of the region, except for the eastern part of the region, characterized by tundra climate (ET) and barren land cover. Fig. 9.6 shows the duration of drought events in the El Maule region, based on SPI-6 (as representative of wet/dry seasons) for the 2001–16 period. The Maule region was almost drought-free from 2001 to 2007. However, drought duration has increased since 2008 onwards, with 2011 being the year most affected by drought, reaching 6–9 months. Finally, Fig. 9.7 shows the duration of

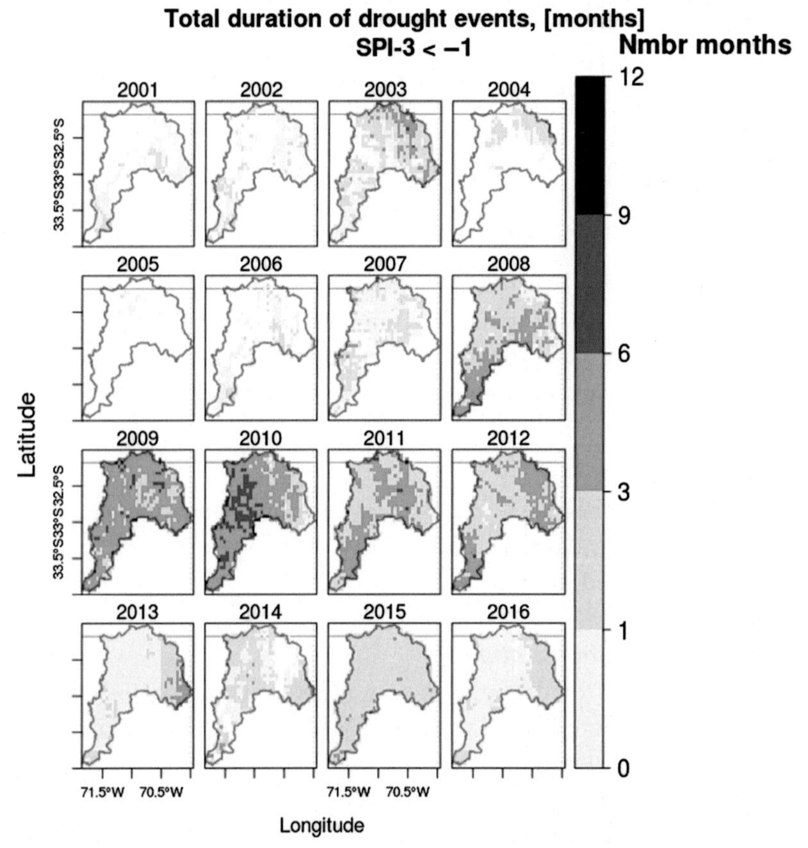

Figure 9.5 Total duration of drought events in months using the SPI-3 over Valparaíso Region.

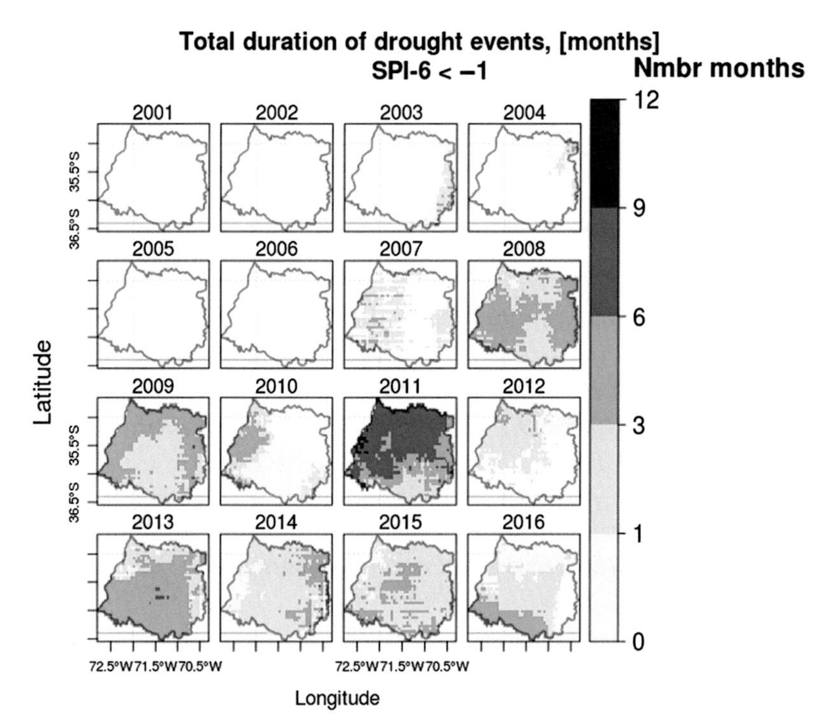

Figure 9.6 Total duration of drought events in months using the SPI-6 over El Maule Region.

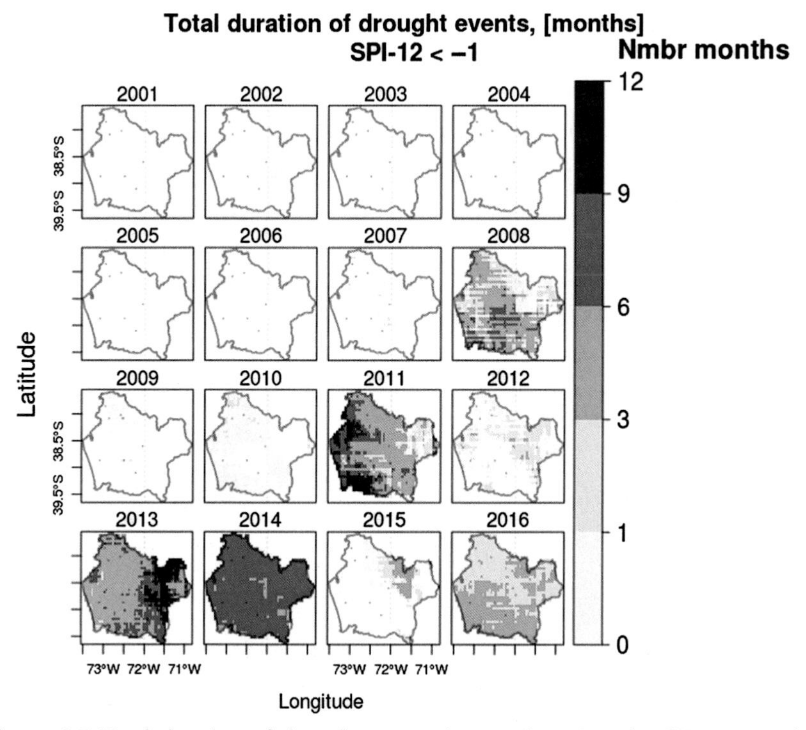

Figure 9.7 Total duration of drought events in months using the SPI-12 over the Araucanía Region.

drought events in the Araucanía region, based on SPI-12 (as a proxy of hydrological drought) for the 2001–16 period. The 2001–07 period was drought-free, but years 2008, 2011, 2013, 2014, and 2016 were affected by several months under drought.

9.4.4.2 Megadrought severity

The total severity of drought events identified with SPI-3, -6, -9, and -12 was computed for the Valparaíso, Maule, and Araucanía regions. In general, all the regions show a spatial pattern of drought severity similar to the one of drought duration, and an increment in drought severity for all timescales for 2008 onwards. Fig. 9.8 shows the severity of drought events

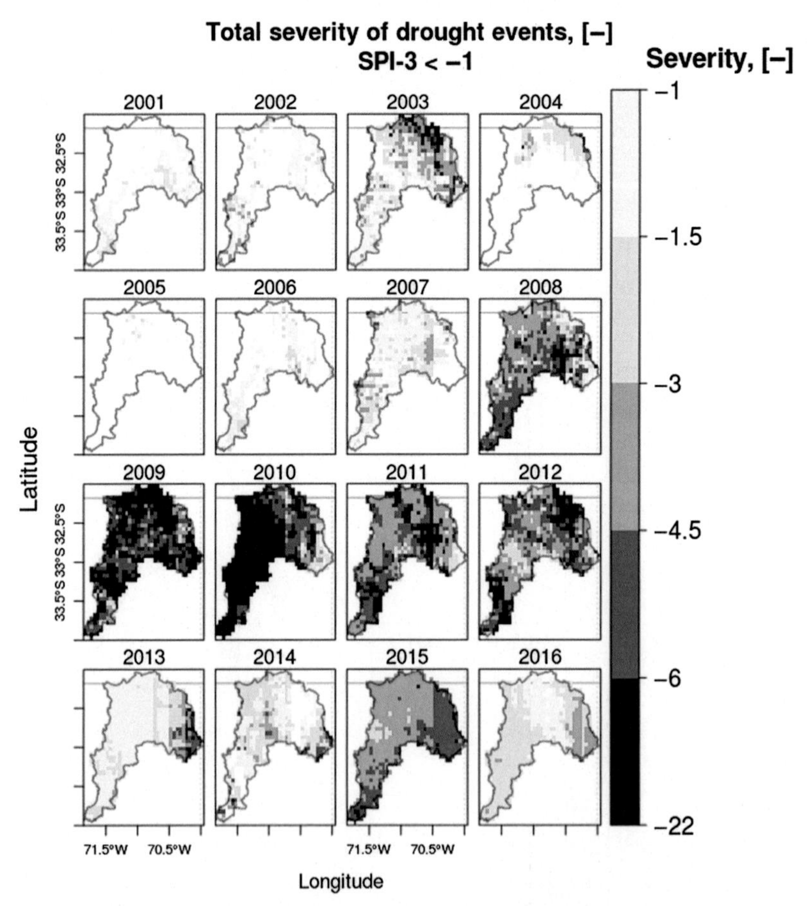

Figure 9.8 Total severity of drought events in the Valparaíso region [dimensionless], based on SPI-3.

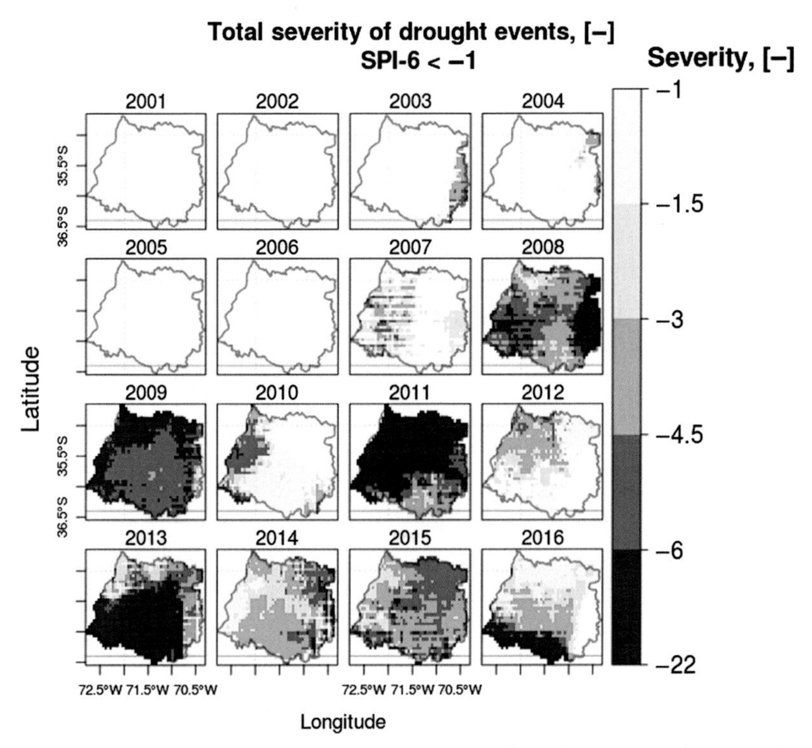

Figure 9.9 Total severity of drought events in the El Maule region, [dimensionless], based on SPI-6.

for the Valparaíso region, based on SPI-3 for the 2001–16 period. The most severe droughts are observed in the 2008–12 and 2014–15 periods when SPI at low timescales (3–6) was used. However, when longer timescales (9–12 months) were used, the most severe droughts are concentrated in the period 2010–12, with extreme severity all over the region. These results show that SPI was able to detect the megadrought period over the Valparaíso Region. Fig. 9.9 shows the severity of drought events for the El Maule region, based on SPI-6 for the 2001–16 period. In general, the period 2001–2007 was almost not affected by droughts, while from 2008 onwards most of the region was under severe drought conditions, particularly in years 2008, 2009, 2011, and 2013. When using SPI-12, most of the region was under strong drought conditions in the period 2011–15. Finally, Fig. 9.10 shows the severity of drought events for the Araucanía region, based on SPI-12 for the 2001–16 period. In agreement with Fig. 9.7, it shows that the period 2001–07 was drought-free, while years 2008, 2011, 2013, 2014, and 2016 were affected

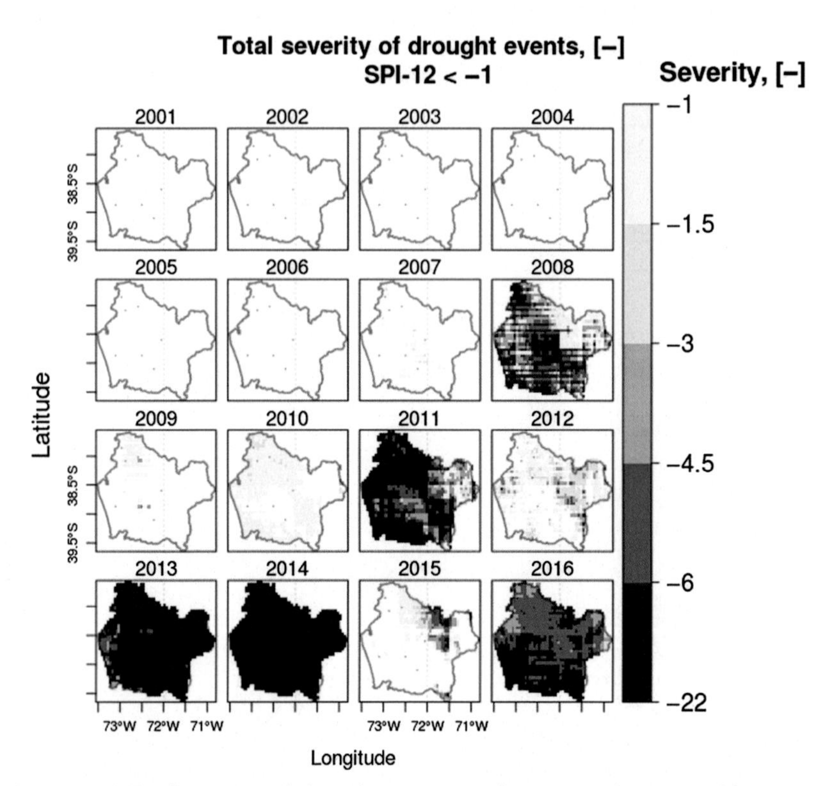

Figure 9.10 Total severity of drought events in the Araucanía region, [dimensionless], based on SPI-12.

by severe droughts. At lower timescales (3 and 6 months), the years 2009, 2010, 2013, 2015, and 2016 presented the most severe conditions.

9.4.5 Discussion

As described in the Results section, the spatial extent, severity, and duration of drought events in the studied areas vary when the SPI is computed at different timescales. However, for all regions, drought duration and severity have been increasing since 2008.

Figs. 9.11 and 9.12 compare, respectively, the relative drought duration and mean annual severity for the 2001–09 reference period against the 2010–16 megadrought period (Garreaud et al., 2017). Valparaíso shows values up to 40% of relative drought duration during the megadrought, mainly in the northeastern part characterized by high-elevated areas. The Maule region also shows high relative duration values (up to

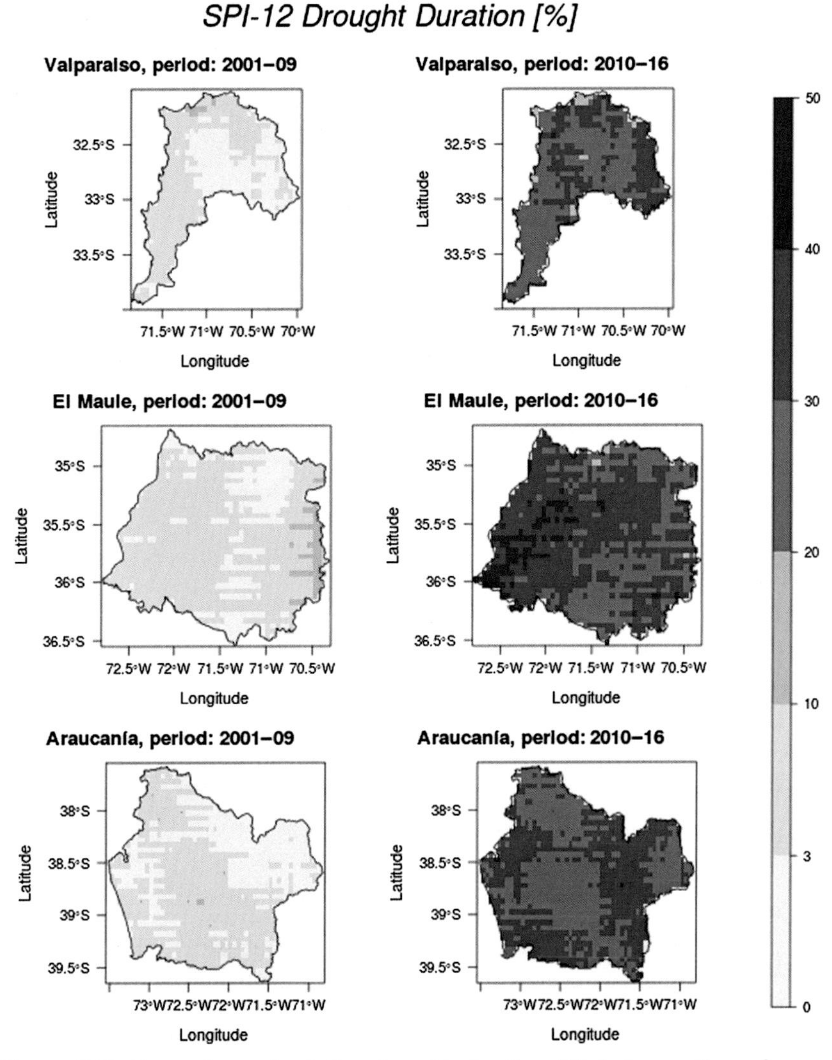

Figure 9.11 Comparison of relative drought duration [%] between the reference period 2001–09 and the megadrought period 2010–16, using SPI-12.

50%) in the western region, characterized by the presence of exotic plantations and crops. On the other hand, the highest severity in the Valparaíso region during the megadrought is observed in the central valley, in contrast to the reference period 2001–09 where the highest severity was found in the western part. The Maule region also shows an increment in drought severity in the western area, while during the reference period 2001–09 the highest annual severity was found in the eastern

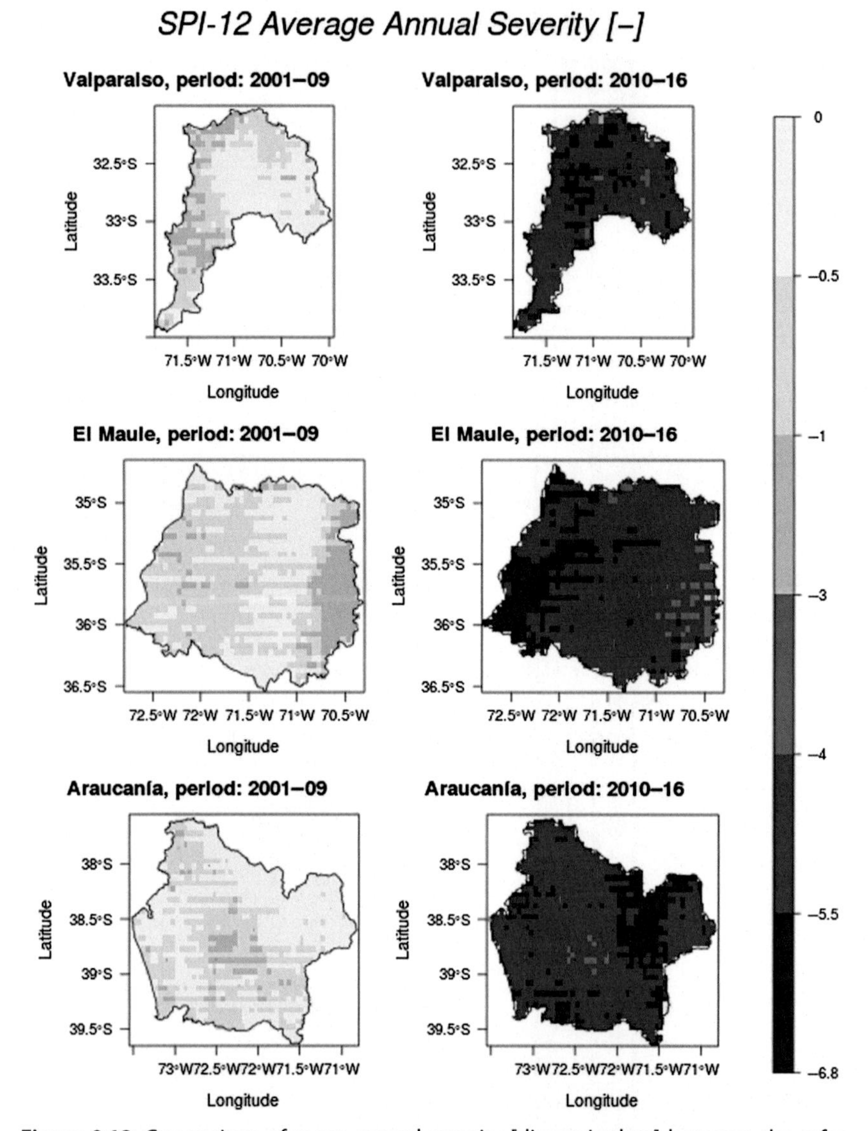

Figure 9.12 Comparison of mean annual severity [dimensionless] between the reference period 2001–09 and the megadrought period 2010–16, using SPI-12.

part of the region. Finally, the Araucanía region also presents a general increase in drought duration and severity during the megadrought period. The highest durations and severities were observed in the central valley during the period 2001–09, while during the megadrought the highest values were found in the western mountainous parts of the region.

The increase in duration and severity of drought events during the megadrought is due to an important reduction in the monthly amounts of precipitation falling over the three study areas (Garreaud et al., 2017). However, duration and severity of drought events vary among the different land cover types present in the study areas (not shown here).

9.5 Conclusions

Knowledge about the causes and characteristics of drought events is required to move towards a proactive risk management in data scarce regions. However, developing countries, such as those in LAC, usually do not have the dense network of meteorological stations that would allow a reliable characterization of the spatiotemporal variability of key meteorological variables necessary to describe the spatial and temporal patterns of drought events. Therefore, in this chapter, we evaluated the suitability of a state-of-the-art satellite-based precipitation estimate (CHIRPSv2) to identify the spatial extension, duration, and severity of the megadrought that has affected Chile during the last decade. This evaluation was carried out over three severely affected regions of Chile (Valparaíso, El Maule, and La Araucanía) based on the SPI at 3-, 6-, 9-, and 12-month temporal scales. The main findings were:

- Satellite-based data sets are a promising technology for drought monitoring in developing countries such as LAC. These data sets have long temporal coverage and relatively high spatial resolution, which provides spatiotemporal values of key hydrological variables. However, a specific assessment is still recommended for each study area in order to select a suitable data set for analyzing drought.
- The spatial extension of drought events varies along different years and depends also on the timescale used for carrying out the analysis. Different timescales might be used for monitoring different types of drought, with short timescales (1–3) usually most suited for agricultural droughts and long timescales (9–12) usually better for monitoring the long-term impacts of droughts on water resources.
- The SPI values computed at different timescales were able to identify the spatial extent, duration, and severity of the Chilean megadrought for the period 2010–16, in particular in the Valparaíso, Maule, and

Araucanía regions. In general, an increase in drought duration and severity is observed among the study areas since 2008 onwards, compared to the previous 2001–09 period.
- This study confirms that the Chilean megadrought already reported by Garreaud et al. (2017) for the period 2010–15 was still ongoing during the year 2016.
- The impacts of droughts presented different patterns for the main land cover types present in the study areas. Therefore, land classification must be carefully considered when characterizing a drought event and its impacts.
- The increase in severity and duration of drought events during the megadrought period was mostly driven by a decrease in precipitation, in agreement with previous studies. However, different factors such as temperature, land cover, and vegetation water requirements might play an important role in the evaluation of the spatiotemporal characteristics of droughts.

Acknowledgments

Mauricio Zambrano-Bigiarini thanks the CONICYT-Fondecyt 11150861 and the CONICYT-Fondap 15110009 for their support in the production of this chapter.

References

AghaKouchak, A., Farahmand, A., Melton, F.S., Teixeira, J., Anderson, M.C., Wardlow, B.D., et al., 2015. Remote sensing of drought: progress, challenges and opportunities. Rev. Geophys. 53, 452–480.

Barker, L.J., Hannaford, J., Chiverton, A., Svensson, C., 2016. From meteorological to hydrological drought using standardised indicators. Hydrol. Earth Syst. Sci. 20, 2483–2505.

Beguería, S., Vicente-Serrano, S.M., 2017. SPEI: calculation of the Standardised Precipitation-Evapotranspiration Index. R Package Cersion 1.7. Available at <https://CRAN.R-project.org/package = SPEI>.

Boisier, J.P., Rondanelli, R., Garreaud, R.D., Muñoz, F., 2016. Anthropogenic and natural contributions to the Southeast Pacific precipitation decline and recent megadrought in central Chile. Geophys. Res. Lett. 43, 413–421.

Byun, H.R., Wilhite, D.A., 1999. Objective quantification of drought severity and duration. J. Clim. 12, 2747–2756.

CIPER, 2018. La naturaleza poltica de la sequía en Petorca. <https://ciperchile.cl/2018/04/27/la-naturaleza-politica-de-la-sequia-en-petorca/> (online; last accessed Jul 2018).

CONICYT, 2007. The wine and vine grape production sector in Chile. In: Technical Report. Conicyt. Santiago, Chile. Available online on <http://www.conicyt.cl/documentos/dri/ue/Vino_Wine_BD.pdf>.

CR2, 2015. Report to the Nation: The Central Chile Mega-Drought [in Spanish]. In: Technical Report. Center for Climate and Resilience Research (CR2). Santiago, Chile. Available online at <http://www.cr2.cl/wp-content/uploads/2015/11/informe-megasequia-cr21.pdf> (last visited 10.08.17).

Chang, T.J., Kleopa, X.A., 1991. A proposed method for drought monitoring. J. Am. Water Resour. Assoc. 27, 275.

Charvériat, C., 2000. Natural disasters in Latin America and the Caribbean: an overview of risk. SSRN Electron. J, IDB Working Paper No. 364.

R. Core Team, 2018. R: A Language and Environment for Statistical Computing. Vienna, Austria. https://www.R-project.org/.

Crausbay, S.D., Ramirez, A.R., Carter, S.L., Cross, M.S., Hall, K.R., Bathke, D.J., et al., 2017. Defining ecological drought for the twenty-first century. Bull. Am. Meteorol. Soc. 98, 2543–2550.

Dai, A., 2011. Drought under global warming: a review. Wiley Interdiscip. Rev.: Clim. Change 2, 45.

Dai, A., 2013. Increasing drought under global warming in observations and models. Nat. Clim. Change 3, 52–58.

El Desconcierto, 2018. Al borde de la sequía total: Reportaje reveló que Cabildo podría convertirse en la primera comuna del país sin agua. <https://goo.gl/nwrFW6> (online; last accessed May 2018).

Dracup, John A., Lee, Kil Seong, Paulson Jr., Edwin G., 1980. On the definition of droughts. Water Resour. Res. 16, 297–302.

Fick, S.E., Hijmans, R.J., 2017. Worldclim 2: new 1-km spatial resolution climate surfaces for global land areas. Int. J. Climatol. 37, 4302–4315.

Flint, L.E., Flint, A.L., Mendoza, J., Kalansky, J., Ralph, F.M., 2018. Characterizing drought in California: new drought indices and scenario-testing in support of resource management. Ecol. Process. 7.

Funk, C., Peterson, P., Landsfeld, M., Pedreros, D., Verdin, J., Shukla, S., et al., 2015. The climate hazards infrared precipitation with stations-a new environmental record for monitoring extremes. Sci. Data 2, 150066.

Gao, Bc, 1996. NDWI-A normalized difference water index for remote sensing of vegetation liquid water from space. Remote Sens. Environ. 58, 257–266.

Garreaud, R.D., Alvarez-Garreton, C., Barichivich, J., Boisier, J.P., Christie, D., Galleguillos, M., et al., 2017. The 2010–2015 megadrought in central Chile: impacts on regional hydroclimate and vegetation. Hydrol. Earth Syst. Sci. 21, 6307–6327.

Gibbs, W.J., Maher, J.V., 1967. Rainfall deciles as drought indicators. In: Bulletin (Australia. Bureau of Meteorology); no.48, Melbourne: Bureau of Meteorology. Available online on <https://trove.nla.gov.au/work/21297477>.

Gillette, H.P., 1950. A creeping drought under way. Water Sewage Works 104–105.

Glantz, M.H., Katz, R.W., 1977. When is a drought a drought? Nature 267, 192–193.

Guha-Sapir, D., Hargitt, D., Hoyois, P., 2004. Thirty years of natural disasters 1974–2003: The numbers. Presses univ. de Louvain.

Guttman, N.B., 1999. Accepting the standardized precipitation index: a calculation algorithm. JAWRA J. Am. Water Resour. Assoc. 35, 311–322.

Hayes, M., Svoboda, M., Wall, N., Widhalm, M., 2011. The Lincoln declaration on drought indices: universal meteorological drought index recommended. Bull. Am. Meteorol. Soc. 92, 485–488.

Hayes, M.J., Svoboda, M.D., Wilhite, D.A., Vanyarkho, O.V., 1999. Monitoring the 1996 drought using the Standardized Precipitation Index. Bull. Am. Meteorol. Soc. 80, 429–438.

Hayes, M.J., Svoboda, M.D., Wardlow, B.D., Anderson, M.C., Kogan, F., 2012. Drought monitoring: Historical and current perspectives. In: Wardlow, B.D., Anderson, M.C., Verdin, J.P. (Eds.), Remote Sensing of Drought: Innovative Monitoring Approaches. CRC Press/Taylor & Francis.

Heim Jr, R.R., 2002. A review of twentieth-century drought indices used in the United States. Bull. Am. Meteorol. Soc. 83, 1149–1165.

Hijmans, R.J., 2017. raster: Geographic Data Analysis and Modeling. In: R Package Version 2.6-7. <https://CRAN.R-project.org/package = raster>.
Hosking, J.R.M., Wallis, J.R., 1995. A comparison of unbiased and plotting-position estimators of L moments. Water Resour. Res. 31, 2019–2025.
Hu, Q., Willson, G.D., 2000. Effects of temperature anomalies on the Palmer Drought Severity Index in the central United States. Int. J. Climatol. 20, 1899–1911.
JRC, 2011. Product Fact Sheet: SPI – Europe. In: Technical Report. European Commission, Joint Research Centre, DESERT Action, LMNH Unit. Available on <http://edo.jrc.ec.europa.eu/documents/factsheets/factsheet_spi.pdf> (Last visited: October 2016).
Jarvis, A., Reuter, H., Nelson, A., Guevara, E., 2008. Hole-filled seamless SRTM data V4, International Centre for Tropical Agriculture (CIAT). <http://srtm.csi.cgiar.org> (online; last accessed May 2012).
Jiménez Cisneros, B., Oki, T., Arnell, N., Benito, G., Cogley, J., Döll, P., et al., 2014. Climate change 2014: impacts, adaptation, and vulnerability. Part A: Global and sectoral aspects. Contribution of Working Group II to the Fifth Assessment Report of the Intergovernmental Panel on Climate Change. Cambridge University Press, Cambridge, United Kingdom and New York, pp. 229–269, chapter Freshwater resources.
Kallis, G., 2008. Droughts. Annu. Rev. Environ. Resour. 33, 85.
Keyantash, J., Dracup, J.A., 2002. The quantification of drought: an evaluation of drought indices. Bull. Am. Meteorol. Soc. 83, 1167–1180.
Kogan, F.N., 1990. Remote sensing of weather impacts on vegetation in non-homogeneous areas. Int. J. Remote Sens. 11, 1405–1419.
Liu, H.Q., Huete, A., 1995. A feedback based modification of the NDVI to minimize canopy background and atmospheric noise. IEEE Trans. Geosci. Remote Sens. 33, 457–465.
Lloyd-Hughes, B., 2014. The impracticality of a universal drought definition. Theor. Appl. Climatol. 117, 607–611.
Masud, M.B., Khaliq, M.N., Wheater, H.S., 2017. Future changes to drought characteristics over the Canadian Prairie Provinces based on NARCCAP multi-RCM ensemble. Clim. Dynam. 48, 2685–2705.
McKee, T.B., Doesken, N.J., Kleist, J., 1993. The relationship of drought frequency and duration to time scales. In: Proceedings of the 8th Conference on Applied Climatology, pp. 179–183.
El Mercurio, 2018. Crisis del agua en Petorca: "los requerimientos hídricos no se condicen con la disponibilidad". <https://goo.gl/bVDiiM> (online; last accessed Jul 2018).
Meroni, M., Rembold, F., Fasbender, D., Vrieling, A., 2017. Evaluation of the Standardized Precipitation Index as an early predictor of seasonal vegetation production anomalies in the Sahel. Remote Sens. Lett. 8, 301–310.
Mishra, A.K., Singh, V.P., 2010. A review of drought concepts. J. Hydrol. 391, 202–216.
Mishra, A.K., Ines, A.V.M., Das, N.N., Prakash Khedun, C., Singh, V.P., Sivakumar, B., et al., 2015. Anatomy of a local-scale drought: application of assimilated remote sensing products, crop model, and statistical methods to an agricultural drought study. J. Hydrol. 526, 15–29.
Nalbantis, I., Tsakiris, G., 2009. Assessment of hydrological drought revisited. Water Resour. Manage. 23, 881.
Narasimhan, B., Srinivasan, R., 2005. Development and evaluation of Soil Moisture Deficit Index (SMDI) and Evapotranspiration Deficit Index (ETDI) for agricultural drought monitoring. Agric. For. Meteorol. 133, 69–88.
Naumann, G., Alfieri, L., Wyser, K., Mentaschi, L., Betts, R.A., Carrao, H., et al., 2018. Global changes in drought conditions under different levels of warming. Geophys. Res. Lett.

Palmer, W., 1965. Meteorological drought. Research Paper No. 45. Department of Commerce, Washington, DC.

Palmer, W., 1968. Keeping track of crop moisture conditions, nationwide: the New Crop Moisture Index. Weatherwise 21, 156.

Prudhomme, C., Giuntoli, I., Robinson, E.L., Clark, D.B., Arnell, N.W., Dankers, R., et al., 2014. Hydrological droughts in the 21st century, hotspots and uncertainties from a global multimodel ensemble experiment. Proc. Natl. Acad. Sci. U.S.A 111, 3262–3267.

Quiring, S.M., Ganesh, S., 2010. Evaluating the utility of the Vegetation Condition Index (VCI) for monitoring meteorological drought in Texas. Agric. For. Meteorol. 150, 330–339.

Rubel, F., Brugger, K., Haslinger, K., Auer, I., 2017. The climate of the European Alps: Shift of very high resolution Köppen–Geiger climate zones 1800–2100. Meteorologische Zeitschrift 26, 115–125.

Rubilar, R., Blevins, L., Toro, J., Vita, A., Muoz, F., 2008. Early response of pinus radiata plantations to weed control and fertilization on metamorphic soils of the Coastal Range, Maule Region, Chile. Bosque (Valdivia) 29, 74–84.

Shafer, B.A., Dezman, L.E., 1982. Development of a Surface Water Supply Index (SWSI) to Assess the Severity of Drought Conditions in Snowpack Runoff Areas. Reno, Nevada.

Stagge, J.H., Tallaksen, L.M., Gudmundsson, L., Van Loon, A.F., Stahl, K., 2015. Candidate distributions for climatological drought indices (SPI and SPEI). Int. J. Climatol. 35, 4027–4040.

Stahl, K., 2001. Hydrological Drought – A Study Across Europe (Ph.D. thesis). Geosciences Faculty, Albert-Ludwigs-University of Freiburg.

Tannehill, I.R., 1947. Drought, its Causes and Effects, Vol. 64. Princeton University Press, New Jersey.

Tsakiris, G., Loukas, A., Pangalou, D., Vangelis, H., Tigkas, D., Rossi, G., et al., 2007. Chapter 7. Drought characterization. Options Méditerranéennes 58, 85–102.

Tucker, C.J., 1979. Red and photographic infrared linear combinations for monitoring vegetation. Remote Sens. Environ. 8, 127–150.

Van Loon, A.F., 2015. Hydrological drought explained. WIREs Water 2, 359.

Van Loon, A.F., Van Lanen, H.A.J., 2013. Making the distinction between water scarcity and drought using an observation-modeling framework. Water Resour. Res. 49, 1483–1502.

Van Loon, A.F., Gleeson, T., Clark, J., Van Dijk, A.I.J.M., Stahl, K., Hannaford, J., et al., 2016a. Drought in the anthropocene. Nat. Geosci. 9, 89–91.

Van Loon, A.F., Stahl, K., Di Baldassarre, G., Clark, J., Rangecroft, S., Wanders, N., et al., 2016b. Drought in a human-modified world: reframing drought definitions, understanding, and analysis approaches. Hydrol. Earth Syst. Sci. 20, 3631–3650.

Vicente-Serrano, S.M., 2006. Differences in spatial patterns of drought on different time scales: an analysis of the Iberian Peninsula. Water Resour. Manage. 20, 37–60.

Vicente-Serrano, S.M., López-Moreno, J.I., 2005. Hydrological response to different time scales of climatological drought: an evaluation of the Standardized Precipitation Index in a mountainous Mediterranean basin. Hydrol. Earth Syst. Sci. 9, 523–533.

Vicente-Serrano, S.M., Beguería, S., López-Moreno, J.I., 2010. A multiscalar drought index sensitive to global warming: the Standardized Precipitation Evapotranspiration Index. J. Clim. 23, 1696–1718.

WMO, 2012. Standardized Precipitation Index User Guide (M. Svoboda, M. Hayes and D. Wood). WMO-No. 1090. World Meteorological Organization. Geneva. Available online on <http://www.wamis.org/agm/pubs/SPI/WMO_1090_EN.pdf> (last accessed October 2016).

WMO-GWP, 2016. Handbook of Drought Indicators and Indices (M. Svoboda and B.A. Fuchs). WMO-No. 1173. Integrated Drought Management Programme (IDMP), Integrated Drought Management Tools and Guidelines Series 2. World Meteorological Organization (WMO) and Global Water Partnership (GWP). Geneva. Available online on <http://library.wmo.int/pmb_ged/wmo_1173_en.pdf> (last accessed October 2016).

Wang, G., 2005. Agricultural drought in a future climate: results from 15 global climate models participating in the IPCC 4th assessment. Clim. Dyn. 25, 739–753.

Wilhite, D.A., 2000. Chapter 1. Drought as a Natural Hazard: Concepts and Definitions. Paper 69. volume 1. Drought Mitigation Center Faculty Publications, New York, vol 1 edition.

Wilhite, D.A., Glantz, M.H., 1985. Understanding: the drought phenomenon: the role of definitions. Water Int. 10, 111–120.

Woldemeskel, F.M., Sivakumar, B., Sharma, A., 2013. Merging gauge and satellite rainfall with specification of associated uncertainty across Australia. J. Hydrol. 499, 167–176.

Yevjevich, V.M., 1967. An objective approach to definitions and investigations of continental hydrologic droughts. In: Hydrology Papers (Colorado State University); no. 23.

Zambrano, F., Lillo-Saavedra, M., Verbist, K., Lagos, O., 2016. Sixteen years of agricultural drought assessment of the BioBío region in Chile using a 250 m resolution Vegetation Condition Index (VCI). Remote Sens. 8, 530.

Zambrano, F., Wardlow, B., Tadesse, T., Lillo-Saavedra, M., Lagos, O., 2017. Evaluating satellite-derived long-term historical precipitation datasets for drought monitoring in Chile. Atmospheric Res. 186, 26–42.

Zambrano-Bigiarini, M., hydroTSM: time series management, analysis and interpolation for hydrological modelling. In: R Package Version 0.5-1. <https://doi.org/10.5281/zenodo.83964>.

Zambrano-Bigiarini, M., Nauditt, A., Birkel, C., Verbist, K., Ribbe, L., 2017. Temporal and spatial evaluation of satellite-based rainfall estimates across the complex topographical and climatic gradients of Chile. Hydrol. Earth Syst. Sci. 21, 1295–1320.

Zargar, A., Sadiq, R., Naser, B., Khan, F.I., 2011. A review of drought indices. Environ. Rev. 19, 333–349.

Zhang, F., Zhang, Lw, Wang, Xz, Hung, Jf, 2013. Detecting agro-droughts in southwest of China using MODIS satellite data. J. Integr. Agric. 12, 159–168.

Zhao, Y., Feng, D., Yu, L., Wang, X., Chen, Y., Bai, Y., et al., 2016. Detailed dynamic land cover mapping of Chile: accuracy improvement by integrating multi-temporal data. Remote Sens. Environ. 183, 170.

CHAPTER TEN

Recent advances in remote sensing of precipitation and soil moisture products for riverine flood prediction

Stefania Camici[1], Wade T. Crow[2] and Luca Brocca[1]
[1]Research Institute for Geo-Hydrological Protection of the National Research Council (CNR-IRPI), Perugia, Italy
[2]USDA ARS Hydrology and Remote Sensing Laboratory, Beltsville, MD, United States

10.1 Introduction

Every year, extensive economic and social damages are caused by floods worldwide (e.g., Willner et al., 2018). This issue is compounded by the recent intensification of the frequency and magnitude of extreme events (e.g., heavy precipitation) likely due to climate change (e.g., Hirabayashi et al., 2013; Camici et al., 2014; Wake, 2014; Alfieri et al., 2017). Future projections of general and regional circulation models suggest further increases in extreme precipitation frequency in high-emission scenarios. This needs to be carefully considered, knowing that, on one hand, models tend to be on the low side of observations (Hirsch and Archfield, 2015), and, on the other hand, there is an increase in the vulnerability and exposure of assets to flood risk (Jongman et al., 2014). The impact of flood risk on society is well known and the need for developing new mitigation approaches is evident (e.g., Pollner et al., 2010; Aghakouchak et al., 2014).

Remote sensing provides different datasets and products useful for improving our capability of monitoring and predicting floods (e.g., Li et al., 2016; Brocca et al., 2017b). Throughout this chapter the term flood will be used in the context of simulating and forecasting (some hours or days ahead) river discharge at a given cross-section. Specifically, the use of remote sensing, coupled with simplified and/or physically based rainfall-runoff models, for simulating the flood hydrograph is explored. The term

Extreme Hydroclimatic Events and Multivariate Hazards in a Changing Environment.
DOI: https://doi.org/10.1016/B978-0-12-814899-0.00010-9

flood will not refer to flooded areas and the use of remote sensing, e.g., optical and Synthetic Aperture Radar (SAR) images, for monitoring inundated areas will be discussed in Chapter 12. Similarly, coastal flooding due to storm surge is not considered here. Two major hydrological variables for the prediction of floods, that is, precipitation and soil moisture are discussed in this chapter. Certainly, remote sensing has been providing information also on other variables that can be used for flood prediction, such as evapotranspiration, land surface temperature, and river discharge (e.g., Silvestro et al., 2013; Tarpanelli et al., 2017). However, these variables will not be treated in this chapter.

Two articles review the use of satellite precipitation (Maggioni and Massari, 2018) and satellite precipitation and soil moisture (Li et al., 2016) for flood modeling and forecasting. Specifically, Maggioni and Massari (2018) reviewed a number of studies using different satellite precipitation products (SPPs) for flood prediction. The authors describe the source of errors and uncertainty of SPPs and review their performance in flood prediction across continents. Maggioni and Massari (2018) concluded that current SPPs are still inadequate for operational flood prediction and highlighted two possible research directions for improving their use in flood prediction: (1) identify the conditions in which SPPs are successful; and (2) explore the possibility of merging SPPs with ancillary information, like soil moisture (Chen et al., 2014; Ciabatta et al., 2016; Massari et al., 2018). Li et al. (2016) mainly focused on the potential use of SPPs and satellite soil moisture products (SSMPs) for operational flood forecasting. While providing an overview of different SPPs and SSMPs employed in the context of floods, the authors highlighted a number of challenges and opportunities to improve their use in flood simulation. For instance, the correction of the time-varying bias between remote sensing products and ground/modeling observations/states, the need of improving modeling structure, and the potential of merging multiple satellite products are among the challenges to be addressed for a better use of remote sensing in operational flood forecasting. The interested reader should refer to these review papers for a comprehensive overview of the recent scientific literature on the topic. This chapter benefits from these review articles and highlights the most recent studies employing SPPs and SSMPs for flood forecasting. The major challenges to be addressed for overcoming the current limitations of using satellite products for flood forecasting will also be highlighted.

Table 10.1 lists the currently (freely) available SPPs and SSMPs. The list is by no means complete and is only meant to provide an overview of

Table 10.1 List of the satellite products for precipitation and soil moisture most frequently used for flood applications.

Product name	References	Space coverage	Period	Space/time sampling
Satellite precipitation products				
TMPA 3B42RT	Huffman et al. (2007)	± 50°	1997–2017	0.25°/3 h
GPM IMERG	Hou et al. (2014)	± 60°	2014–ongoing	0.1°/30 min
CMORPH v0.x	Joyce et al. (2004)	± 60°	2011–ongoing	0.073°/3 h
PERSIANN CCS	Sorooshian et al. (2000)	± 60°	2003–ongoing	0.04°/1 h
GSMaP-NRT	Ushio et al. (2009)	± 60°	2008–ongoing	0.1°/1 h
H SAF H03	Mugnai et al. (2013)	Europe/Africa	2010–ongoing	8 km/15 min
SM2RAIN-CCI	Ciabatta et al. (2018)	Global (land)	1998–2015	0.25°/1 day
SM2RAIN-ASCAT	Brocca et al. (2017b)	Global (land)	2007–ongoing	12.5 km/1 day
Satellite soil moisture products				
SMOS	Kerr et al. (2001)	Global (land)	2010–ongoing	0.25°/2–3days
ESA CCI SM	Dorigo et al. (2017)	Global (land)	1978–ongoing	0.25°/1 day
SMAP	Entekhabi et al. (2010)	Global (land)	2015–ongoing	9 km/2–3 days
AMSR2-JAXA	Koike et al. (2004)	Global (land)	2012–ongoing	0.25°/1 day
AMSR2-LPRM	Owe et al. (2008)	Global (land)	2012–ongoing	0.25°/1 day
ASCAT	Wagner et al. (2013)	Global (land)	2007–ongoing	12.5 km/1 day
SMOS-BEC	Piles et al. (2014)	Spain	2010–ongoing	1 km/2–3 days
SENTINEL-1	Bauer-Marschallinger et al. (2018a)	Europe	2014–ongoing	500 m/3–6 days
SCATSAR	Bauer-Marschallinger et al. (2018b)	Europe	2014–ongoing	500 m/1 day

All the products are freely available. For some of the products, different spatiotemporal resolutions and coverage are available. The datasets currently available with the better spatial and temporal resolution are listed (for the acronyms see text).

the most widely used satellite products in the scientific literature. Specifically, SPPs are available starting from 1998, with nowadays 30-min temporal and 0.1° spatial sampling through the last Global Precipitation Measurement (GPM) mission, as thoroughly discussed in Chapter 1 of this book. The three most widely used SPPSs are: (1) the Tropical Rainfall Measurement Mission (TRMM) Multisatellite Precipitation Analysis (TMPA); (2) the Climate Prediction Center (CPC) MORPHing technique (CMORPH); and (3) the Precipitation Estimation from Remotely Sensed Information using Artificial Neural Network (PERSIANN). SSMPs have been available since 1978 with daily temporal resolution and mostly ~25 km spatial sampling. Currently, four quasi-operational coarse resolution SSMPs are available: (1) the Soil Moisture Active and Passive (SMAP) mission; (2) the Advanced Microwave Scanning Radiometer 2 (AMSR2); (3) the Soil Moisture and Ocean Salinity (SMOS) mission; and (4) the Advanced SCATterometer (ASCAT).

A large number of precipitation and soil moisture satellite products is available (for free) with global (or quasi-global) coverage and over long time periods. Therefore, there is a strong need to fully understand what their role for improving flood prediction might be. The next two sections underline the main challenges and future research directions aimed at fully exploiting the spatially distributed nature of satellite observations.

10.2 Satellite precipitation products

Precipitation is the most important variable driving the generation of floods and its correct estimation in space and time is critical for obtaining reliable predictions of river discharge. Satellite observations might support the estimation of precipitation, especially in poorly gauged regions.

10.2.1 Challenges

Among the different scientific articles evaluating the hydrological performance of SPPs (see reviews by Serrat-Capdevila et al., 2014 and Maggioni and Massari, 2018), several studies compare the performances of different products in order to select the best one to be used for flood prediction. This section discusses some challenges related to this approach, particularly with the use of SPPs for near real-time flood applications (where near real-time is defined here as a timeliness of less than 1 day).

10.2.1.1 Gauge-corrected versus satellite-only products

Gauge-corrected and satellite-only products are two different kinds of data that have different applications. As gauge-corrected products are not available in near real-time, satellite-only products represent a valid option for operational applications. It is important to underline the inherent difference between these two kinds of products in order to avoid misunderstandings when used by the stakeholders. Some satellite produces are bias corrected using gauge observations. Comparisons between gauge-corrected and satellite-only products with respect to ground-based rain-gauge observations would be unfair if the reference dataset was the same used in the bias correction process. In other words, the gauge-corrected satellite product and the benchmark would not be independent and the superior performances of the gauge-corrected products with respect to the satellite-only ones would be obvious.

10.2.1.2 Differences in gauge correction approaches

Gauge correction of satellite products has been performed through very different approaches. An illustrative example is the availability of three different CMORPH V1.0 products: (1) the raw satellite-only precipitation product (CMORPH RAW); (2) the bias corrected product (CMORPH CRT); and (3) the satellite-gauge blended product (CMORPH BLD). CMORPH RAW is obtained by using satellite observations only. CMORPH CRT is generated through adjusting the RAW product against the CPC unified daily gauge analysis using the probability density function matching bias correction method. CMORPH CRT is further combined with the gauge analysis through an optimal interpolation (OI) technique to generate CMORPH BLD (Joyce et al., 2004). Therefore, CMORPH CRT uses raingauge to adjust RAW precipitation at annual (or monthly) scale, but the correction is made once and the product is potentially available in near real-time. CMORPH BLD requires daily or monthly gauge analyses, that is, BLD product is not available in near real-time, and the product is corrected more heavily than CRT towards rain gauges. In other words, CMORPH CRT is mostly coming from satellite observations, while the opposite is valid for CMORPH BLD (closely related to raingauge observations). Other SPPs (e.g., TMPA and

PERSIANN) follow similar algorithms. The difference in SPPs is of paramount importance in flood prediction.

10.2.1.3 Near real-time versus climatic applications

Different applications require different type of products, that is, for flood and landslide simulation near real-time products are needed, whereas climate studies require long-term and high quality products. As mentioned above, the correction of satellite products with raingauge might prevent their use in near real-time applications, as the gauge-corrected products are usually available with a delay greater than 15 days. Therefore, it is meaningless to test gauge-corrected products for flood (and landslide) applications for which the availability of near real-time observations is mandatory. Conversely, climatic studies could benefit from the exploitation of the high quality gauge-corrected products. Further investigations on how to fully take advantage of satellite observations in order to effectively improve near real-time predictions and climatic applications are encouraged.

10.2.2 Future directions

Based on the above challenges, an effort should be made to investigate the potential benefit of using satellite-only precipitation observations for flood prediction and forecasting. Indeed, this kind of product is surely independent from raingauge observations and they are potentially available in near real-time to be used in an operational context. The following research directions are suggested in this respect.

10.2.2.1 Added-value of satellite precipitation products to high-quality ground-based observations

Most of the studies published in the scientific literature have compared the performance of satellite-only and satellite-corrected precipitation products with respect to high-quality ground-based raingauge observations in reproducing discharge (e.g., Behrangi et al., 2011; Alazzy et al., 2017). However, in an operational context, it is expected that high-quality, near real-time observations were available. Therefore, the questions to be addressed should be: can SPPs provide an added value to high-quality ground-based observations? how to optimally merge SPPs and high-quality ground-based observations for improving flood prediction? Even high-quality observations are affected by errors and SPPs might be used to further improve their quality for flood forecasting. To answer this research question, satellite-only products should be used to avoid interdependency with gauge data and the final target should be the optimal representation of river discharge. This topic has been rarely investigated (e.g., Ciabatta et al., 2016), but it is expected to have a major impact in the context of an operational flood warning systems.

10.2.2.2 Performance metrics for assessing the quality of satellite precipitation products in estimating floods

Classical continuous (e.g., root mean square error and Pearson's correlation) and categorical (e.g., probability of detection) scores are frequently used for assessing the quality of SPPs. In the context of floods, SPPs are often evaluated for different ranges of precipitation (e.g., greater than 15 mm/day) with the target of their evaluation for high rainfall events reproduction and, hence, for flood prediction (e.g., Jiang et al., 2018). However, a detailed study comparing the performance in terms of rainfall and floods over a large number of basins has never been carried out. For this purpose, we have compared the performance of rainfall products and flood simulations in 975 basins throughout Europe. Specifically, we analyzed the performance in terms of rainfall (against the ground-based European daily high-resolution gridded datasets of precipitation, E-OBS, Haylock et al., 2008) and discharge (against stream gauge observations) of three different SPPs: TMPA 3B42RT, CMORPH-RAW, and PERSIANN. The "Modello idrologico Semi-Distribuito di tipo continuo" (MISDc, Brocca et al., 2011) rainfall-runoff model was used for predicting river discharge from rainfall and air temperature observations. Rainfall performance scores (R: Pearson's correlation; KGE: Kling-Gupta efficiency; RMSE: root mean square error; and BIAS) were used as predictors to explain the performance in terms of flood simulations. Results have revealed that a single statistical score is not enough for assessing the quality of SPPs for flood prediction (Fig. 10.1) and a combination of scores should be preferred. Specifically, for the basins and products analyzed in this example, the best performances in terms of floods are obtained for rainfall products with the following range of scores computed against E-OBS dataset: BIAS lower than 20%, R higher than 0.6, and KGE higher than 0.5. This kind of analysis helps to identify which product, climate condition, and morphological setting would be more favorable to obtain improvements in flood prediction through SPPs.

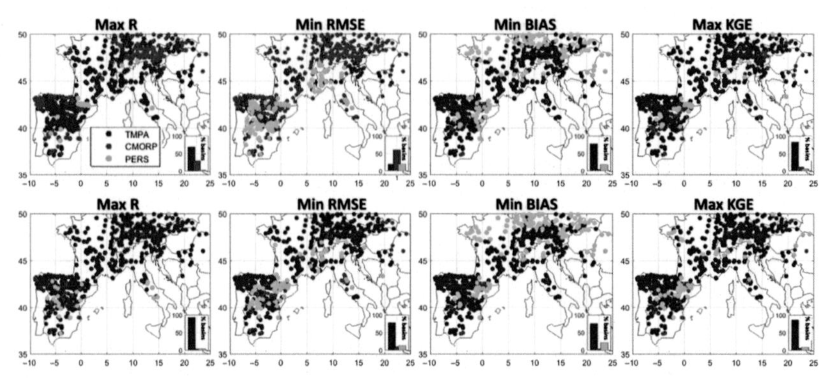

Figure 10.1 Best performing SPP in terms of rainfall (upper panels) and discharge (lower panels). Each dot has a different color as a function of the best SPP (blue = TMPA, red = CMORPH, green = PERSIANN) for the different performance scores. Except for the BIAS, the other scores do not show a consistent spatial pattern for rainfall and discharge, thus suggesting that a combination of scores should be considered *R*, Pearson's correlation; *RMSE*, root mean square error; *KGE*, Kling–Gupta efficiency; *SPPs*, satellite precipitation products.

10.2.2.3 Spatial assessment of satellite precipitation products

The major contribution of SPPs for flood simulation is not related to their capability of reproducing the temporal variability of precipitation, but rather the spatial variability. Due to time-varying bias frequently observed in SPPs, their "temporal quality" is usually found to be unsatisfactory. State-of-the-art SPPs are based on instantaneous measurements collected at the satellite overpass time. The huge temporal variability of rainfall prevents the possibility of obtaining accurate rainfall accumulations based on instantaneous measurements, unless a very good temporal coverage is available. This aspect is expected to be the main reason for the superior performance of gauge-based rainfall products (more stable in time) with respect to SPPs usually obtained in the scientific literature (e.g., Maggioni and Massari, 2018). However, similarly to what is usually done with meteorological radar rainfall estimates, SPPs can be used to spatially distribute the point information provided by rain gauges. This aspect can be investigated either by using a high-density ground-based network of rainfall stations as benchmark, or (more interestingly) through a spatially distributed network of river discharge stations. The application of a distributed rainfall-runoff model to multiple discharge stations could reveal, above all for larger basins, the added value of SPPs in reproducing rainfall spatial patterns. If merging ground-based and SPPs provides improved performance in terms of floods, one can expect a better assessment of rainfall spatial variability. To our knowledge, this topic has not been investigated in detail yet.

10.3 Satellite soil moisture

Soil moisture is the main state variable of rainfall-runoff models and it must be properly assessed for predicting the flood response of a basin to a given rainfall event. For low antecedent soil moisture conditions, even a large rainfall event might have no effect in terms of floods. Differently, medium to high antecedent soil moisture conditions could result in catastrophic flood conditions, even for moderate to high rainfall events (e.g., Brocca et al., 2009; Massari et al., 2014b; Crow et al., 2017).

10.3.1 Challenges

Two different approaches have been employed for using SSMPs in flood prediction. The first approach relies on the use of SSMPs for directly assessing the initial soil moisture conditions before the occurrence of a flood event (Brocca et al., 2009; Massari et al., 2014a; Crow et al., 2017). The second approach is based on data assimilation: SSMPs are assimilated into a rainfall-runoff model and the model performance with and without the use of SSMPs is assessed (e.g., Brocca et al., 2012; Massari et al., 2015; Chen et al., 2014; and Table 4 in Li et al., 2016).

10.3.1.1 Antecedent wetness conditions through satellite soil moisture products

The value of SSMPs as a representation of basin antecedent wetness condition (AWC) is potentially limited by multiple factors. Most notably, SSMPs: (1) lack vertical sampling of the soil column beyond the first few centimeters; (2) are restricted to relatively coarser spatial resolutions; and (3) are subject to significant retrieval errors over dense vegetation. Objective assessments of SSMPs are needed to clarify the impact of these limitations in flood forecasting. Given large remaining challenges in the implementation of effective land data assimilation practices (see discussion below), these assessments are best provided by direct comparisons of SSMPs and AWC (and/or storm-scale rainfall response).

Such comparisons provide an objective means to evaluate the flood forecasting value of various soil moisture (or soil moisture proxy) data products. For example, Crow et al. (2017) sampled the rank correlation between pre-storm AWC estimates, acquired from a range of soil moisture products, and the storm-scale runoff coefficients (RC), defined as the fraction of rainfall volume converted into fast runoff volume during a storm event. For large rainfall events (>25 mm/day) within a series of medium-scale (2000–10,000 km^2) basins within the south-central United States, Crow et al. (2017) demonstrated that pre-storm 9-km surface (0–5 cm) soil moisture estimates extracted from the SMAP Level 4 Surface and Root-Zone (SMAP L4SM) soil moisture product provide a better predictor of storm-scale RC than surface soil moisture products derived from: (1) older SSMPs; (2) surface soil moisture products derived from land surface modeling; and (3) pre-storm stream gauge measurements (Fig. 10.2). This type of assessment provides an objective way to prioritize the implementation of various SSMPs in flood forecasting systems. In particular, it stresses that, despite their limitations, SSMPs have a viable role in operational flood forecasting efforts.

In addition, SSMPs can be used to evaluate the relationship between AWC and RC predicted by hydrologic models. Crow et al. (2018) demonstrated that the observed correlation between AWC (acquired from the SMAP L4SM) and RC (acquired from high-quality rain and stream flow gauges) is commonly stronger than the correlation between comparable soil moisture and runoff predictions acquired from a large-scale hydrologic model. This suggests that many existing models are underrepresenting the role of AWC on subsequent basin-scale runoff response, and therefore, neglecting a viable source of flood predictability.

10.3.1.2 Data assimilation of satellite soil moisture products: a complex recipe?

The challenges in using SSMPs in riverine flood prediction have been already summarized in Massari et al. (2015) and Li et al. (2016). Particularly, we believe that the full characterization of the spatial/temporal variability of the error in SSMPs can benefit their use in rainfall-runoff modeling. Several previous studies considered a constant SSMP error in time and/or space (e.g., Brocca et al., 2012), which could be a difficult assumption to relax. Similarly, it is difficult to quantify the modeling error. For example, in an Ensemble Kalman Filter (EnKF, Evensen, 2009) approach, the model error is usually obtained by randomly perturbing the

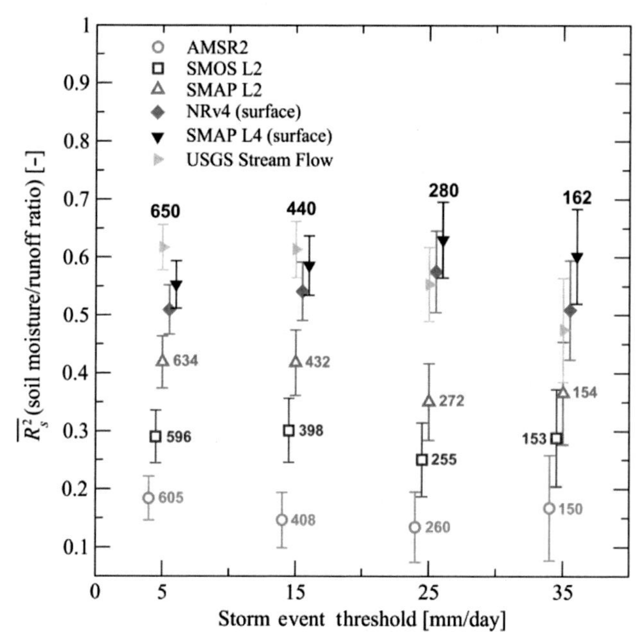

Figure 10.2 For 16 medium-scale (2000–10,000 km2) basins located in the south-central United States, the rank correlation (R^2) between various pre-storm proxies and storm-scale RCs (i.e., the fraction of rainfall volume converted into runoff). "NRv4" represents a surface soil moisture product derived (solely) from a land surface modeling product and "USGS" results are based on the use of pre-storm streamflow observations as a soil moisture proxy. Results are binned based to the daily rainfall intensity threshold used to define an event. Error bars represent 95% confidence intervals associated with temporal sampling uncertainty. Numbers represent the total number of events sampled during the 2015–17 study period. *Adapted from Crow, W. T., Chen, F., Reichle, R.H., Liu, Q., 2017. L band microwave remote sensing and land data assimilation improve the representation of prestorm soil moisture conditions for hydrologic forecasting. Geophys. Res. Lett., 44 (11), 5495–5503.*

model forcing data, states, and/or parameters. However, the entity, type, and distribution of those perturbations are often subjective.

A recent example of SSMP assimilation, obtained from ASCAT for flood modeling in the Mediterranean region has been published by Loizu et al. (2018), where two different rainfall-runoff models (MISDc and TOPLATS, Brocca et al., 2011; Peters-Lidard et al., 1997) have been applied to two catchments in Spain and Italy. Fig. 10.3 shows the summary of the results for the two catchments and models in terms of Nash–Sutcliffe Efficiency (NSE) as a function of the assumed observation error in the satellite observations. When using MISDc, a slight

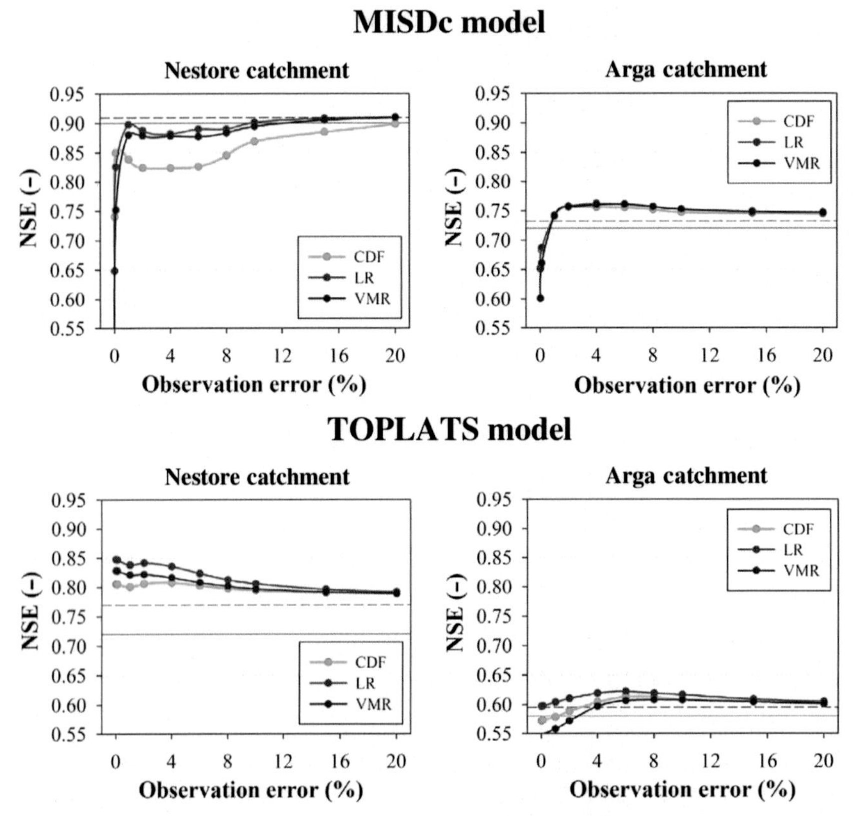

Figure 10.3 NSE as a function of the assumed error in satellite soil moisture observations for the Nestore (Italy) and Arga (Spain) catchments and by using the MISDc and TOPLATS models; results are shown for three different rescaling techniques (CDF: CDF matching; LR: linear rescaling; VMR: variance matching rescaling). The solid horizontal line indicates the model results in the validation period, and the dotted line indicates the results of the open loop. For MISDc model, a slight improvement in the results can be obtained only for the Arga catchment as the model performance for Nestore are very good and difficult to improve. For TOPLATS, the model performances are generally lower, and a larger benefit in the assimilation is obtained. *NSE*, Nash–Sutcliffe Efficiency. *Adapted from Loizu, J., Massari, C., Alvarez-Mozos, J., Tarpanelli, A., Brocca, L., Casali, J., 2018. On the assimilation set-up of ASCAT soil moisture data for improving streamflow catchment simulation. Adv. Water Resour. 111, 86–104.*

improvement can be obtained in the Spanish catchment, whereas the model performance in the Italian catchment is very good to start with and thus difficult to improve. The TOPLATS performance is generally poorer and benefits from data assimilation (i.e., in the Italian catchment NSE increases from 0.71 to 0.85 after the assimilation of SSMP). These results, together with those highlighted by Massari et al. (2015) and Li et al. (2016), clearly underline the need of continuing and advancing research activities on this topic to provide guidelines for the optimal use of SSMPs in a data assimilation framework for flood modeling (see also Brocca et al., 2017a).

10.3.2 Future directions

10.3.2.1 Multiple basins and products assessment

Mostly due to limitations in the availability of in situ discharge observations and the short record of some of the SSMPs (e.g., SMAP was launched in 2015), several past studies using SSMPs for flood prediction have considered a single product for a few (<15) river basins, or different products (e.g., SMOS, SMAP, and ASCAT) for a single basin. To systematically assess the quality and the potential of SSMPs, a comprehensive evaluation over multiple basins with multiple SSMPs is required. Additionally, the use of multiple hydrological models and data assimilation techniques should be tested. Some light should be shed on the contrasting results frequently obtained in the scientific literature (Massari et al., 2015, Brocca et al., 2017b), providing guidelines for the optimal exploitation of SSMPs for flood prediction.

10.3.2.2 New high resolution satellite soil moisture products

The recent launch of Sentinel-1 satellites, with currently two satellites in orbit (launched in April 2014 and April 2016), opens new challenges in the use of SSMPs in flood prediction. For the first time in history, satellite-based soil moisture observations at high spatial resolution (e.g., 500 m, Bauer-Marschallinger et al., 2018a) and good revisit time (3 to 6 days in Europe) have become available. The added value of this new kind of soil moisture measurements is still under investigation, as the first large-scale Sentinel-1 soil moisture products have become available only in 2018. We believe that the 1-km Sentinel-1 SSMPs have a large potential in enhancing our understanding of hydrological model accuracy in reproducing soil moisture spatial variability. Studies published so far on this topic have obtained very unreliable and unsatisfactory results (Brocca et al., 2017a).

10.3.2.3 Integration of satellite soil moisture and precipitation

Most past studies have considered the use of SPPs and SSMPs separately. Some authors have recently demonstrated that the integration of SPPs with soil moisture information obtained from satellite products improved the quality of the final precipitation estimates over land (e.g., Crow et al., 2011; Pellarin et al., 2013; Brocca et al., 2014; Bauer-Marschallinger et al., 2018b; Bhuiyan et al., 2018). Based on this concept, scientists have started integrating SPPs and SSMPs for improving flood prediction.

Chen et al. (2014) used SSMPs (ASCAT and SMOS) for correcting both model states (through data assimilation, EnKF) and rainfall (through SMART algorithm, Crow et al., 2011). They applied such an approach to real-world data in 13 basins in the central United States by using the real-time version of TMPA (TMPA-RT) as rainfall forcing and the Sacramento hydrological model (SAC, Burnash et al., 1973) for rainfall-runoff transformation. Results showed that the rainfall correction was able to improve streamflow predictions during high-flow periods, whereas the state correction was more effective under low-flow conditions. Alvarez-Garreton et al. (2016) applied the same approach and methods to four large basins in Australia, while Massari et al. (2014b) performed a similar analysis by using in situ precipitation and soil moisture data for a small basin in southern France. Ciabatta et al. (2016) merged soil moisture derived rainfall data, through SM2RAIN algorithm (Brocca et al., 2014), with in situ and satellite rainfall (TMPA-RT) to obtain a higher quality rainfall product to be used for flood prediction in four basins in Italy. Results revealed that the use of SSMPs improves both rainfall estimation (better agreement with in situ rainfall data) and flood simulation (improved performance of discharge simulations, see also Camici et al., 2018). Specifically, the improved performances were ascribed to the better stability over time of the soil moisture corrected precipitation product with respect to the original TMPA-3B42RT.

A recent study by Massari et al. (2018) compared the performance of state correction (through data assimilation) versus rainfall correction (through SM2RAIN derived rainfall) in flood simulations. Fig. 10.4 shows the box plot of discharge performances, in terms of ANSE (NSE for high-flow conditions), across 15 basins located in the Mediterranean area by using MISDc model forced with TMPA-RT as SPP. The state correction slightly improves the results, while rainfall correction provides a much larger benefit for high-flow simulations (close to the simulation obtained with E-OBS ground-based dataset). Results are different if low

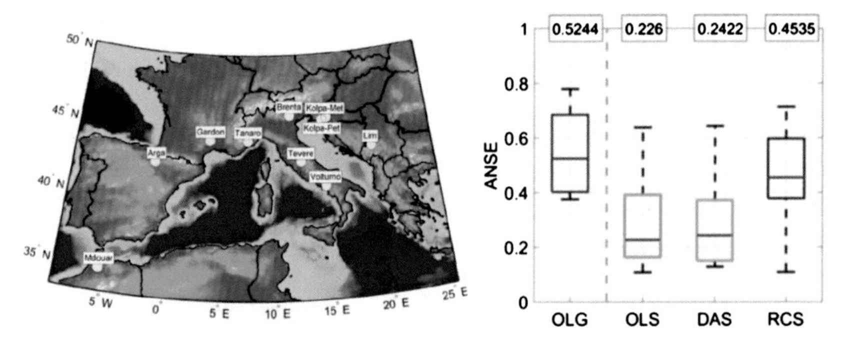

Figure 10.4 Left: Location of the investigated catchments in the Mediterranean area. Right: Summary of the results in terms of ANSE (Nash–Sutcliffe efficiency for high-flow conditions) for all the investigated basins during the validation period. Red box plots refer to the results obtained by forcing MISDc with E-OBS datasets (OLG), the number in the square boxes represent the median values. Results are shown for the simulation forced with TMPA-RT (OLS), with ASCAT soil moisture assimilation (DAS) and with ASCAT rainfall correction (RCS). *Adapted from Massari, C., Camici, S., Ciabatta, L., Brocca, L. (2018). Exploiting satellite-based surface soil moisture for flood forecasting in the Mediterranean area: state update versus rainfall correction. Remote Sens., 10(2), 292.*

flow simulations are considered, with better performance of state correction, even though the differences are small.

These studies clearly highlight the benefits of integrating SPPs and SSMPs for flood prediction. Specifically, it has been found that the integration is able to improve rainfall estimation by exploiting the more stable temporal bias of soil moisture derived rainfall product and the higher skill of state-of-the-art SPPs in reproducing high-intensity rainfall events (e.g., Chiaravalloti et al., 2018). This topic needs further investigations.

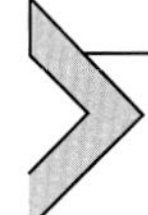

10.4 Towards fully exploiting satellite soil moisture and precipitation products

This chapter briefly reviewed the use of SPPs and SSMPs for flood prediction. Starting from the recent review papers by Maggioni and Massari (2018) and Li et al. (2016), we have identified the challenges and future research directions for fully exploiting satellite products in flood simulations. Overall, the main challenge that should be addressed by hydrologists is the optimal use of observations. Hydrologists are traditionally used to work at small to medium spatial scales (<500–1000 km^2) and only recently "global hydrology" has been emerging as a field of study. At small scales, the temporal dimension of observations, e.g., point-scale hourly rainfall time series, is predominant. Generally, there are few measurements in space (e.g., <20–30 rain gauges) and a lot of measurements in time (1 year equal to 8760 hourly observations). Conversely, satellite observations are more informative in the spatial dimension with respect to the temporal dimension. By taking Sentinel-1 soil moisture data as an example (Bauer-Marschallinger et al., 2018b), for a basin of 1000 km^2, 4000 values are available in each image (spatial resolution of 500 m) and around 100 images per year. Therefore, Sentinel-1 soil moisture observations will provide very detailed information on the soil moisture spatial pattern and less detail on the soil moisture temporal dynamics. As traditional hydrology is not used to working with this kind of data, an overhaul in the modeling tools and techniques for maximizing the use of satellite information in flood modeling is expected. For instance, Beck et al. (2017) evaluated the performance of eight SPPs over 9053 catchments worldwide by using a lumped rainfall-runoff model, thus not considering the capability of the products in reproducing rainfall spatial variability.

A second important aspect that has been underexplored so far is the use of SPPs and SSMPs in operational flood forecasting, that is, for predicting discharge hours and days ahead. A single study employing SSMPs has been published by Wanders et al. (2014), which should be followed up by further research.

Finally, scientists and end-users should be made aware of the availability and potential of satellite observations. People that need such data are often simply unaware of their existence. Software and tools that facilitate the downloading and analysis of satellite products should be developed and openly distributed. At the same time, there is a need for promoting collaboration and exchange within the remote sensing community (data developers) and the different communities of data users (e.g., hydrologists, agronomists, ecologists) in order to fully exploit satellite datasets in real-world applications. Capacity building is fundamental in order to globalize societal applications of satellite datasets and to build consensus on key questions and recommendations (Brocca et al., 2017b).

Acknowledgments

The authors would like to acknowledge the support of the European Space Agency (ESA) STSE WACMOS-MED project (contract n° 4000114770/15/I-SBo), the ESA STSE SMOS + rainfall project (contract n° 4000114738/15/I-SBo), and partly by the Italian Civil Protection Department.

References

Aghakouchak, A., Feldman, D., Stewardson, M.J., Saphores, J.D., Grant, S., Sanders, B., 2014. Australia's drought: lessons for California. Science 343 (6178), 1430–1431.

Alazzy, A.A., Lü, H., Chen, R., Ali, A.B., Zhu, Y., Su, J., 2017. Evaluation of satellite precipitation products and their potential influence on hydrological modeling over the Ganzi river basin of the Tibetan Plateau. Adv. Meteorol. 2017.

Alfieri, L., Bisselink, B., Dottori, F., Naumann, G., Roo, A., Salamon, P., et al., 2017. Global projections of river flood risk in a warmer world. Earth's Future 5 (2), 171–182.

Alvarez-Garreton, C., Ryu, D., Western, A.W., Crow, W.T., Su, C.H., Robertson, D. R., 2016. Dual assimilation of satellite soil moisture to improve streamflow prediction in data-scarce catchments. Water Resour. Res. 52 (7), 5357–5375.

Bauer-Marschallinger, B., Naeimi, V., Cao, S., Paulik, C., Schaufler, S., Stachl, T., et al., 2018a. Towards global soil moisture monitoring with Sentinel-1: harnessing assets and overcoming obstacles. submitted to IEEE Trans. Geosci. Remote Sens.

Bauer-Marschallinger, B., Paulik, C., Hochstöger, S., Mistelbauer, T., Modanesi, S., Ciabatta, L., et al., 2018b. Soil moisture from fusion of scatterometer and SAR: closing the scale gap with temporal filtering. Remote Sensing.

Beck, H.E., Vergopolan, N., Pan, M., Levizzani, V., van Dijk, A.I.J.M., Weedon, G., et al., 2017. Global-scale evaluation of 22 precipitation datasets using gauge observations and hydrological modeling. Hydrol. Earth Syst. Sci. 21, 6201–6217.

Behrangi, A., Khakbaz, B., Jaw, T.C., AghaKouchak, A., Hsu, K., Sorooshian, S., 2011. Hydrologic evaluation of satellite precipitation products over a mid-size basin. J. Hydrol. 397 (3-4), 225–237.

Bhuiyan, M.A.E., Nikolopoulos, E.I., Anagnostou, E.N., Quintana-Seguí, P., Barella-Ortiz, A., 2018. A nonparametric statistical technique for combining global precipitation datasets: development and hydrological evaluation over the Iberian Peninsula. Hydrol. Earth Syst. Sci. 22 (2), 1371.

Brocca, L., Melone, F., Moramarco, T., Morbidelli, R., 2009. Antecedent wetness conditions based on ERS scatterometer data. J. Hydrol. 364 (1-2), 73–87.

Brocca, L., Melone, F., Moramarco, T., 2011. Distributed rainfall-runoff modelling for flood frequency estimation and flood forecasting. Hydrol. Process. 25 (18), 2801–2813.

Brocca, L., Moramarco, T., Melone, F., Wagner, W., Hasenauer, S., Hahn, S., 2012. Assimilation of surface and root-zone ASCAT soil moisture products into rainfall-runoff modelling. IEEE Trans. Geosci. Remote Sens. 50 (7), 2542–2555.

Brocca, L., Ciabatta, L., Massari, C., Moramarco, T., Hahn, S., Hasenauer, S., et al., 2014. Soil as a natural rain gauge: estimating global rainfall from satellite soil moisture data. J. Geophys. Res. 119 (9), 5128–5141.

Brocca, L., Ciabatta, L., Massari, C., Camici, S., Tarpanelli, A., 2017a. Soil moisture for hydrological applications: open questions and new opportunities. Water 9 (2), 140.

Brocca, L., Crow, W.T., Ciabatta, L., Massari, C., de Rosnay, P., Enenkel, M., et al., 2017b. A review of the applications of ASCAT soil moisture products. IEEE J. Selected Topics Appl. Earth Observ. Remote Sens. 10 (5), 2285–2306.

Burnash, R.J.C., Ferral, R.L., McGuire, R.A., 1973. A generalized streamflow simulation system—Conceptual modeling for digital computers. Tech. Rep., Joint Federal State River Forecast Center. Sacramento, CA, 204 pp.

Camici, S., Brocca, L., Melone, F., Moramarco, T., 2014. Impact of climate change on flood frequency using different climate models and downscaling approaches. J. Hydrol. Eng. 19 (8), 04014002.

Camici, S., Ciabatta, L., Massari, C., Brocca, L., 2018. How reliable are satellite precipitation estimates for driving hydrological models: a verification study over the Mediterranean area. J. Hydrol. 563, 950–961.

Chen, F., Crow, W.T., Ryu, D., 2014. Dual forcing and state correction via soil moisture assimilation for improved rainfall–runoff modeling. J. Hydrometeorol. 15 (5), 1832–1848.

Chiaravalloti, F., Brocca, L., Procopio, A., Massari, C., Gabriele, S., 2018. Assessment of GPM and SM2RAIN-ASCAT rainfall products over complex terrain in southern Italy. Atmosph. Res. 206, 64–74.

Ciabatta, L., Brocca, L., Massari, C., Moramarco, T., Gabellani, S., Puca, S., et al., 2016. Rainfall-runoff modelling by using SM2RAIN-derived and state-of-the-art satellite rainfall products over Italy. Int. J. Appl. Earth Observ. Geoinform. 48, 163–173.

Ciabatta, L., Massari, C., Brocca, L., Gruber, A., Reimer, C., Hahn, S., et al., 2018. SM2RAIN-CCI: a new global long-term rainfall data set derived from ESA CCI soil moisture. Earth Syst. Sci. Data 10, 267–280.

Crow, W.T., van Den Berg, M.J., Huffman, G.J., Pellarin, T., 2011. Correcting rainfall using satellite-based surface soil moisture retrievals: The Soil Moisture Analysis Rainfall Tool (SMART). Water Resour. Res. 47 (8).

Crow, W.T., Chen, F., Reichle, R.H., Liu, Q., 2017. L band microwave remote sensing and land data assimilation improve the representation of prestorm soil moisture conditions for hydrologic forecasting. Geophys. Res. Lett. 44 (11), 5495–5503.

Crow, W.T., Chen, F., Reichle, R.H., Xia, Y., Liu, Q., 2018. Exploiting soil moisture, precipitation, and streamflow observations to evaluate soil moisture/runoff coupling in land surface models. Geophys. Res. Lett. 45.

Dorigo, W., Wagner, W., Albergel, C., Albrecht, F., Balsamo, G., Brocca, L., et al., 2017. ESA CCI soil moisture for improved earth system understanding: state-of-the art and future directions. Remote Sens. Environ. 203, 185–215.

Entekhabi, D., Njoku, E.G., Neill, P.E., Kellogg, K.H., Crow, W.T., Edelstein, W.N., et al., 2010. The soil moisture active passive (SMAP) mission. Proc. IEEE 98 (5), 704–716.

Evensen, G., 2009. Data Assimilation: The Ensemble Kalman Filter. Springer Science & Business Media.

Haylock, M.R., Hofstra, N., Klein Tank, A.M.G., Klok, E.J., Jones, P.D., New, M., 2008. A European daily high-resolution gridded data set of surface temperature and precipitation for 1950–2006. J. Geophys. Res.: Atmospheres 113 (D20).

Hirabayashi, Y., Mahendran, R., Koirala, S., Konoshima, L., Yamazaki, D., Watanabe, S., et al., 2013. Global flood risk under climate change. Nat. Climate Change 3 (9), 816–821.

Hirsch, R.M., Archfield, S.A., 2015. Flood trends: not higher but more often. Nat. Clim. Change 5 (3), 198–199.

Hou, A.Y., Kakar, R.K., Neeck, S., Azarbarzin, Aa, Kummerow, C.D., Kojima, M., et al., 2014. The global precipitation measurement mission. Bull. Am. Meteorol. Soc. 95, 701–722.

Huffman, G.J., Adler, R.F., Bolvin, D.T., Gu, G., Nelkin, E.J., Bowman, K.P., et al., 2007. The TRMM multisatellite precipitation analysis (TMPA): quasi-global, multiyear, combined-sensor precipitation estimates at fine scales. J. Hydrometeorol. 8 (1), 38–55.

Jiang, Q., Li, W., Wen, J., Qiu, C., Sun, W., Fang, Q., et al., 2018. Accuracy evaluation of two high-resolution satellite-based rainfall products: TRMM 3B42V7 and CMORPH in Shanghai. Water 10 (1), 40.

Jongman, B., Hochrainer-Stigler, S., Feyen, L., Aerts, J.C., Mechler, R., Botzen, W.W., et al., 2014. Increasing stress on disaster-risk finance due to large floods. Nat. Clim. Change 4 (4), 264–268.

Joyce, R.J., Janowiak, J.E., Arkin, P.A., Xie, P., 2004. CMORPH: a method that produces global precipitation estimates from passive microwave and infrared data at high spatial and temporal resolution. J. Hydromet. 5, 487–503.

Kerr, Y.H., Waldteufel, P., Wigneron, J.-P., Martinuzzi, J., Font, J., Berger, M., 2001. Soil moisture retrieval from space: the Soil Moisture and Ocean Salinity (SMOS) mission. IEEE Trans. Geosci. Remote Sens. 39, 1729–1735.

Koike, T., Nakamura, Y., Kaihotsu, I., Davaa, G., Matsuura, N., Tamagawa, K., et al., 2004. Development of an advanced microwave scanning radiometer (AMSR-E) algorithm for soil moisture and vegetation water content. Proc. Hydraul. Eng. 48, 217–222.

Li, Y., Grimaldi, S., Walker, J.P., Pauwels, V., 2016. Application of remote sensing data to constrain operational rainfall-driven flood forecasting: a review. Remote Sens. 8 (6), 456.

Loizu, J., Massari, C., Alvarez-Mozos, J., Tarpanelli, A., Brocca, L., Casali, J., 2018. On the assimilation set-up of ASCAT soil moisture data for improving streamflow catchment simulation. Adv. Water Resour. 111, 86–104.

Maggioni, V., Massari, C., 2018. On the performance of satellite precipitation products in riverine flood modeling: a review. J. Hydrol. 558, 214–224.

Massari, C., Brocca, L., Barbetta, S., Papathanasiou, C., Mimikou, M., Moramarco, T., 2014a. Using globally available soil moisture indicators for flood modelling in Mediterranean catchments. Hydrol. Earth Syst. Sci. 18, 839–853.

Massari, C., Brocca, L., Moramarco, T., Tramblay, Y., Didon Lescot, J.-F., 2014b. Potential of soil moisture observations in flood modelling: estimating initial conditions and correcting rainfall. Adv. Water Resour. 74, 44–53.

Massari, C., Brocca, L., Tarpanelli, A., Moramarco, T., 2015. Data assimilation of satellite soil moisture into rainfall-runoff modelling: a complex recipe? Remote Sens. 7 (9), 11403–11433.

Massari, C., Camici, S., Ciabatta, L., Brocca, L., 2018. Exploiting satellite-based surface soil moisture for flood forecasting in the Mediterranean area: state update versus rainfall correction. Remote Sens. 10 (2), 292.

Mugnai, A., Casella, D., Cattani, E., Dietrich, S., Laviola, S., Levizzani, V., et al., 2013. Precipitation products from the hydrology SAF. Natural Hazards Earth Syst. Sci. 13 (8), 1959–1981.

Owe, M., de Jeu, R.A.M., Holmes, T., 2008. Multisensor historical climatology of satellite-derived global land surface moisture. J. Geophys. Res. 113, F01002.

Pellarin, T., Louvet, S., Gruhier, C., Quantin, G., Legout, C., 2013. A simple and effective method for correcting soil moisture and precipitation estimates using AMSR-E measurements. Remote Sens. Environ. 136, 28–36.

Peters-Lidard, C.D., Zion, M.S., Wood, E.F., 1997. A soil-vegetation-atmosphere transfer scheme for modeling spatially variable water and energy balance processes. J. Geophys. Res. D Atmos. 102, 4303–4324.

Piles, M., Sánchez, N., Vall-llossera, M., Camps, A., Martinez-Fernandez, J., Martinez, J., et al., 2014. A downscaling approach for SMOS land observations: evaluation of high-resolution soil moisture maps over the Iberian Peninsula. IEEE J. Sel. Topics Appl. Earth Observ. 7 (9), 3845–3857.

Pollner, J., Kryspin-Watson, J., Nieuwejaar, S., 2010. Disaster Risk Management and Climate Change Adaptation in Europe and Central Asia. World Bank, Washington, DC.

Serrat-Capdevila, A., Valdes, J.B., Stakhiv, E., 2014. Water management applications for satellite precipitation products: synthesis and recommendations. J. Am. Water Resour. Assoc. 50, 509–525.

Silvestro, F., Gabellani, S., Delogu, F., Rudari, R., Boni, G., 2013. Exploiting remote sensing land surface temperature in distributed hydrological modelling: the example of the Continuum model. Hydrol. Earth Syst. Sci. 17 (1), 39–62.

Sorooshian, S., Hsu, K.L., Gao, X., Gupta, H.V., Imam, B., Braithwaite, D., 2000. Evaluation of PERSIANN system satellite–based estimates of tropical rainfall. Bull. Am. Meteorol. Soc. 81 (9), 2035–2046.

Tarpanelli, A., Amarnath, G., Brocca, L., Massari, C., Moramarco, T., 2017. Discharge estimation and forecasting by MODIS and altimetry data in Niger-Benue River. Remote Sens. Environ. 195, 96–106.

Ushio, T., Sasashige, K., Kubota, T., Shige, S., Okamoto, K.I., Aonashi, K., et al., 2009. A Kalman filter approach to the global satellite mapping of precipitation (GSMaP) from combined passive microwave and infrared radiometric data. J. Meteorol. Soc. Jpn. Ser. II 87, 137–151.

Wagner, W., Hahn, S., Kidd, R., Melzer, T., Bartalis, Z., Hasenauer, S., et al., 2013. The ASCAT soil moisture product: a review of its specifications, validation results, and emerging applications. Meteorologische Zeitschrift 22 (1), 5–33.

Wake, B., 2014. Hydrology: the rains are coming. Nat. Clim. Change 4 (3), 171.

Wanders, N., Karssenberg, D., Roo, A.D., De Jong, S.M., Bierkens, M.F.P., 2014. The suitability of remotely sensed soil moisture for improving operational flood forecasting. Hydrol. Earth Syst. Sci. 18 (6), 2343–2357.

Willner, S.N., Otto, C., Levermann, A., 2018. Global economic response to river floods. Nat. Clim. Change . Available from: https://doi.org/10.1038/s41558-018-0173-2.

Further reading

Brocca, L., Massari, C., Ciabatta, L., Moramarco, T., Penna, D., Zuecco, G., et al., 2015. Rainfall estimation from in situ soil moisture observations at several sites in Europe: an evaluation of SM2RAIN algorithm. J. Hydrol. Hydromech. 63 (3), 201–209.

CHAPTER ELEVEN

On the potential of altimetry and optical sensors for monitoring and forecasting river discharge and extreme flood events

Angelica Tarpanelli[1] and Jérôme Benveniste[2]
[1]Research Institute for Geo-Hydrological Protection, National Research Council, Perugia, Italy
[2]European Space Agency (ESA-ESRIN), Directorate of Earth Observation Programmes, EO Science, Applications and Climate Department, Frascati, Italy

11.1 Introduction

Flood is one of the most catastrophic natural disaster causing casualties, fatalities, displacement of people, and damage to the environment, industrial settlements, infrastructures, artistic and historical sites (WMO, 2011; European Commission, 2007). As a result of climate change, an increase of the frequency and magnitude of these phenomena is expected (National Academies of Sciences, 2016), that along with the increasing world population will bring more people to live in highly flood-prone areas. Accurate knowledge of flood extents and the humans at risk in these areas is crucial in the emergency phase in order to plan relief efforts and, successively, to assess the extension of the damaged area. The final objective is to take adequate and coordinated measures to reduce and to mitigate the impact of flood episodes (European Commission, 2007).

A reliable assessment of flood extent can be carried out through field data (i.e., in situ surveys, photographs, videos, point measurements), GPS surveys, aerial photographs, and satellite remote sensing. This latter provides a unique opportunity to monitor and forecast floods on a global scale. Recently, satellite imagery has been largely used for damage estimation through the acquisition of a snapshot of the flood extent (Serpico et al., 2012; Martinis et al., 2009; Mason et al., 2012; Pulvirenti et al., 2011; Wang et al., 2005; see also Section 2.9 for further details). In particular, images from Synthetic Aperture Radar (SAR) sensors are used for

Extreme Hydroclimatic Events and Multivariate Hazards in a Changing Environment.
DOI: https://doi.org/10.1016/B978-0-12-814899-0.00011-0

detecting flooded areas in all-weather conditions, day and night. The high-resolution products provided by the latest generation of SAR imagers, such as TerraSAR-X and the COSMO-SkyMed constellation, are very useful for flood detection also during emergency operations (Martinis et al., 2015).

The cost of the data products and the revisit time of these sensors limit their applicability and move the interest towards free-access data products from higher revisit time satellites, such as the new Copernicus missions, Sentinel-1 (10 m of spatial resolution and a revisit time of about 6 days, at mid-latitudes, with the constellation of two satellites, 1A and 1B) (Torres et al., 2012). Alternatively, optical sensors are used mostly because of the easy detection of flooded areas edges, especially in urban areas, where SAR images exhibit drawbacks due to shadowing, layover, and foreshortening. Even if optical sensors cannot penetrate clouds, they have been mainly used to observe post-flood inundation extent and they have advantages of a relatively high repeat cycle (advanced very high resolution radiometer or moderate resolution imaging spectroradiometer, MODIS), that make them suitable for hydrological studies. For both SAR and optical sensors, it is difficult to have one (or more) image during a flood event, and it is highly improbable to acquire those images during the maximum extension of the inundation. On the other hand, floodplains may remain inundated for 2–5 days, enlarging the possibility of monitoring the event.

Floods are monitored by measuring the water volume flowing in a unit time, i.e., the river discharge. River discharge is provided by the measurements of surface water level and flow velocity, along with the topographical survey of the cross-section. These amounts are available thanks to hydrological monitoring networks that include stream gauges for observing water surface elevation and velocimeters for river flow velocity. Such instruments are not uniformly distributed around the globe and the time periods of acquisition have variable length. Due to installation and maintenance costs, long and up-to-date time-series are difficult to obtain, especially in developing countries, where the economic conditions are a deterrent for the design of river gauge networks and their installation.

Thanks to the recent increasing availability of remote sensing products and the densification of constellations that have augmented the temporal observability, the use of satellite radar altimetry is becoming an interesting alternative for measuring water surface of medium-large rivers. As described in Section 1.3, the potential of altimetry for the monitoring of rivers has been demonstrated and is now enhanced with the new SAR Altimeters (also called Delay-Doppler), leading to improved resolutions and accuracy.

This chapter discusses the use of satellite altimetry measurements and optical sensors for monitoring and forecasting floods. The last section, which covers the potential and limits of these applications, also describes the benefits that will derive from future altimeter and optical missions for flood forecasting and inland water applications. While Section 2.3 describes the advantages and the limitations of using rainfall and soil moisture satellite products in flood forecasting, here the focus is on the utility of satellite-based observations of water level and river optical characteristics for flood modeling and forecasting.

11.2 Use of altimetry for in land water

Radar altimetry measures the two-way travel time of a radar pulse between the satellite antenna and the Earth's surface at the nadir of the spacecraft. This measurement is performed through the recording of the pulse bounced back by the reflecting facets within the altimeter footprint. The radar pulse does not penetrate water, but is reflected by the water surface and received by the sensor onboard the satellite. The return echoes and the two-way travel time are recorded by the tracker (onboard), which is used to keep the receiving window in the appropriate position to gather the full echo waveform. An accurate estimate of the (uncorrected) range between the satellite and target on the ground is provided by the retracker (processing on ground), which yields a fine correction to locate the position of the retracking point of the echo waveform. The (uncorrected) range obtained by the retracker is then corrected for the propagation of the pulse through the atmosphere (dry and wet tropospheric correction) and the ionosphere (ionospheric correction) and for solid tidal effects on the Earth (the solid tide correction) (Calmant et al., 2008). Finally, the difference between the satellite height relative to a reference ellipsoid (orbit) and the distance derived from the radar measurement (range) gives the water surface height with respect to the reference ellipsoid (e.g., WGS84). To supply data to hydrologists, a further step is performed to remove the geoid height from the ellipsoidal height to obtain orthometric heights. For more detail on satellite altimetry, the reader is referred to Section 1.3.

Altimetry missions were originally designed for monitoring the ocean and successively for mapping ice sheets and sea ice. So far, no altimetry mission has been dedicated to the monitoring of inland water, although satellites have been recording continuously over the continents. The ocean retracker is inefficient over continental water, where the echo waveforms have very variable shapes. Thus, a myriad of retrackers were designed to exploit these measurements over rivers, lakes, and wetlands.

The first studies referred to water level monitoring of the Great Lakes in the United States and the African Great Lakes by using altimetry data from Seasat, Geosat, and Topex/Poseidon (Brooks 1982; Morris and Gill 1994; Birkett 1995; Cazenave et al., 1997; Mercier et al., 2002). Successively, a number of studies extended the application of altimetry to rivers, that is, Amazon, Negro, Ob, Mekong, Gange, Brahmaputra, and Po, to cite a few (Frappart et al., 2006; Leon et al., 2006; Kouraev et al., 2004; Birkinshaw et al., 2010; Tarpanelli et al., 2013a; Domeneghetti et al., 2014). Due to the size of the footprint, which is larger than the water bodies of small-medium rivers (width of 40–800 m), the surrounding topography often contaminates the returned radar signal. Therefore, the height measurement is more accurate for wide rather than narrow rivers, but it also depends on the kind of altimeter sensor onboard the satellite (conventional or synthetic aperture radars). For example, in the Amazon River, the largest river in the world, Koblinsky et al. (1993) found a root mean square error (RMSE) of about 70 cm with Geosat, and Frappart et al. (2006) and Santos da Silva et al. (2010) obtained an RMSE of about 30 cm with Envisat. With the SAR altimeters not only the accuracy is enhanced, but, thanks to the high along-track resolution, narrower rivers can be monitored (Schneider et al., 2018). Nevertheless, the processing of satellite altimetry measurements for small-medium rivers remains challenging due to the large antenna footprint even in SAR mode in the cross-track direction (which has the same footprint width as conventional altimetry). To process the convoluted echo waveforms, dedicated retrackers are designed to seek for the power coming from the water surface in the waveforms nourished from echoes from multiple targets (Sulistioadi et al., 2015; Biancamaria et al., 2017).

Another limitation of radar altimetry over inland water is due to the spatial and temporal sampling. Space-time resolutions are linked to the satellite orbit, in particular its repeat period. The intertrack distance at the equator depends on the mission repeat cycle: 10 days for Topex/Poseidon and Jason-1/2/3 missions, leading to an intertrack of 315 km, 27 days for the Sentinel-3A/3B missions, leading to an intertrack of 104 km with one satellite and 53 km with Sentinel-3A and -3B working in tandem, and 35 days for ERS-1/2, Envisat, and SARAL/AltiKa, leading to an intertrack of 80 km. The location where the satellite track intersects the river reach is called virtual station (VS). Each mission exhibits a pseudo repeat cycle where the global coverage is uniform, but the measurement is not always made at the exact same point but nearby; for Sentinel-3 this pseudo repeat cycle is 4 days. This revisit time is not fully compatible with the dynamics of rivers, although flood events occur in a period from some hours to some days depending on the climatic regime of the river. For example, in the case of flash floods, radar altimetry fails to catch the rapid increase of water level and the short duration events. Flash floods can occur under different conditions, related to natural extreme events, such as volcano eruptions, glacial dam lake outburst, hurricane and tropical storm, or anthropological causes, such as a dam failure. On the contrary, for river regimes driven by the monsoon effect (long and seasonal rainfall), the duration of the flood can last some days and in these cases altimetry can monitor the water variation with a good sampling. In all the other cases, the revisit time can represent a drawback for the estimation and forecast of high peak values, typical of extreme events.

Radar altimetry over inland water has been applied to the estimation of flooded areas (Sanyal and Lu, 2005), to the calibration of the parameters of hydrological and hydraulic models (Domeneghetti et al., 2014; Schneider et al., 2018), and to the estimation of river discharge by rating curve or multimission approaches (Bjerklie et al., 2003; Sichangi et al., 2016; Tarpanelli et al., 2017). Some examples are provided in the following sections with a particular focus on the evaluation of flood extreme events.

11.3 Flood modeling and forecasting using altimetry

With the purpose of assessing and predicting flood events and their impact, the development of flood inundation models has become a worldwide objective, especially considering the increasing number of flood occurrences. In the last 30 years, progress in flood inundation modeling has improved the quality of flood risk mapping, damage assessment, real-time flood forecasting, and water resources planning (Merz et al., 2010; Arduino et al., 2005).

Flood inundation models can be grouped in two classes (Teng et al., 2017): (1) hydrodynamic models including one-, two-, and three-dimensional methodologies (abbreviated as 1D, 2D, and 3D, respectively) that simulate flow routing within the channel through physical laws of fluid motion with different degrees of complexity; and (2) empirical and statistical methods based on ground measurements, surveys, and remote sensing observations (e.g. Schumann et al., 2009; Smith, 1997). The next sections describe how altimetry water level has been used in inundation models and the main performance indices for evaluating the goodness of models and their forecasting capabilities.

11.3.1 Performance indices

The performance of inundation models is often quantified through the Pearson coefficient of correlation (R), the RMSE, the Nash–Sutcliffe (NS) efficiency, and the coefficient of persistence (PC). The correlation coefficient expresses the linear relationship between predicted and observed values and ranges between -1 (inverse linear correlation) and $+1$ (direct linear correlation), where 0 corresponds to no correlation. The RMSE represents the square root of the second sample moment of the residuals (or differences) between predicted and observed values and it ranges from 0 (perfect fit) to $+\infty$ (lower RMSE is better than higher). NS efficiency (Nash and Sutcliffe, 1970) is a performance index specifically used in hydrological applications because of its sensitivity to extreme values. It is defined as follows:

$$\mathrm{NS} = 1 - \frac{\sum_1^T \left(Q_{\mathrm{sim}}^t - Q_{\mathrm{obs}}^t\right)^2}{\sum_1^T \left(Q_{\mathrm{obs}}^t - \overline{Q_{\mathrm{obs}}}\right)^2} \tag{11.1}$$

where $\overline{Q_{\mathrm{obs}}}$ is the mean of observed discharges, Q_{sim} is the modeled discharge, and Q_{obs} is the observed discharge at time t. NS can range from $-\infty$ to 1, with 1 corresponding to a perfect match of modeled discharge to the observed data. NS equal to 0 indicates that the model predictions are as accurate as the mean of the observed data, whereas an efficiency less than 0 occurs when the observed mean is a better predictor than the model. Finally, the PC compares the prediction of the model forecast Q^t_{sim} with the observation at $t-$ TL, Q^{t-TL}_{obs}, where TL is the wave travel time, assuming that the forecast coincides with the most recent observed value. It is defined as

$$PC = 1 - \frac{\sum_1^T \left(Q^t_{obs} - Q^t_{sim}\right)^2}{\sum_1^T \left(Q^t_{obs} - Q^{t-TL}_{obs}\right)^2} \quad (11.2)$$

where PC values lower than 0 denote bad performances, while values greater than 0 indicate good to perfect (PC = 1) forecast.

Additionally, the model performance in forecasting extreme events is also measured through the following categorical indices:

- Probability of detection (POD), defined as total number of correct event forecasts (hits) divided by the total number of events observed. The optimal value of POD is 1.
- False alarm ratio (FAR) defined as the number of false alarms (an event forecast that is associated with no event observed) divided by the total number of event forecasts. The optimal value of FAR is 0.
- Threat score (TS) defined as the number of events successfully estimated over the total of hit, missed, and false events. The optimal value of TS is 1.

11.3.2 Hydrodynamic models and data assimilation

Altimetry-derived water levels, as traditional in situ river stage measurements, have several uses in hydraulic modeling. Not only they are integrated with flood modeling through the calibration of parameters and validation of results, but they can also improve model skills by data assimilation approaches (Bates et al.,1997; Martinis et al., 2009; Matgen et al., 2011; Mure-Ravaud et al.,2016; Pulvirenti et al., 2011; Sanyal and Lu, 2005; Schumann et al., 2009; Smith, 1997; Ticehurst et al., 2015; Yan et al., 2014).

Often model parameters are not physically related to measurable quantities and it is quite common that such parameters are used to represent the properties of large areas, even if their variability across the area is evident. Thus, calibration should tune parameters to minimize the misfit between simulated and observed states. For hydrological models, calibration is performed by comparing simulated and measured discharge, whereas for hydraulic models, both discharge and water level can be used.

Validation is carried out by running the model with input data from an independent period, different from the one used for parameter calibration.

For example, Domeneghetti et al. (2014) used water level time-series of four VSs from two satellites, ERS-2 and Envisat, for calibrating a quasi-2D hydraulic model of the Po River in Italy along a stretch of 140 km. They considered 16 years of simulations from 1995 to 2011, splitting the period into two: the first half is used for calibrating the Manning roughness coefficient and the second half for validating their results. The analysis showed that satellite data can effectively help to assess the hydrometric regime for rivers that are at least 400 m wide, such as the Po River. Moreover, their validation was extended to the evaluation of high flow conditions, during the flood event of October 2000. They compared the river surface water elevation obtained by several configurations of the hydraulic model. The Manning roughness coefficient values were calibrated by using in situ stations only, satellite data only, and a combination of in situ and satellite data. The comparison of these several configurations showed that the model calibrated with in situ observed data only underestimated water surface elevation, whereas with the use of satellite data, it overestimated water surface elevation. The integration of both in situ and radar altimetry data enhanced the model ability of reproducing the maximum water profile slope, with a very limited error (about 1.26%) with respect to the slope of the observed marks.

By using the long-repeat orbit of 369 days of Cryosat-2, the ESA ice mission, Schneider et al. (2018) analyzed a longer stretch of the same Po River. The unprecedented potential of the CryoSat sensor is demonstrated by the possibility to calibrate the Manning coefficient each 10 km along the entire river, thanks to the short distance between satellite tracks (~7 km). This level of detail in the roughness distribution is not common, even for ground networks. Typically, only a ground station near the outlet is available and the calibration of the roughness coefficient provides only one average value, which has to be used for the full river stretch. The consequence is a possible over/underestimation of the high flow that cannot be correctly predicted by this average value. On the contrary, the advantage to have measurements of water level with a high spatial distribution, as in the case of CryoSat-2, largely improves our capability to map reliable extreme flood events.

Even if models are highly performing and well calibrated, the estimation of discharge is still subject to large uncertainties due to errors in the model structure, forcing data, parameterization, and initial conditions.

A possible solution to reduce the uncertainty is to assimilate observations into hydrological and/or hydraulic models. Data assimilation is the process through which a model state is updated using independent observations in order to obtain the best estimate of the current state of the system. The current state of the hydrological system is important in real-time applications, especially for flood forecasting. In such cases, sequential data assimilation, in which an update is performed whenever a new measurement becomes available is preferable.

In the last years, radar altimetry water levels have been successfully used to update states in hydrological and hydraulic models (Andreadis et al., 2007; Getirana et al., 2009; Paiva et al., 2013; Michailovsky et al., 2012). Studies in the Amazon (Getirana et al., 2009; Paiva et al., 2013), Brahmaputra (Michailovsky et al., 2013), and Zambezi (Michailovsky et al., 2012) rivers provide an example of the improvement that can be obtained by integrating satellite data with flood models. Andreadis et al. (2007) showed that assimilating virtual wide swath altimeter data (i.e., surface water ocean topography, SWOT) to a raster-based hydrodynamic models with known bathymetry could improve modeled river depth and discharge. They applied the Ensemble Kalman Filter to assimilate virtual water levels coming from a SWOT simulator, in the LISFLOOD-FP 2D model built for the Ohio River. They found lower errors (10% relative error against the 23% for the simulation without assimilations, called open-loop simulations) in the simulated river discharges.

Michailovsky et al. (2013) assimilated radar altimetry data from six Envisat VSs into a Muskingum routing model of the main reach of the Brahmaputra River driven by the outputs of a calibrated Budyko type rainfall-runoff model. They used an extended Kalman filter to update the routed water volumes. They found a significant improvement in terms of NS efficiency for daily discharge: from 0.78 in the open-loop simulation to 0.84 in the assimilation run. The application of radar altimetry in hydrodynamic models is deeply affected by the uncertainty in remotely sensed water stages. Domeneghetti et al. (2015) analyzed the influence of the error of ERS-2 and Envisat water levels in the calibration of the quasi-2D hydraulic model in the Po River and found that Envisat outperforms ERS-2 in estimating the roughness coefficient. Uncertainty in remote sensing data has become a major topic in current research and it is expected that the accuracy of measurements will improve with new satellite missions, consequently enhancing the model accuracy.

11.3.3 Simplified and empirical models

With the availability of more and more satellite data, traditional approaches for flood modeling and discharge estimation underwent adaptations to better suit these new data (Birkinshaw et al., 2010; Smith and Pavelsky, 2008). The well-known empirical formulas, based on the hydraulic relationships between river characteristics, have been revised and adjusted to the use of satellite observations. A meaningful example is provided by the studies of Bjerklie et al. (2003; 2005). They reviewed a series of traditional formulas and validated the relationships with more than 1000 flow measurements in more than 100 rivers in the United States and New Zealand, including four measurements in the Amazon River. Results showed that existing satellite-based sensors are quite accurate for measuring water-surface width, water level, and slope, and provide estimates of discharge with an average uncertainty of less than 20%. Radar altimetry is well suited to be included in these formulas, provided that the cross-section is known. These equations consider the water stage, that is, the difference between the water surface level and the bottom of the river. Radar altimetry provides water heights and not the water depths, therefore without a topographical survey of the river in the specific section, radar altimetry measurements are useless. Past studies sought to estimate river bathymetry by relating it to morphological characteristics (Mersel et al., 2013; Domeneghetti., 2016) or using the entropy theory (Moramarco et al, 2013).

Tarpanelli et al. (2013a) compared the performance in terms of river discharge between one of the empirical formulas of Bjerklie et al. (2003) and a simplified routing model, the rating curve model (RCM; Moramarco et al., 2005). Starting from the knowledge of the discharge recorded at an upstream section and the stages observed at another downstream section of a river reach, RCM is a simple method able to estimate the flow condition in the downstream section, taking into account the lateral inflow contributions. Tarpanelli et al. (2013a) applied both approaches to two sections of the Po River, by using water level information derived by two different satellites, ERS-2 and Envisat. The RCM method showed NS values higher than 0.73 and an average relative error of 23% in the estimation of discharge. Worse performance was observed with the empirical formulas proposed by Bjerklie et al. (2003), especially in the overestimation of high flows. The knowledge of river discharge at the upstream section in the RCM is the main reason for the better performance with respect to the empirical formulas. However, if well calibrated and if the information of river velocity was available, the empirical formulas would provide better results.

Sichangi et al. (2016) used two simplified empirical formulas to express river discharge: the first assuming that river discharge is proportional to the river stage (the rating curve) and the second assuming that river discharge is proportional to the river stage and river width. They retrieved the stage from Envisat and Jason-2 and the river width from MODIS. They calibrated the parameters of the empirical formulas by using observed discharge at 14 gauge stations from eight of the world's major rivers. They found that river discharge can be satisfactorily expressed as a function of width and stage. Errors ranged between 0.60 and 0.97 in terms of NS, with worse performance in the artic rivers (Lena and Yenisey Rivers), which are frozen from November to April and where satellite-derived discharge was always higher than the observed values. A comparison with the Bjerklie et al. (2003) empirical formulas confirms their overestimation of river discharge, especially for extreme high values. Despite the good results obtained by Sichangi et al. (2016), temporal resolution remains a major limitation, especially due to the satellite altimetry water stage. Section 11.3.4 presents some studies that attempt to solve this major challenge.

Based on the definition of river discharge as the product of the flow velocity times the flow area, Tarpanelli et al. (2015) combined the information coming from altimetry and optical sensors to estimate the Po River flow. The flow area was computed by using water surface elevation from Envisat and the bathymetry estimated through the entropy theory (Moramarco et al. 2013). For flow velocity, the approach of Tarpanelli et al. (2013b) was adopted, where the near infrared bands of MODIS images were used to distinguish the land pixel (dry) from the water pixel (wet). The ratio between their reflectances was used to derive the variation of river flow velocity (or river discharge). Then, the product of the flow area (as a function of altimetry-derived water level) and the flow velocity by MODIS provided long time-series of river discharge. The same methodology could be applied to observations from different satellites, like medium-resolution imaging spectrometer (MERIS) and ocean and land color instrument. In this study case, the high flows were underestimated mostly because the optical sensor was not able to see through clouds during the flood and the lower water levels present during post-flood conditions (i.e., when the sky turned to be clear) contaminated the final estimates. This procedure has been applied to the cross-section of Pontelagoscuro, in the Po River. Further tests are necessary to validate this methodology and to provide a consistent proof of the possible merging of satellite observations and physically based models.

11.3.4 Flood forecasting models

One of the measures to mitigate flood disaster is represented by early warning systems, based on real-time flood forecasting models. Flood forecasting systems can be implemented considering precipitation and streamflow data in a rainfall-runoff model to forecast discharge and water level hours or days ahead (depending on the size of the basin). In this context, altimetry (and more in general satellite data), if combined with weather, hydrological, and hydrodynamic forecast methods, can mitigate the negative effects of a flood. Among the examples of flood forecasting available in the literature (Hirpa et al., 2013), this section summarized two case studies: one in Nigeria (Tarpanelli et al., 2017) and one in Ganges-Brahmaputra river basin (Biancamaria et al., 2011).

By using accurate stage and/or discharge predictions with an appropriate forecast lead-time, Tarpanelli et al. (2017) predicted discharges in Nigeria, by using Envisat and Jason-2 water level time-series. Flood forecasting in the Makurdi section, along the Benue River, is carried out by considering the rating curve between the discharge at time t, $Qd(t)$, and the water level at an upstream section $\mathrm{hu}(t - \Delta t)$, where Δt is the wave travel time and coincides with the forecast lead-time. Results showed that the proposed approach performed very well, with NS equal to 0.99, and PC equal to 0.59, especially for high flows (PC = 0.79) and during the flood event (PC = 0.55). Unfortunately, the long revisit time limited the model performance. Although the water level time-series included five satellite tracks for improving the sampling, they could not guarantee a good estimation of the flood peak. In this regard, the information from altimetry data was compared to the one obtained from the optical sensors, i.e., MODIS. The capability of the optical sensors to observe the river once a day allowed dense information to be transferred to a downstream Section 4 days in advance. Fig. 11.1 shows 4-day ahead forecasted discharges at Makurdi gauged station using MODIS imagery from AQUA and Terra satellites (daily product) and radar altimetry from ENVISAT. This suggests the high potential of an approach able to combine altimetry and optical information, which will be further discussed in Section 11.4.

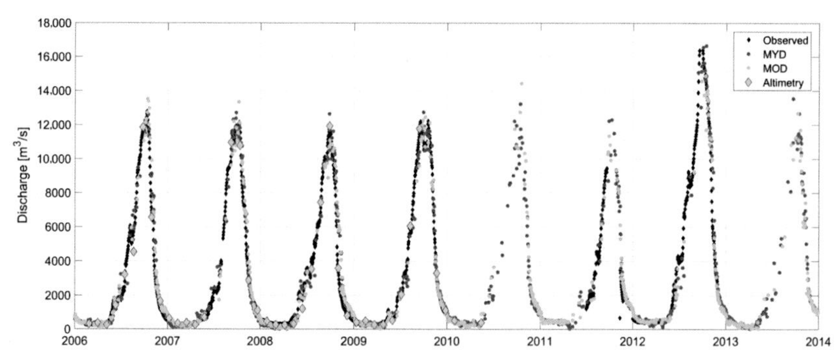

Figure 11.1 (A) 4-day ahead forecasted discharges at Makurdi gauged station by using MODIS from Terra (MOD) and AQUA (MYD) satellite and radar altimetry from ENVISAT. Closer view of the 2012 flood event for daily (B) and 8-day (C) MODIS products and for altimetry (D). *MODIS*, Moderate resolution imaging spectroradiometer.

A good example of the benefit provided by altimetry is represented by the study of Biancamaria et al. (2011) that investigated the transboundary system of the Ganges-Brahmaputra River basin and showed how to overcome political constrains between riparian nations with the use of Topex/Poseidon satellite altimetry. Most of the water flowing in Bangladesh comes from the countries upstream, mainly India. Hydrological data are not shared among countries and the forecasting activities in Bangladesh are limited to a maximum 3-day lead-time. Biancamaria et al. (2011) extended the lead-time, by using Topex/Poseidon satellite altimetry measurements in India and in situ measurements in Bangladesh. They showed the possibility to forecast water elevation anomalies in the Ganges and Brahmaputra Rivers, during the critical monsoon season from June to September, with an RMSE of about 0.40 m for lead-times up to 5 days and RMSE between 0.60 and 0.80 m for lead-times of 10 days. This study demonstrated the potential of altimetry in transboundary river management and in flood forecasting operational activities.

11.4 Merging altimetry with other remote sensing data for extreme event estimation

We have seen that the main limitation of altimetry-derived water level is the revisit time due to the orbit repeat period of the satellite. Current and past missions are characterized by frequency of observations of 10, 27, and 35 days. Such frequency of sampling is too long for catching floods in most of the rivers. The multimission approach seems the best way to solve this issue. There are two ways to merge this information: altimetry multimission approach and multisatellite approach.

The merging of multisatellite altimeter data is carried out by considering all altimetry tracks available along a river and by transferring the water level information towards a specific location through the hydraulic concept of wave travel time. This is what Tourian et al. (2016) proposed in their study along the Po, Mississippi, Congo, and Danube rivers. Although the temporal resolution is improved to ~3 days, the capability to simulate high flow seems insufficient, especially in the Po River, where most of the peak values are missing. Some improvements to the approach were introduced in a follow-up study by Tourian et al. (2017) that uses a different merging technique for deriving daily water level and river discharge time-series by using the rating curve and quantile matching approach in the Niger River.

The merging of multisatellite includes the joint exploitation of radar altimeter observations with other satellite sensors, such as optical sensors. Beside the empirical formulas used to estimate flows, which can be considered as a possible way to merge satellite data, another method has been recently introduced: the machine learning approach. Based on the knowledge of input and output data for a training period, machine learning is becoming very popular thanks to the increasing computing power offered by cloud computing facilities and by currently available platforms for the management and storage of large amounts of data.

Tarpanelli et al. (2018) merged data from different satellite optical sensors and altimeters for estimating daily river discharge in the Niger River and in Po River through an artificial neural networks (ANN) technique. They combined altimetry measurements from Envisat, Jason-2, ERS-2, Topex/Poseidon, and Cryosat-2 and optical images from MODIS and MERIS (with an attempt also with Landsat data). ANN offers an easy but effective way of combining input data from different sources. Two independent periods were considered: one for training the network and one for validating river discharge against in situ observations. The use of Landsat underlined the importance of temporal resolution with respect to spatial resolution for the evaluation of extreme events. Landsat has a long revisit time (14–16 days), which caused the missing of most peak values, as shown in Fig. 11.2.

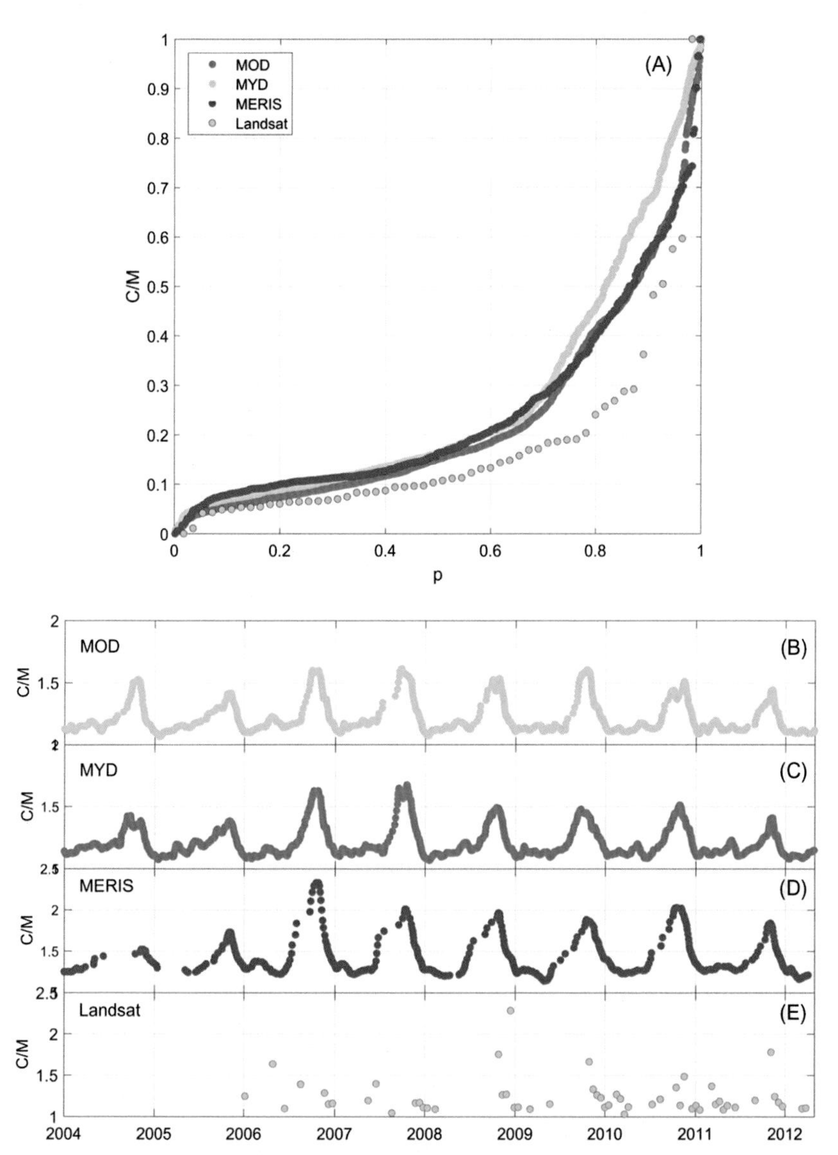

Figure 11.2 Time-series of C/M* derived by different optical products for Niger River at Lokoja: MODIS-Terra (MOD), MODIS-AQUA (MYD), MERIS, and Landsat. *MODIS*, Moderate resolution imaging spectroradiometer; *MERIS*, medium-resolution imaging spectrometer.

With respect to the single use of each sensor, the multimission approach, which involves optical sensors and altimetry, was found to be the most reliable tool to estimate river discharge with relative RMSEs of 0.12% and 0.27% and NS of 0.98 and 0.83 for the Niger and Po Rivers,

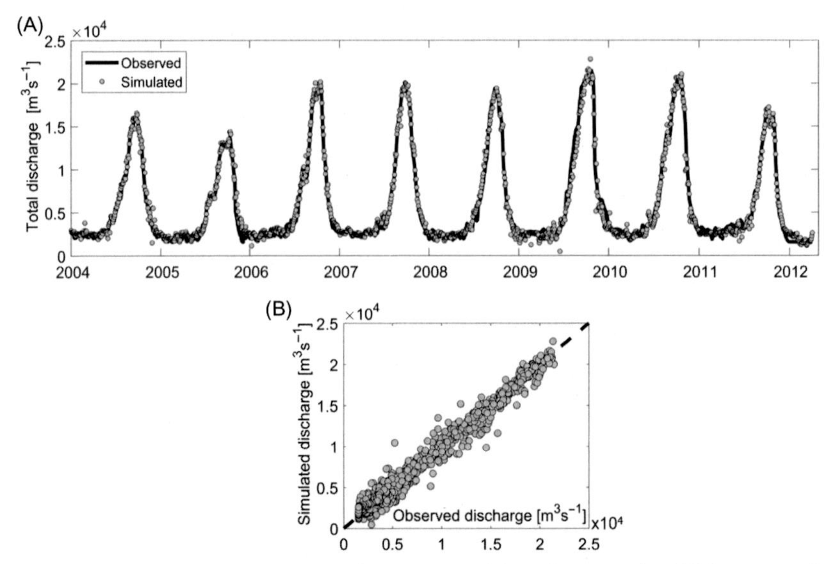

Figure 11.3 Comparison in terms of time-series (A) and scatterplot (B) between the discharge observed in situ and simulated by merging all the products through ANN, for the Niger River. *ANN,* Artificial neural network.

respectively. Fig. 11.3 shows the simulated and in situ time-series for the Niger River in terms of time-series and scatterplot. The merged product followed very well the in situ time-series. Considering thresholds ranging between 5000 and 20,000 m^3/s, median value of POD was 0.95, of FAR was 0.04 and of TS was 0.9. The high values of the categorical performance demonstrated the peculiarity of the approach to reproduce the high flows, and hence the extreme events.

11.5 Conclusions

Floods are fast and often unpredictable phenomena. Traditional monitoring networks, when available and dense enough, remain the best option for monitoring these natural events. Satellite altimetry can be used in several applications to support the evaluation of floods (when and where in situ observations are not available or dense enough): from the calibration of the hydrodynamic model parameters to the assimilation of those data into models, as well as the evaluation of empirical formulas to forecast flood peaks. Most operational applications still have limitations in the temporal resolution of the predictions. The satellite revisit time is often not compatible with the fast river dynamics. However, techniques have been developed to deal with this constraint and to benefit from the added information from satellites.

The advent of new sensors with improved technology (SAR Altimeters) on CryoSat-2, Sentinel-3A and -3B, already in orbit and in 2020 on Sentinel-6/Jason-CS, allows to decrease the uncertainty in the observations, promoting the development of tools for a more accurate flood forecasting.

An interesting aspect to consider is the merging of satellite information. Different sensing technologies allow the observation of different physical characteristics that can be used to improve the estimation of river discharges. Moreover, the increasing amount of satellites orbiting around the Earth should facilitate the observation of extreme events when they occur. Further studies need to address the use of radar altimetry together with other sensors (e.g., optical, SAR, and/or thermal) for enhancing flood monitoring. Machine learning can be efficiently used to merge data of different natures and the current availability of superclusters and cloud platforms assists the technical operations to manage large amounts of data.

Despite the limitation in space and time, all the examples reported in this chapter show the large potential of radar altimetry, especially if specifically designed for inland water. In 2021, National Aeronautics and Space Administration, NASA and Centre National d'Études Spatiales with contributions from the Canadian and UK Space Agencies, have planned the first global two-dimensional survey of inland water by launching the SWOT mission. The scientific community has already been working on algorithms to estimate discharge from these new altimetry observations. Differently from the current radar altimetry, SWOT has a wide swath technology, based on the interferometry principle (Fu et al., 2009) that with repeated measurements of elevation at high resolution will provide observations of most rivers and lakes in the world.

References

Andreadis, K.M., Clark, E.A., Lettenmaier, D.P., Alsdorf, D.E., 2007. Prospects for river discharge and depth estimation through assimilation of swath-altimetry into a raster-based hydrodynamics model. Geophys. Res. Lett. 34 (10).

Arduino, G., Reggiani, P., Todini, E., 2005. Recent advances in flood forecasting and flood risk assessment. Hydrol. Earth Syst. Sci. 9 (4), 280–284.

Bates, P.D., Horritt, M.S., Smith, C.N., Mason, D., 1997. Integrating remote sensing observations of flood hydrology and hydraulic modelling. Hydrol. Process 11 (14), 1777–1795.

Biancamaria, S., Frappart, F., Leleu, A.S., Marieu, V., Blumstein, D., Desjonquères, J.D., et al., 2017. Satellite radar altimetry water elevations performance over a 200 m wide river: evaluation over the Garonne River. Adv. Space Res. 59 (1), 128–146. Available from: https://doi.org/10.1016/j.asr.2016.10.008.

Biancamaria, S., Hossain, F., Lettenmaier, D.P., 2011. Forecasting transboundary river water elevations from space. Geophys. Res. Lett. 38 (11).

Birkett, C.M., 1995. The contribution of Topex/Poseidon to the global monitoring of climatically sensitive lakes. J. Geophys. Res. 100 (C12), 25179–25204. Available from: https://doi.org/10.1029/95JC02125.

Birkinshaw, S.J., O'Donnell, G.M., Moore, P., Kilsby, C.G., Fowler, H.J., Berry, P.A.M., 2010. Using satellite altimetry data to augment flow estimation techniques on the Mekong River. Hydrol. Process 24, 3811–3825. Available from: https://doi.org/10.1002/hyp.7811.

Bjerklie, D.M., Dingman, S.L., Vorosmarty, C.J., Bolster, C.H., Congalton, R.G., 2003. Evaluating the potential for measuring river discharge from space. J. Hydrol. 278, 17–38.

Bjerklie, D.M., Moller, D., Smith, L.C., Dingman, S.L., 2005. Estimating discharge in rivers using remotely sensed hydraulic information. J. Hydrol. 309 (1-4), 191–209.

Brooks, R.L., 1982. Lake Elevation From Satellite Radar Altimetry From a Validation Area in Canada. Report. Geoscience Research Corporation, Salibury, MD.

Calmant, S., Seyler, F., Cretaux, J.F., 2008. Monitoring continental surface waters by satellite altimetry. Surv. Geophys. 29, 247–269. Available from: https://doi.org/10.1007/s10712-008-9051-1.

Cazenave, A., Bonnefond, P., DoMinh, K., 1997. Caspian sea level from Topex/Poseidon altimetry: level now falling. Geophys. Res. Lett. 24, 881–884. Available from: https://doi.org/10.1029/97GL00809.

Domeneghetti, A., 2016. On the use of SRTM and altimetry data for flood modeling in data-sparse regions. Water Resour. Res. 52 (4), 2901–2918.

Domeneghetti, A., Tarpanelli, A., Brocca, L., Barbetta, S., Moramarco, T., Castellarin, A., et al., 2014. The use of remote sensing-derived water surface data for hydraulic model calibration. Remote Sens. Environ. 149, 130–141. Available from: https://doi.org/10.1016/j.rse.2014.04.007.

Domeneghetti, A., Castellarin, A., Tarpanelli, A., Moramarco, T., 2015. Investigating the uncertainty of satellite altimetry product for hydrodynamic modelling. Hydrol. Process 29 (23), 4908–4918. Available from: https://doi.org/10.1002/hyp.10507.

European Commission, 2007. Directive 2007/60/EC of the European Parliament and of the Council of 23 October 2007 on the Assessment and Management of Flood Risks.

Frappart, F., Calmant, S., Cauhope, M., Seyler, F., Cazenave, A., 2006. Preliminary results of ENVISAT RA-2-derived water levels validation over the Amazon basin. Remote Sens. Environ. 100, 252–264. Available from: https://doi.org/10.1016/j.rse.2005.10.027.

Fu, L.L., Alsdorf, D., Rodríguez, E., Morrow, R., Mognard, N., Lambin, J., et al., 2009. The SWOT (surface water and ocean topography) mission: spaceborne radar interferometry for oceanographic and hydrological applications. In: Hall, J., Harrison, D.E., Stammer, D. (Eds.), Proceedings of OceanObs'09: Sustained Ocean Observations and Information for Society, Vol. 2. September 21–25, 2009. ESA Publication WPP, Venice, Italy, p. 306. <https://doi.org/10.5270/OceanObs09.cwp.33>.

Getirana, A.C.V., Bonnet, M.P., Calmant, S., Roux, E., Rotunno, O.C., Mansur, W.J., 2009. Hydrological monitoring of poorly gauged basins based on rainfall-runoff modeling and spatial altimetry. J. Hydrol. 379, 205–219.

Hirpa, F.A., Hopson, T.M., De Groeve, T., Brakenridge, G.R., Gebremichael, M., Restrepo, P.J., 2013. Upstream satellite remote sensing for river discharge forecasting: application to major rivers in South Asia. Remote Sens. Environ. 131, 140–151.

Koblinsky, C.J., Clarke, R.T., Brenner, A.C., Frey, H., 1993. Measurement of river level variations with satellite altimetry. Water Resour. Res. 29 (6), 1839–1848. Available from: https://doi.org/10.1029/93WR00542.

Kouraev, A.V., Zakharova, E.A., Samain, O., Mognard, N.M., Cazenave, A., 2004. Ob' river discharge from TOPEX/Poseidon satellite altimetry (1992–2002). Remote Sens. Environ. 93, 238–245. Available from: https://doi.org/10.1016/j.rse.2004.07.007.

Leon, J.G., Calmant, S., Seyler, F., Bonnet, M.P., Cauhope, M., Frappart, F., et al., 2006. Rating curves and estimation of average water depth at the upper Negro River based on satellite altimeter data and modeled discharges. J. Hydrol. 328, 481–496. Available from: https://doi.org/10.1016/j.jhydrol.2005.12.006.

Martinis, S., Twele, A., Voigt, S., 2009. Towards operational near real-time flood detection using a split-based automatic thresholding procedure on high resolution TerraSAR-X data. Nat. Hazard. Earth Syst. Sci. 9 (2), 303–314. Available from: https://doi.org/10.5194/nhess-9-303-2009.

Martinis, S., Kersten, J., Twele, A., 2015. A fully automated TerraSAR-X based flood service. ISPRS J. Photogramm. 104, 203–212.

Mason, D., Davenport, I., Neal, J., Schumann, G., Bates, P.D., 2012. Near real-time flood detection in urban and rural areas using high resolution synthetic aperture radar images. IEEE Trans. Geosci. Remote Sens. 50 (8), 3041–3052. Available from: https://doi.org/10.1109/TGRS.2011.2178030.

Matgen, P., Hostache, R., Schumann, G., Pfister, L., Hoffmann, L., Savenije, H.H.G., 2011. Towards an automated SAR-based flood monitoring system: lessons learned from two case studies. Phys. Chem. Earth 36 (7-8), 241–252.

Mercier, F., Cazenave, A., Maheu, C., 2002. Interannual lake level fluctuations (1993–1999) in Africa from Topex/Poseidon: connections with ocean–atmosphere interactions over the Indian ocean. Glob. Planet. Change 32, 141–163. Available from: https://doi.org/10.1016/S0921-8181(01)00139-4.

Mersel, M.K., Smith, L.C., Andreadis, K.M., Durand, M.T., 2013. Estimation of river depth from remotely sensed hydraulic relationships. Water Resour. Res. 49, 1–15.

Merz, B., Hall, J., Disse, M., Schumann, A., 2010. Fluvial flood risk management in a changing world. Nat. Hazard. Earth Sys. 10 (3), 509.

Michailovsky, C.I., McEnnis, S., Berry, P.A.M., Smith, R., Bauer-Gottwein, P., 2012. River monitoring from satellite radar altimetry in the Zambezi River basin. Hydrol. Earth Syst. Sci. 16 (7), 2181–2192.

Michailovsky, C.I., Milzow, C., Bauer-Gottwein, P., 2013. Assimilation of radar altimetry to a routing model of the Brahmaputra River. Water Resour. Res. 49, 4807–4816. Available from: https://doi.org/10.1002/wrcr.20345.

Moramarco, T., Barbetta, S., Melone, F., Singh, V.P., 2005. Relating local stage and remote discharge with significant lateral inflow. J. Hydrol. Eng. 10 (1), 58–69.

Moramarco, T., Corato, G., Melone, F., Singh, V.P., 2013. An entropy-based method for determining the flow depth distribution in natural channels. J. Hydrol. 497, 176–188.

Morris, C.S., Gill, S.K., 1994. Variation of Great Lakes waters from geosat altimetry. Water Resour. Res. 30, 1009–1017. Available from: https://doi.org/10.1029/94WR00064.

Mure-Ravaud, M., Binet, G., Bracq, M., Perarnaud, J.J., Fradin, A., Litrico, X., 2016. A web based tool for operational real-time flood forecasting using data assimilation to update hydraulic states. Environ. Modell. Softw. 84, 35–49.

Nash, J.E., Sutcliffe, J.V., 1970. River flow forecasting through conceptual models, Part I: A discussion of principles. J. Hydrol. 10 (3), 282–290.

National Academies of Sciences, 2016. Engineering, and Medicine. Attribution of Extreme Weather Events in the Context of Climate Change. The National Academies Press, Washington, DC. Available from: https://doi.org/10.17226/21852.

Paiva, R.C.D., Collischonn, W., Buarque, D.C., 2013. Validation of a full hydrodynamic model for large-scale hydrologic modelling in the Amazon. Hydrol. Process 27, 333–346. Available from: https://doi.org/10.1002/hyp.842.

Pulvirenti, L., Chini, M., Pierdicca, N., Guerriero, L., Ferrazzoli, P., 2011. Flood monitoring using multi-temporal COSMO-SkyMed data: image segmentation and signature interpretation. Remote Sens. Environ. 115 (4), 990–1002. Available from: https://doi.org/10.1016/j.rse.2010.12.002.

Santos da Silva, J., Calmant, S., Seyler, F., Rotunno Filho, O.C., Cochonneau, G., Mansur, W.J., 2010. Water levels in the Amazon basin derived from the ERS 2 and ENVISAT radar altimetry missions. Remote Sens. Environ. 114, 2160–2181. Available from: https://doi.org/10.1016/j.rse.2010.04.020.

Sanyal, J., Lu, X.X., 2005. Remote sensing and GIS-based flood vulnerability assessment of human settlements: a case study of Gangetic West Bengal, India. Hydrol. Process 19 (18), 3699–3716.

Schneider, R., Tarpanelli, A., Nielsen, C., Madsen, H., Bauer-Gottwein, P., 2018. Evaluation of multi-mode Cryosat-2 altimetry data over the Po River against in situ data and a hydrodynamic model. Adv. Water Resour. 112, 17–26. Available from: https://doi.org/10.1016/j.advwatres.2017.11.027.

Schumann, G., Bates, P.D., Horritt, M.S., Matgen, P., Pappenberger, F., 2009. Progress in integration of remote sensing–derived flood extent and stage data and hydraulic models. Rev. Geophys. 47 (4).

Serpico, S.B., Dellepiane, S., Boni, G., Moser, G., Angiati, E., Rudari, R., 2012. Information extraction from remote sensing images for flood monitoring and damage evaluation. Proc. IEEE 100 (10), 2946–2970.

Sichangi, A.W., Wang, L., Yang, K., Chen, D., Wang, Z., Li, X., et al., 2016. Estimating continental river basin discharges using multiple remote sensing data sets. Remote Sens. Environ. 179, 36–53.

Smith, L.C., 1997. Satellite remote sensing of river inundation area, stage, and discharge: a review. Hydrol. Process 11 (10), 1427–1439.

Smith, L.C., Pavelsky, T.M., 2008. Estimation of river discharge, propagation speed, and hydraulic geometry from space: Lena River, Siberia. Water Resour. Res. 44 (3).

Sulistioadi, Y.B., Tseng, K.H., Shum, C.K., Hidayat, H., Sumaryono, M., Suhardiman, A., et al., 2015. Satellite radar altimetry for monitoring small rivers and lakes in Indonesia. Hydrol. Earth Syst. Sci. 19 (1), 341–359. Available from: https://doi.org/10.5194/hess-19-341-2015.

Tarpanelli, A., Santi, E., Tourian, M.J., Filippucci, P., Amarnath, G., Brocca, L., 2018. Daily river discharge estimates by merging satellite optical sensors and radar altimetry through artificial neural network. IEEE Trans. Geosci. Remote Sens. In press.

Tarpanelli, A., Barbetta, S., Brocca, L., Moramarco, T., 2013a. River discharge estimation by using altimetry data and simplified flood routing modeling. Remote Sens. 5 (9), 4145–4162. Available from: https://doi.org/10.3390/rs5094145.

Tarpanelli, A., Brocca, L., Lacava, T., Melone, F., Moramarco, T., Faruolo, M., et al., 2013b. Toward the estimation of river discharge variations using MODIS data in ungauged basins. Remote Sens. Environ. 136, 47–55. Available from: https://doi.org/10.1016/j.rse.2013.04.010.

Tarpanelli, A., Brocca, L., Barbetta, S., Faruolo, M., Lacava, T., Moramarco, T., 2015. Coupling MODIS and radar altimetry data for discharge estimation in poorly gauged river basins. IEEE. J. Sel. Top. Appl. 8 (1), 141–148. Available from: https://doi.org/10.1109/JSTARS.2014.2320582.

Tarpanelli, A., Amarnath, G., Brocca, L., Massari, C., Moramarco, T., 2017. Discharge estimation and forecasting by MODIS and altimetry data in Niger-Benue River. Remote Sens. Environ. 195, 96–106. Available from: https://doi.org/10.1016/j.rse.2017.04.015.
Teng, J., Jakeman, A.J., Vaze, J., Croke, B.F., Dutta, D., Kim, S., 2017. Flood inundation modelling: a review of methods, recent advances and uncertainty analysis. Environ. Model. Softw. 90, 201–216.
Ticehurst, C., Dutta, D., Karim, F., Petheram, C., Guerschman, J.P., 2015. Improving the accuracy of daily MODIS OWL flood inundation mapping using hydrodynamic modelling. Nat. Hazard. 78 (2), 803–820. Available from: https://doi.org/10.1007/s11069-015-1743-5.
Torres, R., Snoeij, P., Geudtner, D., Bibby, D., Davidson, M., Attema, E., et al., 2012. GMES Sentinel-1 mission. Remote Sens. Environ. 120, 9–24. Available from: https://doi.org/10.1016/j.rse.2011.05.028.
Tourian, M.J., Tarpanelli, A., Elmi, O., Qin, T., Brocca, L., Maramarco, T., et al., 2016. Spatiotemporal densification of river water level time series by multimission satellite altimetry. Water Resour. Res. 52. Available from: https://doi.org/10.1002/2015WR017654.
Tourian, M.J., Schwatke, C., Sneeuw, N., 2017. River discharge estimation at daily resolution from satellite altimetry over an entire river basin. J. Hydrol. 546, 230–247. Available from: https://doi.org/10.1016/j.jhydrol.2017.01.009.
WMO W., 2011. Manual on flood forecasting and warning. WMO, No. 1072.
Wang, Y., Colby, J., Mulcahy, K., 2005. An efficient method for mapping flood extent in a coastal floodplain using Landsat TM and DEM data. Int. J. Remote Sens. 23 (18), 3681–3696. Available from: https://doi.org/10.1080/01431160110114484.
Yan, K., Tarpanelli, A., Balint, G., Moramarco, T., Di Baldassarre, G., 2014. Exploring the potential of radar altimetry and SRTM topography to support flood propagation modeling: the Danube Case Study. J. Hydrol. Eng. 20 (2). Available from: https://doi.org/10.1061/(ASCE)HE.1943-5584.0001018.

CHAPTER TWELVE

Inundation mapping by remote sensing techniques

Xinyi Shen and Emmanouil Anagnostou
Department of Civil and Environmental Engineering, University of Connecticut, Mansfield, CT, United States

12.1 Introduction

Near real-time (NRT) inundation mapping during flood events is vital to support rescue and damage recovery decisions and to facilitate rapid assessment of property loss and damage. Both multispectral and microwave sensors are used for inundation mapping with different levels of difficulty and capacity. Via water indices, extracting water surfaces is straightforward and reliable, but observations are limited by cloud and unclear weather conditions, which are common during severe floods. Via the detecting of scattering mechanisms, synthestic aperture radar (SAR) can map inundation during almost any weather condition due to its penetration ability. As an active sensor, it works diurnally as well. However, the retrieval algorithms using SAR data are difficult to design and, until today, pure automated algorithms that require zero human interference are rare. With the growing remote sensing data availability and the development of retrieval techniques, automation and reliability of SAR data are expected to emerge soon. This chapter focuses on evaluating existing inundation mapping algorithms using SAR data developed from 1980 until today. Given that data, study area, and validation method vary from study to study, we primarily used four criterion: automation, robustness, applicability, and accuracy.

12.2 Principles

The change in surface roughness is the key to detect inundation using SAR data. Where the ground surface is covered with water, its low roughness exhibits almost ideal reflective scattering that has strong contrast

Extreme Hydroclimatic Events and Multivariate Hazards in a Changing Environment.
DOI: https://doi.org/10.1016/B978-0-12-814899-0.00012-2

to the scattering of natural surfaces in dry conditions. Two opposite changes of backscattering intensity can be caused due to different scattering mechanisms: dampening and enhancement.

In the case of an open flood where water is not obscured by vegetation or buildings, dampening occurs because most of the scattering intensity is concentrated to the forward direction, whereas the back direction is very weak. In vegetated areas, different scattering mechanisms can happen depending on the vegetation structure and submerged status. For fully submerged locations, scattering dampening occurs as in an open flood, while for partially or not submerged locations, scattering enhancement is observed because the dihedral scattering dominates. The backscattering components of vegetation (with a canopy layer) are derived in the enhanced Michigan microwave canopy scattering (enhanced MIMICS) model by Shen et al. (2010), from the first-order approximation of the vectorized radiative transfer (VRT) equation.

As shown in Fig. 12.1, the vegetation–ground system can be conceptualized by three layers from the top to the bottom: the canopy layer (C), trunk layer (T), and the ground rough surface (G). The G–C–G term, for instance, indicates the wave scattered by the ground surface after it penetrates the canopy, that is then scattered by the canopy back to the ground, and finally scattered back to the sensor after penetrating the canopy. The Ds term indicates that the scattering process on ground surface only occurs once between the two penetrations of canopy. A trunk is usually modeled by a cylindrical structure whose length is much longer than the diameter. Term 4a (T–G–C) in Fig. 12.1 represents the

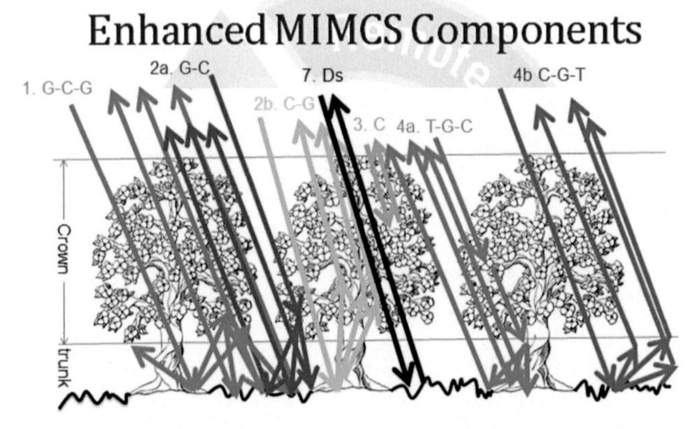

Figure 12.1 Backscattering components from vegetation with a canopy layer.

scattering wave that penetrates the canopy layer first, scattered forwarding by the trunk, scattering back to the antenna's direction and finally penetrates the trunk and canopy layer. Term 4b (C−G−T) represents the scattering wave in the reverse order. Under physical optical approximation, its forward scattering (to all azimuth directions) dominates. As a result, in terms 4a and 4b, only the forward scattering from the trunk is retained. If G is rough, the ground scattering energy is distributed to the upper hemisphere and the forward direction is not very strong. When flooded, the ground surface is close to ideally smooth and the forward scattering is significantly enhanced, which in turn, significantly enhances the 4a and 4b terms.

In natural areas, both open flood and beneath-vegetation flood can occur whereas in urban areas both the open-flood dampening effects and the L-shaped caused enhancement can be observed. Smooth inundated surfaces together with vertical buildings, when illuminated by the radar antenna, can form an L-shaped corner scatterer. In addition, more areas can be shadowed by buildings than by terrain, which indicates larger uncertainty in flood conditions in urban areas.

12.3 Error sources

Unfortunately, water is not the only surface that exhibits close-to-specular scattering. Smooth surfaces on the scale of the measuring wavelength and shadowed areas share almost identical scattering properties with water surfaces. We refer to them as water-like surfaces hereafter. Water-like surfaces vary slightly with frequency. At L-band (wavelength $\approx$ 23 cm) for example, even pasture can be identified as a water-like surface, if the grass height is short. At C-band (wavelength $\approx$ 5 cm), bare soil can be identified as a type of water-like surface, whereas at X-band (wavelength $\approx$ 3 cm), the aforementioned surfaces may be separated from water-like surfaces. If the objective is to detect open water automatically, water-like surfaces create "false alarms" or "over detections."

Noise-like speckles are encountered for almost all SAR applications. They are not real noise. As the distances between the elementary scatterers and the radar receiver vary due to the random location of scatterers, the received waves from each scatterer, although coherent in frequency, are no longer coherent in phase. A strong signal is received, if wavelets are added relatively constructively, while a weak signal is received if the waves are out of phase (Lee and Pottier, 2009). As a result, homogeneous and continuous areas exhibit strong inhomogeneity in SAR images.

The original geometry of SAR images is range-azimuth. In order to be geo-referenced, SAR images are often corrected to ground distance before the generation of high-level products. Limited by the accuracy of input elevation data, the geometric correction algorithm and orbit accuracy, it is common to see location errors at the level of a few pixels.

12.4 Methods

SAR-based inundation mapping methods are usually more complex than the ones based on optical sensors. This is due to processes added to reduce the error propagated from one or more aforementioned error sources. Consequently, one method may consist of several different approaches and it is therefore difficult to uniquely label it.

12.4.1 Supervised methods

Mapping water areas is a matter of classification, which includes supervised and unsupervised approaches. Supervised classification methods require a training set. A training set consists of labeled inundated areas or core water location and their corresponding pixels extracted from SAR data. The use of training sets allows achieving better accuracy without deeply understanding the physics before designing an algorithm (Borghys et al., 2006; Kussul et al., 2008; Pulvirenti et al., 2010; Song et al., 2007; Townsend, 2001; Töyrä et al., 2002; Zhou et al., 2000). However, drawbacks are at least two-fold: they cannot be automated and there is a local dependence, i.e., a well-trained classifier over one area may not work well in another area.

Pulvirenti et al. (2011b) developed an "almost-automatic" fuzzy logic classifier, which uses fixed thresholds determined by a theoretical scattering model to automatically classify pixels or to use human labeled samples. This method has been proven to be fairly accurate, but, on one hand, the former requires a microwave scattering model and introduces modeling uncertainty while, on the other hand, the latter is time-consuming.

Possible approaches to circumvent the drawbacks of supervised methods are the following unsupervised methods: threshold determination, segmentation, and change detection.

12.4.2 Threshold determination

The specular reflective properties of open water drove many efforts (Yamada, 2001; Giustarini et al., 2013; Hirose et al., 2001; Matgen et al., 2011) to determine a threshold below which pixels were identified as water. A single threshold may not hold well in large-area water bodies (Tan et al., 2004) or for the whole swath of SAR images since it suffers from heterogeneity of the environment, such as wind-roughening and satellite system parameters (Martinis et al., 2009). Martinis and Rieke (2015) demonstrate the temporal heterogeneity of the backscattering of permanent water bodies, implying the threshold temporal variability. To address the spatial variability, Martinis et al. (2009) apply a split-based approach (SBA) together with an object oriented (OO) segmentation method (Baatz, 1999). Matgen et al. (2011) introduce a histogram segmentation method and Giustarini et al. (2013) generalize the calibration process. Essentially, a threshold-based approach needs either a bimodal image histogram (Fig. 12.2 in Matgen et al., 2011) or some sample data to initialize the water distribution. To deal with nonbimodal histograms, manually drawing regions of interest (ROI) is the most straightforward solution, although it impedes automation. For automation, SBA (Martinis et al., 2009) ensures that only the splits showing a bimodal histogram (of water and nonwater pixels) are used to derive a global threshold. On the other hand, Lu et al. (2014) loosened the bimodal histogram restriction by initializing the water distribution using a "core flooding area" automatically

Figure 12.2 Workflow of the RAPID system. *RAPID*, Radar produced inundation diary.

derived from change detection, using multitemporal SAR images. More generally, dual-polarization (the common configuration of many active sensors and fully polarized SAR data) can be utilized for threshold segmentation (Shen et al., 2019). Shen et al. (2019) introduce a sophisticated method to identify water samples using water pixels of high probability via a water occurrence map. Pekel et al. (2016) have generated an induction map from 30-year Landsat images, while automatically removing strong scatterers and invalid low-scattering pixels that contaminated the samples.

12.4.3 Segmentation

In contrast to pixel-based threshold-determination, image segmentation techniques, which group connected homogeneous pixels into patches, can provide information at object level (i.e., at a higher level than pixel) and are therefore believed to be more resistant to speckles because they utilize morphological information instead of radiometric information alone. The active contour method (ACM; Horritt, 1999; Horritt et al., 2001) allows a certain amount of backscattering heterogeneity within a water body and incorporates morphological metrics, such as curvature and tension. Martinis et al. (2009) have applied OO segmentation (Baatz, 1999) with SBA to reduce false alarm and speckles. Heremans et al. (2003) compared the ACM and OO and concluded that the latter delineated the water areas more precisely than the former that tended to overestimate the extension of water areas. Pulvirenti et al. (2011a) have developed an image segmentation method consisting of dilution and erosion operators to remove isolated groups of water pixels and small holes in water bodies which are believed to be caused by speckles. Matgen et al. (2011), Giustarini et al. (2013), and Lu et al. (2014) employ a region growing algorithm (RGA) to extend the inundation area from detected water pixels. RGA starts from seeding pixels and then keeps absorbing homogenous pixels from neighbors until no more homogenous pixels exist in neighboring areas.

12.4.4 Change detection

Change detection approaches usually refer to the comparison techniques of pre- and in-flood SAR images in order to detect changes in pixels caused by flooding (Yamada, 2001; Hirose et al., 2001; Lu et al., 2014; Santoro and Wegmüller, 2012; Bazi et al., 2005). A (multiple) prior- or post-flood SAR image(s) is (are) needed, which is referred to as the dry reference hereafter. The principle of change detection techniques is straightforward and should be theoretically effective to overcome the first error source, overdetection on water-like surfaces, as done by Giustarini et al. (2013) and Matgen et al. (2011). But in practice, the aforementioned second and third error sources of geometric error and speckle-caused noise can compromise the effectiveness of change detection. Shen et al. (2019) propose an improved change detection (ICD) method that employs multiple ($\sim$5) pre-flood SAR images and a multicriterion approach to reject false positives caused by water-like bodies and reduce the effect of speckles.

Besides isolating inundated areas from permanent (or more precisely, preexisting) water areas, change detection techniques have been used to initialize water pixel sets before thorough detection (Lu et al., 2014; Martinis et al., 2009), especially when full automation is an objective and no a priori water information is available. Change detection methods can also be used to detect flood beneath vegetation, as described in the next section of this chapter.

12.4.5 Visual inspection/manual editing versus automated process

SAR-derived inundation results are affected by many error sources as discussed in the *Error sources* section, which can hardly be eliminated entirely by any algorithm. Visual inspection and manual editing could help to draw training ROIs (Pulvirenti et al., 2010). However, without full automation, a method can hardly be applied to rapid response of flood disasters, especially during back-to-back flood events, as manual editing requires overwhelming manpower. Studies aiming for automation include Martinis et al. (2009), Horritt (1999), Horritt et al. (2001), Horritt et al. (2003), Matgen et al. (2011), Giustarini et al. (2013), Pulvirenti et al. (2011b), Pulvirenti et al. (2011a), and Pulvirenti et al. (2013).

12.4.6 Open water, closed water versus urban flood

Inundation detection has gained attention in the recent past for vegetated areas, partially submerged wetlands, and urban areas. Theoretically, scattering is enhanced during flood time if a stalk structure exists. Ormsby et al. (1985) have evaluated the backscattering difference caused by flooding under canopy. Martinis and Rieke (2015) have analyzed the sensitiveness of multitemporal/frequency SAR data to different land cover and concluded that X-band could only be used to detect inundation beneath sparse vegetation or forest during the leaf-off period, while L-band, though characterized by better penetration, has a very wide range of backscattering enhancement, which also obscures the reliability of the classification. Townsend (2001) utilized ground data to train a decision tree to identify flooding beneath a forest using Radarsat-1 (http://www.asc-csa.gc.ca/eng/satellites/radarsat1/Default.asp). Horritt et al. (2003) input the enhanced backscattering at C-band and phase difference of co-polarizations (HH-VV) to the ACM to generate water lines from selected known open water (ocean) and coastal dry land pixels. The area between the dry contour and open water is considered as flooded vegetation. Pulvirenti et al. (2010) trained a set of rules using visually interpreted ROI to extract flooded forest and urban areas from COSMO-SkyMed SAR data. Pulvirenti et al. (2013) combined a fuzzy logic classifier (Pulvirenti et al., 2011b) and a segmentation method (Pulvirenti et al., 2011a) to monitor flood evolution in vegetated areas using COSMO-SkyMed (http://www.e-geos.it/cosmo-skymed.html). Given its potential for flood detection under vegetation, most studies adopt a supervised classification, which can hardly be automated. Specifically, the enhanced dihedral scattering of vegetation cannot be considered as a single class because of different vegetation species, structure, and leaf-off and -on conditions. Such heterogeneity makes it difficult to automatically determine a threshold of backscattering enhancement. In other words, the issue in detecting floods beneath vegetation is identifying multiple classes from an image, whose automation is more challenging than identifying a single class.

The challenges in detecting urban inundation share some similarities with identifying floods under vegetation due to vertical structure of many buildings. An additional challenge arises from the asymmetric building structure compared to the symmetric vegetation structure. Knowing the building geometry, orientation, material, and illumination direction is critical for urban flood detection (Ferro et al., 2011) and thus requires more ancillary data that are not always available. Moreover, smooth artificial surfaces and shadowing areas may also create overdetection. Nevertheless, at present, the major challenge of an operational system for urban flood mapping is still data availability. Ultrahigh resolution data ($\sim$1 m) such as TerraSAR-X (https://directory.eoportal.org/web/eoportal/satellite-missions/t/terrasar-x) and COSMO-SkyMed (http://www.e-geos.it/cosmo-skymed.html) are all commercial, which are less easily accessible than free datasets.

12.5 Selected studies to date

Past studies have only partially addressed the operational demands of real-time inundation mapping in terms of automation and accuracy. Often, tedious human intervention to reduce overdetection as well as filtering to reduce underdetection caused by strong scatter disturbances and speckle-caused noise are needed when using satellite SAR data for inundation mapping.

Martinis et al. (2009) have applied SBA (Bovolo and Bruzzone, 2007) to determine the global threshold for binary (water and nonwater) classification. In the SBA, a SAR image is first divided into splits (sub tiles) to determine their individual threshold using KI (Kittler and Illingworth, 1986), global minimum, and quality index. Then only qualified splits showing sufficient water and nonwater pixels are selected to get the global threshold. The OO segmentation (implemented in e-cognition software) is used to segment the image to continuous and nonoverlapping object patches at different scales. Then the global threshold is applied to each object. Eventually, topography is used as an option to fine-tune the results. SBA is employed to deal with the heterogeneity of SAR backscattering from the same object in time and space. The intention of applying OO segmentation algorithm was to reduce false alarms and speckle noise. However, OO was originally designed for high-resolution optical sensors, which has no consideration of noise like speckles and water-like areas. The fine-tune procedure can only deal with flood plains extended from identified water bodies, leaving inundated areas isolated from known water sources.

The ACM, also known as snake algorithm (Horritt, 1999), was the first image segmentation algorithm designed for SAR data to the authors' knowledge. It allows a certain amount of backscattering heterogeneity, while no smoothing across segment boundaries occurs. A smooth contour is favored by the inclusion of curvature and tension constraint. The algorithm spawns smaller snakes to represent multiply connected regions. The snake starts as a narrow strip moving along the course of the river channel, ensuring that it contains only flooded pixels. Overall, it can deal with a low signal-to-noise ratio. Horritt et al. (2003) use ACM to map waterlines under vegetation. They start from known pure ocean pixels to map the active contour of open water and to then map the second active contour, which is the waterline beneath vegetation. Two radar signatures forced the ACM, the enhanced backscattering at C-band and the HH-VV phase difference at L-band. Unlike the OO method which aggregates objects from bottom (pixel level) to top, the segmentation in ACM requires seeding pixels, whose detection is difficult in an automated approach. In addition, similarly to RGA, ACM cannot detect inundated areas isolated from a known water body.

To assess flooding beneath vegetation, Pulvirenti et al. (2011a) have developed an image segmentation method using multitemporal SAR images and utilized a microwave scattering model (Bracaglia et al., 1995) which combines the Matrix Doubling (Ulaby et al., 1986, Fung, 1994) and the integral equation model (Fung, 1994, Fung et al., 1994) to interpret the object backscattering signatures. The image segmentation method dilutes and erodes multitemporal SAR images using different window sizes. Then, an unsupervised classification is carried out at the pixel level. The object patches are formed based on connected pixels of the same class. Eventually, the trend of patch scattering can be compared with canonical values simulated by scattering model to derive the flood map. Pulvirenti et al. (2011b) proposed a fuzzy logic-based classifier to incorporate texture, context, and ancillary data (elevation, land cover, etc.), as an alternative to thresholding and segmentation methods. Eventually, they combined the segmentation and fuzzy logic classifier (Pulvirenti et al., 2013) by using backscattering change and backscattering during flooding as input for densely vegetated areas. The thresholds of the membership functions are fine-tuned with a scattering model. The use of image segmentation reduces the disturbance by speckles; the use of a multitemporal trend takes into account the progress of the water receding process; and the comparison with the scattering model sounds more objective and location independent than a supervised classification or empirical values. However, the adopted image segmentation may reduce the inundation extent detail and does not guarantee the removal of speckles either. Scattering models require detailed vegetation (moisture, height, density, canopy particle orientation and size distribution, trunk height and diameter) and ground soil information (moisture, roughness) as input, which are heterogeneous in space and usually not available.

Towards automation, Matgen et al. (2011) have developed the M2a algorithm to determine the threshold that makes the nonwater pixels (below the threshold) best fit a gamma distribution, a theoretical distribution of any given class in SAR image. They then extend flooded areas using RGA from detected water pixels using a larger threshold, 99% percentile of the "water" backscatter gamma distribution arguing that flood maps resulting from region growing should include all "open water" pixels connected to the seeds. Then they apply a change detection technique on backscattering to reduce within the identified water bodies overdetection caused by water-like surfaces as well as to remove permanent water pixels. Based on the same concept, Giustarini et al. (2013) have developed an iterative approach to calibrate the segmentation threshold, distribution parameter, and region growing threshold. The same segmentation threshold is also applied to the dry reference SAR image to obtain the permanent water area. However, they claim that if the intensity distribution of the SAR image is not bimodal, the automated threshold determination may not work. Lu et al. (2014) have used a changed detection approach to first detect a core flood area that contains more plausible but incomplete collection of flood pixels, and then to derive the statistical curve of the water class to segment water pixels. The major advantage of this approach is that a bimodal distribution is not compulsory. In practice, a nonbimodal distribution often occurs. However, the change detection threshold is difficult to determine and globalize. Their efforts succeeded in automating the process of determining the segmentation threshold yet leaving a number of common errors unaddressed:

1. The argument that flood maps resulting from region growing should include all "open water" pixels connected to the seeds can be untrue and therefore lead to under- and overdetections. If captured after the apex, at times large isolated and scattered flood pockets can develop disconnected from the pre-flooded water bodies due to variability in surface elevation and barriers. Limited by image resolution, narrow water paths or the ones covered by vegetation may appear isolated from the known core water zone as well. Water-like areas can be connected to real water bodies. For example, airports in Boston and San Francisco that are built along the Atlantic and Pacific Ocean, respectively, may be identified as water area because they are water-like (smooth) and connected to real water bodies.
2. Change detection at pixel level can eliminate false alarms caused by water-like surfaces. Both speckles and geometric errors can cause overdetection. In SAR applications, change detection approaches need to be applied with caution. For instance, change of shadow areas can be caused by the change of looking direction. Most studies that use change detection models (including the ones reviewed in this section) compared backscattering of collocated SAR pixels within and outside of the flood period. However, according to (Martinis et al., 2009;

Martinis and Rieke, 2015), backscattering exhibits large heterogeneity in space and time, which indicates that in a single object (water or nonwater), the direct difference in backscattering might mislead to erroneous interpretations. In addition, change detection was carried out from the output of RGA. Consequently, isolated inundation areas could not be detected.

3. Speckle caused error is not properly dealt with.
4. Most approaches had not been tested in large areas and their validation is usually limited to small areas along a river.

12.6 Case study

Shen et al. (2019) developed a radar produced inundation diary (RAPID) system that combined radar polarimetry, SAR statistics, morphological, and machine learning techniques to address the automation and quality problem in NRT inundation mapping. To overcome the common error in SAR-based inundation mapping, RAPID digests multi-source ancillary data including fine-resolution topography, water occurrence, land cover classification (LCC), river width, hydrograph, and water type databases besides SAR data. The RAPID NRT system is designed for the scenario depicted in Fig. 12.2.

Due to the large size of high resolution and global SAR data, overwhelming computational and storage resources are required to process all images blindly. The RAPID system is designed to be triggered by ground flood observation systems, such as US Geological Survey (USGS) flood watch (https://waterwatch.usgs.gov/?id = ww_flood) and Dartmouth Flood Observatory (DFO) river watch (https://floodobservatory.colorado.edu/DischargeAccess.html). Once many stations in a region issue flooding condition, the system will automatically search Sentinel-1 SAR data within the flooding region during flooding and pre-flooding periods. Then, its core algorithm is triggered to generate the inundation maps. The four-step framework of RAPID, depicted in Fig. 12.3, is briefly introduced here: (A) binary classification of water and nonwater at pixel level; (B) morphological processing to reduce over- (primarily) and under-detection at object level; (C) multithreshold compensation to reduce speckle noise; and (D) machine learning-based correction utilizing topography and the knowledge of stream network and water type.

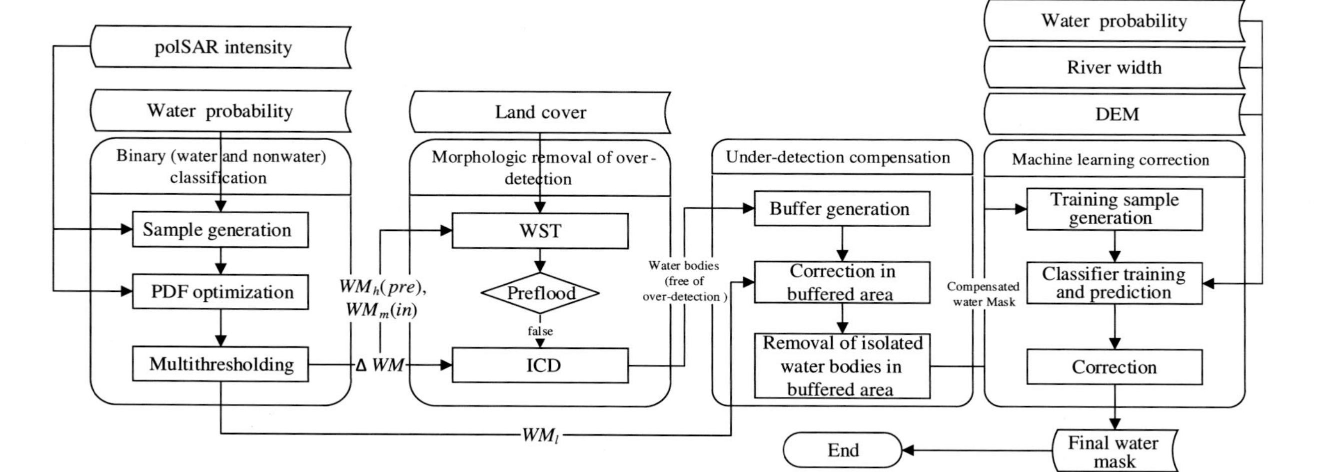

Figure 12.3 Core algorithm of the RAPID system. From columns from the left to the right are step A (binary classification), B (morphological processing consisting of WST and ICD, C (multithreshold compensation), and D (machine-learning-based correction). *ICD*, Improved change detection; *RAPID*, radar produced inundation diary; *WST*, water source tracing.

As pointed out by Lee and Pottier (2009) (Ulaby et al., 1982), noise-like speckle is not real noise, but rather a strong overlap between water and nonwater classes. Consequently, conventional single-threshold methods inevitably cause noisy classification results (Matgen et al., 2004, 2006) and the common practice attempting to filter speckle as noise cannot be the solution. RAPID therefore proposed a multithreshold scheme to reduce speckles. In principle, water sources can be found on a high-resolution LCC map for large water bodies, but may not be found for small water bodies. To prevent overdetection, RAPID applies water source tracing (WST) to map floodplain inundation extended from known water sources. To prevent underdetection of isolated water bodies, RAPID proposes an ICD technique. Unlike traditional change detection on backscattering within the detected mask of RGA (Matgen et al., 2011; Giustarini et al., 2013; Lu et al., 2014), ICD runs on the binary water masks at object level. It does not require a connection between changed water bodies and a known water source and is therefore capable of detecting isolated "islands" of inundation and resistant to speckles. Finally, RAPID employs a machine learning-based approach to reduce the error caused by strong scatterers and remaining speckles. The major limitation of RAPID is that it currently deals only with open floods.

With more and more SAR satellite images open to the public (e.g., Sentinel-1), more and more flood events could be captured since 2016. In this chapter, we demonstrate the inundation mapping capacity using the RAPID system (Shen et al., 2019) with Sentinel-1 SAR in four flood events: Typhoon Nepartak in Yangtze River, July 2016; Hurricane Harvey in Texas, August-September 2017; Vietnam, October 2017; and a flash flood coastal area in Boston on March 3, 2018.

The events and data are described in Table 12.1.

Sentinel-1 level-1 dual-polarized (VH + VV or HV + HH) SAR data for Nepartak in interferometric wide (IW) mode and in IW and (strip map) SM modes for Harvey in Ground Range Detected (GRD) format. After preprocessing the data using the European Space Agency (ESA) toolbox, the Sentinel Application Platform (SNAP), the resulting pixel spacing is 10 m × 10 m. The preprocessing includes four steps:

1. Orbit correction
2. Radiometric calibration
3. Range-Doppler geometric terrain correction
4. Incidence angle normalization

Table 12.1 Data and test events.

Event	Location	Mode	Pre-flood dates	In-flood dates
Nepartak	Hubei, China	IW	May 6, 18, 23, and 30, 2016	Jul. 5, 17, and 22, 2016
Harvey	Texas, USA	IW + SM	Jun. 25, Jul. 18, 24, 30, 31, Aug. 05, 11, 12, 18, 23, 2017	Aug. 29 and 30, and Sept. 4, 5, and 10, 2017
Pre-Damery	Vietnam	IW	Jun. 24, Aug. 23, 2017	Oct. 10, 2017
N/A	Massachusetts, USA	IW	Feb. 07, Feb. 19, 2018	Mar. 03, 2018

Steps 2 and 3 are sometimes referred to as radiometric terrain correction (RTC). For simplicity, we used the algorithm provided by Mladenova et al. (2013) to run step 4.

Table 12.2 provides ancillary data options. Categorized by type, they comprise LCC, water occurrence, hydrographic, and river width products. Of the Landsat-based LCC products, the National Land Cover Database (NLCD) (Homer et al., 2015; Fry et al., 2011; Homer et al., 2007; Vogelmann et al., 2001) is available in the United States at 5-year intervals, and the Finer Resolution Observation and Monitoring of Global Land Cover database (FROM-GLC) (Gong et al., 2013) (http://data.ess.tsinghua.edu.cn/) has been available all over the globe since 2010. In the NLCD taxonomy, water types are coded 90, "water bodies," and 95, "wetland," while highly developed types are 23, "built-up area with medium density," and 24, "built-up area with high density." In the FROM-GLC taxonomy, water types are 50, "wetland," and 60, "water bodies," while the highly developed type is 80, "artificial surfaces." For water occurrence, the only available dataset is produced by Pekel et al. (2016). For river width, two products, the Global River Width Database (Allen and Pavelsky, 2018) for Large Rivers (GRD-LR) (Yamazaki et al., 2014), are available in North America and globally, respectively. For hydrographic data, the National Hydrograph Dataset (NHD) plus v2 dataset (Simley and Carswell Jr, 2009) (http://www.horizon-systems.com/NHDPlus/NHDPlusV2_home.php) is available at 30-m resolution in the United States, while the GRD-LR contains all hydrographic variables at 90-m resolution globally. Since the region for our study was a segment of the Yangtze River, we used FROM-GLC and NARWidth as river width inputs and GRD-LR for LCC, river width, and hydrographic inputs.

Table 12.2 Input data to the radar produced inundation diary (RAPID) kernel algorithm.

Name	Source/Type	Producer	Time span	Coverage	Spatial res.	Revisiting intervals	Needed by step
Sentinel-1	SAR	ESA	Since 2014	Global	3.5/10 m	~2 days	A
NLCD	TM/LCC	USGS	1992–2011	US	30 m	5 years	B
Global30	TM/LCC	Tsinghua Univ.	2010 only	Global	30 m	One time	B
Water Occurrence	TM/water probability	ESA	1984–2015	Global	30 m	Static	A & D
Hydrograph	NHD	Horizon Systems Co.	N/A	US	30 m	Static	D
DEM	STRM	USGS	N/A	Global	30 m	Static	D
NARWidth	TM/River Width	George Allen	N/A	North America	30 m	Static	D
GWD-LR	STRM/River Width and Hydrograph	Dai Yamazaki	N/A	Global	90 m	Static	D
HydroLakes	STRM	WWF	N/A	Global	90 m	Static	D
US Detailed Water bodies	Multiple	ESRI, USGS, USEPA.	N/A	US	4 m	Static	D

12.6.1 Nepartak in Yangtze River, 2016

Typhoon Nepartak, known in the Philippines as Typhoon Butchoy, was the third most intense tropical cyclone worldwide in 2016. Nepartak severely impacted Taiwan and East China, with 86 confirmed fatalities.

Fig. 12.4 illustrates the processing steps for the image obtained on July 17 over a zoomed-in area containing roads and an airport (water-like) and inundated areas. Fig. 12.4C shows an area affected significantly by noise-like speckles and strong scatters, as identified through binary classification (Step A). One can hardly delineate the boundary of flood plain or flooded fields. Due to the synthetic aperture and range compression, strong scatters in the Yangtze River, infrastructure, and vehicles radiate backscattering to surrounding areas along the azimuth and range directions (the cross-shaped underestimated areas). Overdetection is observed over an airport in the bottom right corner of Fig. 12.5C and speckles over most overland areas. The morphological processing (Step B) eliminated most overdetections and slightly reduced underdetection, as shown in Fig. 12.4D. The compensation steps (Step C) further reduced the underdetection caused by the probability density function (PDF) overlapping between water and nonwater classes, as shown in Fig. 12.4(E). The remaining issue—underdetection caused by strong scatterers and speckles—was corrected by the machine learning-based method (Step D), as shown in Fig. 12.4(F). Fig. 12.4(F) shows the quality of the final map.

Results for May 6 and July 17, 2017 are presented in Fig. 12.5. Fig. 12.5D and G show that RAPID accurately captured the inundated area and that false alarms and noise were reduced without sacrificing, respectively, detectability and resolution. Where the LCC map was wrong, preexisting water pixels derived from SAR data could, infrequently, be misidentified as either inundated if they were connected to floodplain or nonwater if they were not connected. Both misidentification can be eliminated by overlaying water masks extracted by revisits of high-resolution optical remote-sensing satellite (such as Landsat or World View).

12.6.2 Harvey in Texas, 2017

Hurricane Harvey is tied with 2005s Hurricane Katrina as the costliest tropical cyclone on record, inflicting US$125 billion in damage, primarily from catastrophic rainfall-triggered flooding in the Houston metropolitan

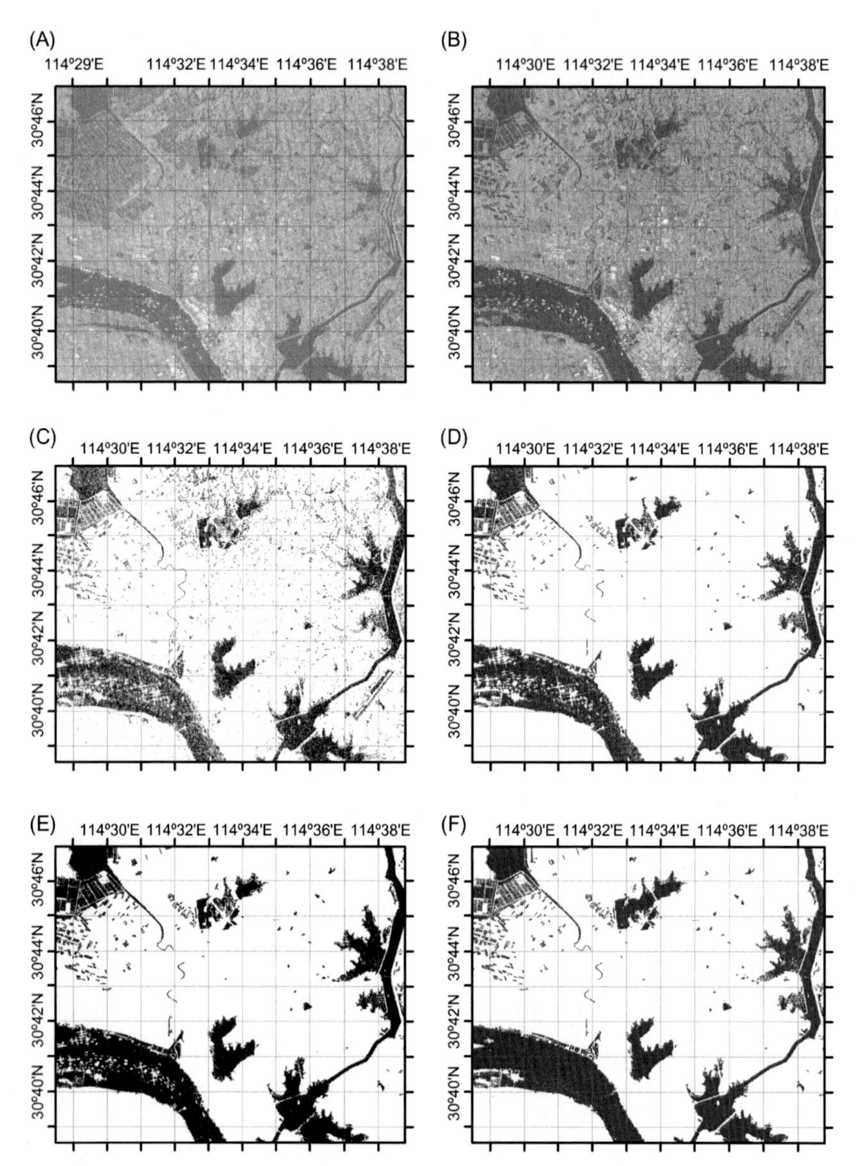

Figure 12.4 Demonstration of the major processing steps (A–D in the Methodology Section) using the water mask in the flooding time, July 17. (A) SAR data on July 17; (B) ESRI map; (C) water mask, WM_h, derived from the binary classification using the optimal threshold, Step A; (D) water mask from Step B, morphological processing; (E) water mask from Step C, compensation; and (F) final water mask from Step D, machine-learning correction.

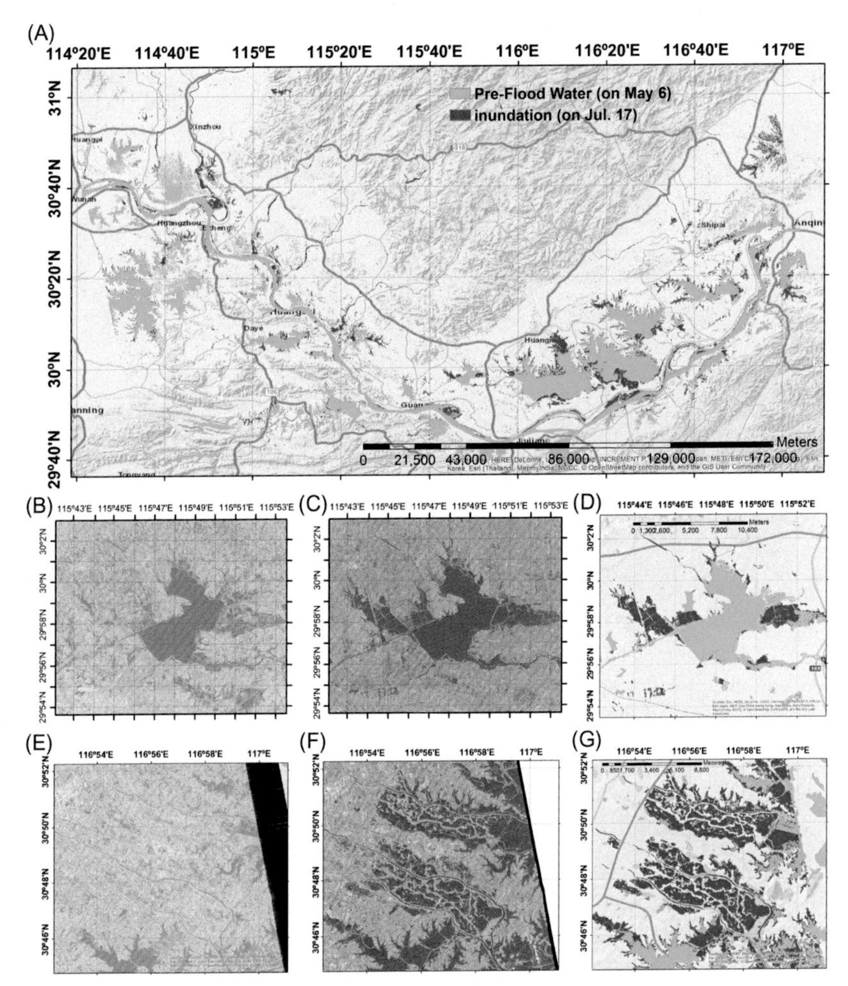

Figure 12.5 Inundation detection of a segment of Yangtze River (centered at Poyang Lake) using July 17, 2016 (B, E, in-flood) vs May 6, 2016 (E, F, pre-flood). (A) is the overview and (D) and (G) are the zoomed-in inundation maps.

area. It was the first major hurricane to make landfall in the United States since Wilma in 2005, ending a record 12-year span in which no hurricanes made landfall at the intensity of a major hurricane throughout the country. In a 4-day period, many areas received more than 1000 mm of rain as the system slowly meandered over eastern Texas and adjacent

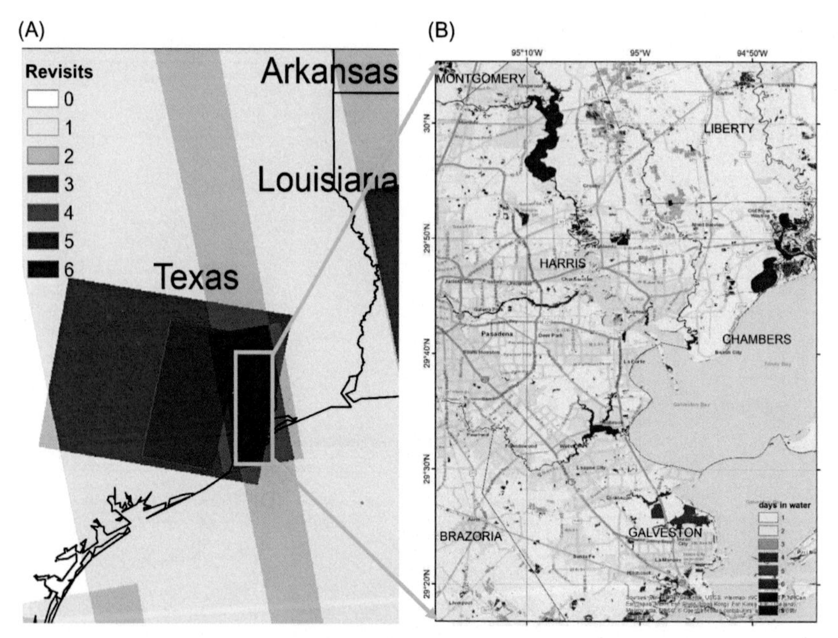

Figure 12.6 During Harvey, (A) Sentinel-1 SAR data availability days and coverage in Texas and (B) the inundated times of the Seabrook area.

waters, causing unprecedented flooding. With peak accumulations of 60.58 in (1539 mm), Harvey was the wettest tropical cyclone on record in the United States. The resulting floods inundated hundreds of thousands of homes which displaced more than 30,000 people and prompted more than 17,000 rescues.

Given the duration of Harvey spanned only from August 25, to September 12, 2017, eight revisits of Sentinel-1 SAR images in the Seabrook area (Fig, 12.6. (A)) were acquired during the event. Fig. 12.6 (B) shows the map of wet times of the common region that is within radar illumination at every Sentinel-1 revisit. Both S1A and S1B satellites data are available on some of the dates from different orbits. Fig. 12.7 shows an extreme inundation day in Texas, August 30, 2017. In Fig. 12.7B we can observe that after the apex, in some areas, inundation formed isolated lowlands from which no water source is visible. As a result, algorithms relying on the connectivity of known water bodies to flooded areas may fail.

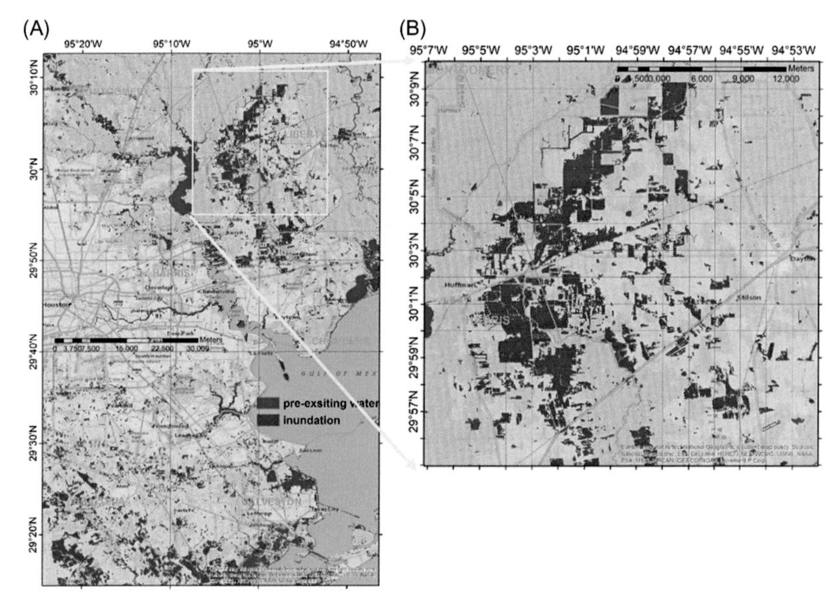

Figure 12.7 Inundation extents on a peak date, August 30, 2017: (A) overview and (B) isolated low land with inundation.

12.6.3 Vietnam, 2017

In Vietnam, a long-lasting flood occurred from July to November, 2017 with peaking in November caused by Typhoon Damrey. The flood map on October 10 was generated in Fig. 12.8. The affected region is characterized by hilly terrain. However, shadow areas did not disturb inundation delineation in hilly areas thanks to the use of ICD (yellow circles in Fig. 12.8).

12.6.4 A flash flood event in coastal area of Massachusetts on March 3, 2018

On March 1–3, 2018, Massachusetts experienced a flash flood caused by coastal storm. Both S1B and S1A passed above the flooded area on March 3 and 6, and images in total were acquired to cover Rhode Island, Massachusetts, New Hampshire, and part of Maine. Snapshots of the flood inundation are reported in Fig. 12.9.

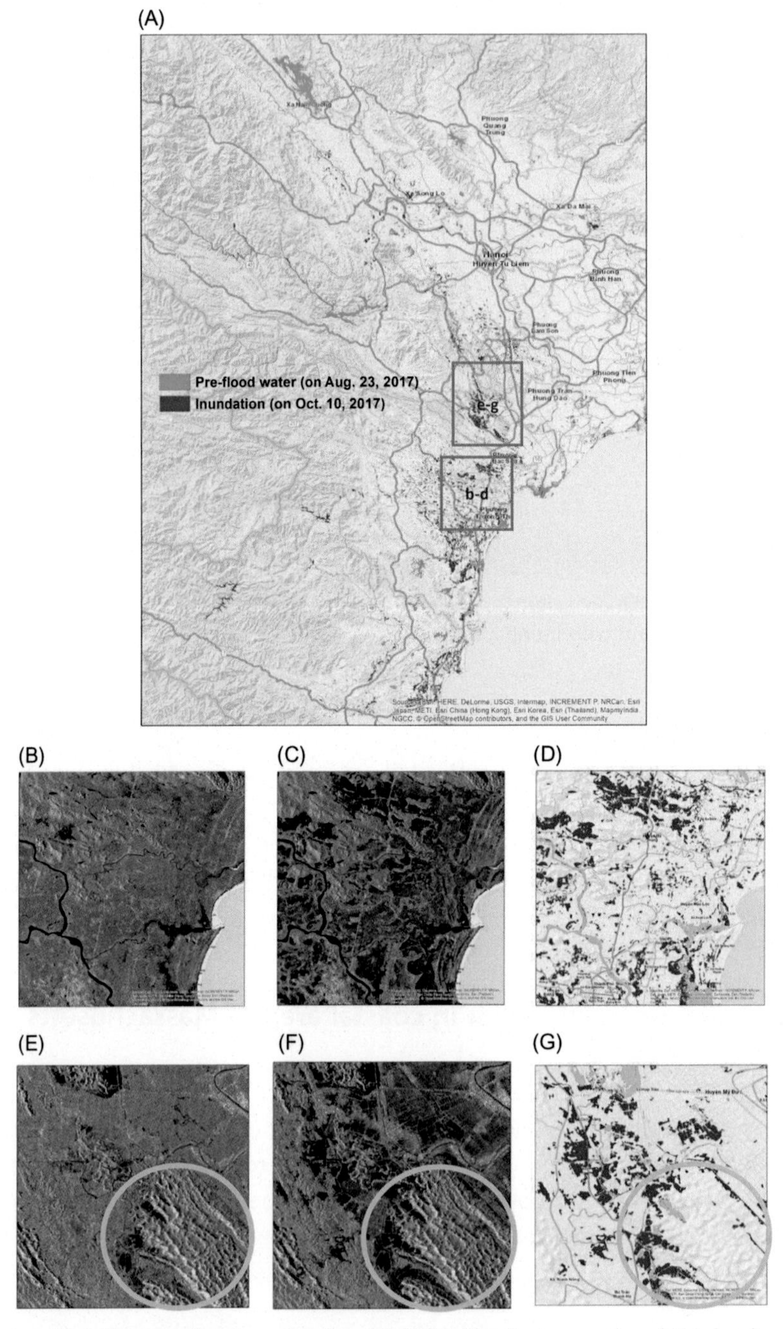

Figure 12.8 Vietnam flood on October 10, 2017. The first second and third columns from (B)–G) are pre-flood, in-flood SAR images, and inundation maps, respectively.

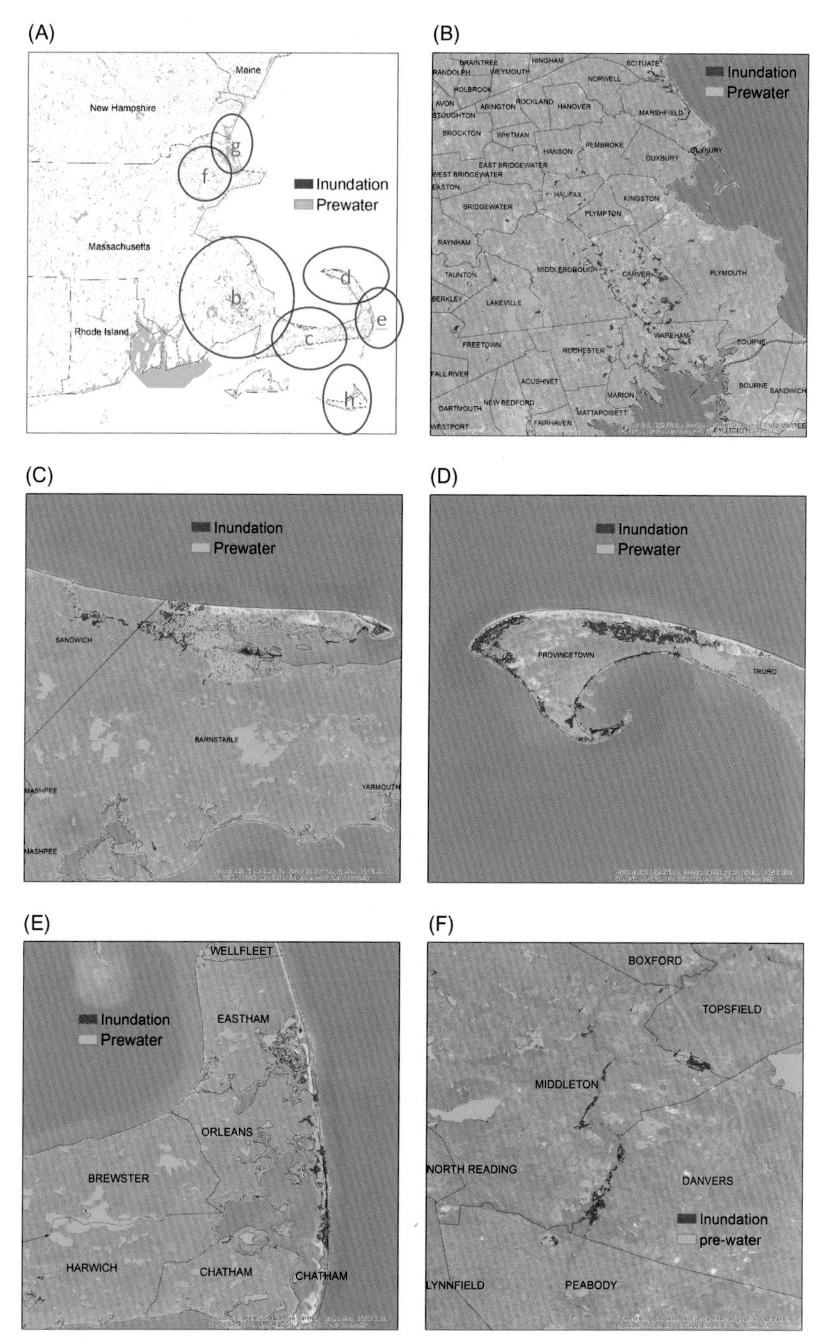

Figure 12.9 Flash flood in Massachusetts on March 3, 2018.

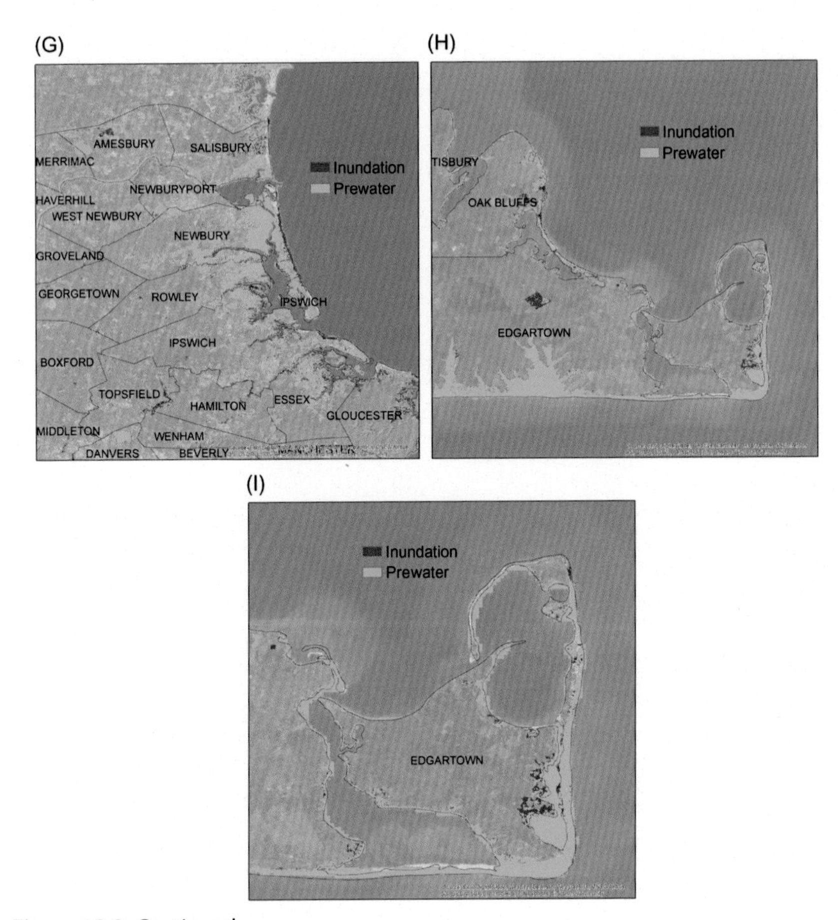

Figure 12.9 Continued.

References

Allen, G.H., Pavelsky, T.M., 2018. Global extent of rivers and streams. Science 361, 585–588.

Baatz, M. Object-oriented and multi-scale image analysis in semantic networks. In: Proc. the 2nd International Symposium on Operationalization of Remote Sensing, Enschede, ITC, Aug. 1999, 1999.

Bazi, Y., Bruzzone, L., Melgani, F., 2005. An unsupervised approach based on the generalized Gaussian model to automatic change detection in multitemporal SAR images. IEEE Trans. Geosci. Remote Sens. 43, 874–887.

Borghys, D., Yvinec, Y., Perneel, C., Pizurica, A., Philips, W., 2006. Supervised feature-based classification of multi-channel SAR images. Pattern Recogn. Lett. 27, 252–258.

Bovolo, F., Bruzzone, L., 2007. A split-based approach to unsupervised change detection in large-size multitemporal images: application to tsunami-damage assessment. IEEE Trans. Geosci. Remote Sens. 45, 1658–1670.

Bracaglia, M., Ferrazzoli, P., Guerriero, L., 1995. A fully polarimetric multiple scattering model for crops. Remote Sens. Environ. 54, 170–179.

Ferro, A., Brunner, D., Bruzzone, L., Lemoine, G., 2011. On the relationship between double bounce and the orientation of buildings in VHR SAR images. IEEE Geosci. Remote Sens. Lett. 8, 612–616.

Fry, J.A., Xian, G., Jin, S., Dewitz, J.A., Homer, C.G., Limin, Y., et al., 2011. Completion of the 2006 national land cover database for the conterminous United States. Photogramm. Eng. Remote Sens. 77, 858–864.

Fung, A.K., 1994. Microwave Scattering and Emission Models and Their Applications. Cambridge, New York.

Fung, A.K., Shah, M.R., Tjuatja, S., 1994. Numerical simulation of scattering from three-dimensional randomly rough surfaces. IEEE Trans. Geosci. Remote Sens. 32, 986–994.

Giustarini, L., Hostache, R., Matgen, P., Schumann, G.J.-P., Bates, P.D., Mason, D.C., 2013. A change detection approach to flood mapping in urban areas using TerraSAR-X. IEEE Trans. Geosci. Remote Sens. 51, 2417–2430.

Gong, P., Wang, J., Yu, L., Zhao, Y., Zhao, Y., Liang, L., et al., 2013. Finer resolution observation and monitoring of global land cover: first mapping results with Landsat TM and ETM + data. Int. J. Remote Sens. 34, 2607–2654.

Heremans, R., Willekens, A., Borghys, D., Verbeeck, B., Valckenborgh, J., Acheroy, M. et al., 2003, Automatic detection of flooded areas on ENVISAT/ASAR images using an object-oriented classification technique and an active contour algorithm. In: Recent Advances in Space Technologies, 2003. RAST'03. International Conference on. Proceedings of IEEE, pp. 311–316.

Hirose, K., Maruyama, Y., D.O. Van, Q., Tsukada, M. & Shiokawa, Y., 2001 Visualization of flood monitoring in the lower reaches of the Mekong River. In: Paper Presented at the 22nd Asian Conference on Remote Sensing, p. 9.

Homer, C., Dewitz, J., Fry, J., Coan, M., Hossain, N., Larson, C., et al., 2007. Completion of the 2001 national land cover database for the counterminous United States. Photogramm. Eng. Remote Sens. 73, 337.

Homer, C., Dewitz, J., Yang, L., Jin, S., Danielson, P., Xian, G., et al., 2015. Completion of the 2011 National Land Cover Database for the conterminous United States—representing a decade of land cover change information. Photogramm. Eng. Remote Sens. 81, 345–354.

Horritt, M., 1999. A statistical active contour model for SAR image segmentation. Image Vis. Comput. 17, 213–224.

Horritt, M.S., Mason, D.C., Luckman, A.J., 2001. Flood boundary delineation from synthetic aperture radar imagery using a statistical active contour model. Int. J. Remote Sens. 22, 2489–2507.

Horritt, M.S., Mason, D.C., Cobby, D.M., Davenport, I.J., Bates, P.D., 2003. Waterline mapping in flooded vegetation from airborne SAR imagery. Remote Sens. Environ. 271–281.

Kittler, J., Illingworth, J., 1986. Minimum error thresholding. Pattern Recogn. 19, 41–47.

Kussul, N., Shelestov, A., Skakun, S., 2008. Grid system for flood extent extraction from satellite images. Earth Sci. Inform. 1, 105.

Lee, J.S., Pottier, E., 2009. Polarimetric Radar Imaging: From Basics To Applications. CRC.

Lu, J., Giustarini, L., Xiong, B., Zhao, L., Jiang, Y., Kuang, G., 2014. Automated flood detection with improved robustness and efficiency using multi-temporal SARdata. Remote Sens. Lett. 5, 240–248.

Martinis, S., Rieke, C., 2015. Backscatter analysis using multi-temporal and multi-frequency SAR data in the context of flood mapping at River Saale, Germany. Remote Sens. 7, 7732–7752.

Martinis, S., Twele, A., Voigt, S., 2009. Towards operational near real-time flood detection using a split-based automatic thresholding procedure on high resolution TerraSAR-X data. Nat. Hazards Earth Syst. Sci. 9, 303–314.

Matgen, P., Henry, J., Pappenberger, F., Pfister, L., De Fraipont, P., Hoffmann, L., 2004. Uncertainty in calibrating flood propagation models with flood boundaries derived from synthetic aperture radar imagery. Proc. 20th Congr. Int. Soc. Photogramm. Remote Sens. Istanbul, Turkey, pp. 352–358.

Matgen, P., EL Idrissi, A., Henry, J.B., Tholey, N., Hoffmann, L., De Fraipont, P., et al., 2006. Patterns of remotely sensed floodplain saturation and its use in runoff predictions. Hydrol. Process. 20, 1805–1825.

Matgen, P., Hostache, R., Schumann, G., Pfister, L., Hoffmann, L., Savenije, H., 2011. Towards an automated SAR-based flood monitoring system: lessons learned from two case studies. Phys. Chem. Earth, A/B/C 36, 241–252.

Mladenova, I.E., Jackson, T.J., Bindlish, R., Hensley, S., 2013. Incidence angle normalization of radar backscatter data. IEEE Trans. Geosci. Remote Sens. 51, 1791–1804.

Ormsby, J.P., Blanchard, B.J., Blanchard, A.J., 1985. Detection of Lowland Flooding Using Active Microwave Systems. American Society of Photogrammetry. 51, 317–328.

Pekel, J.-F., Cottam, A., Gorelick, N., Belward, A.S., 2016. High-resolution mapping of global surface water and its long-term changes. Nature 540, 418–422.

Pulvirenti, L., Pierdicca, N., Chini, M., 2010. Analysis of Cosmo-SkyMed observations of the 2008 flood in Myanmar. Italian J. Remote Sens. 42, 79–90.

Pulvirenti, L., Chini, M., Pierdicca, N., Guerriero, L., Ferrazzoli, P., 2011a. Flood monitoring using multi-temporal COSMO-SkyMed data: image segmentation and signature interpretation. Remote Sens. Environ. 115, 990–1002.

Pulvirenti, L., Pierdicca, N., Chini, M., Guerriero, L., 2011b. An algorithm for operational flood mapping from Synthetic Aperture Radar (SAR) data using fuzzy logic. Nat. Hazards Earth Syst. Sci. 11, 529.

Pulvirenti, L., Pierdicca, N., Chini, M., Guerriero, L., 2013. Monitoring flood evolution in vegetated areas using COSMO-SkyMed data: The Tuscany 2009 case study. IEEE J. Select. Top. Appl. Earth Observ. Remote Sens. 6, 1807–1816.

Santoro, M., Wegmüller, U., 2012. Multi-temporal SAR metrics applied to map water bodies. In: Geoscience and Remote Sensing Symposium (IGARSS), 2012 IEEE International. IEEE, pp. 5230–5233.

Shen, X., Hong, Y., Yuan, W., Chen, S., 2010. A backscattering enhanced canopy scattering model based on MIMICS. In: American Geophysical Union (AGU) 2010 Fall Meeting, December 2010 San Francisco. AGU.

Shen, X., Anagnostou, E.N., Allen, G.H., Brakenridge, R.G., Kettner, A.J., 2019. Near-Real Time Non-obstructed Inundation Mapping by Synthetic Aperture Radar. Remote Sens. Environ. 221, 302–315.

Simley, J.D. & W.J. Carswell Jr. 2009. The national map—hydrography. In: US Geological Survey Fact Sheet.

Song, Y.-S., Sohn, H.-G., Park, C.-H., 2007. Efficient water area classification using Radarsat-1 SAR imagery in a high relief mountainous environment. Photogramm. Eng. Remote Sens. 73, 285–296.

Tan, Q., Bi, S., Hu, J., Liu, Z., 2004. Measuring lake water level using multi-source remote sensing images combined with hydrological statistical data. In: Geoscience and Remote Sensing Symposium, 2004. IGARSS'04. Proceedings. 2004 IEEE International. IEEE, pp. 4885–4888.

Townsend, P.A., 2001. Mapping seasonal flooding in forested wetlands using multi-temporal Radarsat SAR. Photogramm. Eng. Remote Sens. 67, 857–864.

Töyrä, J., Pietroniro, A., Martz, L.W., Prowse, T.D., 2002. A multi-sensor approach to wetland flood monitoring. Hydrol. Process. 16, 1569–1581.

Ulaby, F.T., Moore, R.K., Fung, A.K., 1982. Microwave Remote Sensing: Active and Passive. Artech House Inc, London UK.

Ulaby, F.T., Moore, R.K., Fung, A.K., 1986. Microwave Remote Sensing: Active and Passive. Artech House Inc, London UK.

Vogelmann, J.E., Howard, S.M., Yang, L., Larson, C.R., Wylie, B.K., Van Driel, N., 2001. Completion of the1990s National Land Cover Data Set for the conterminous United States from Landsat Thematic Mapper data and ancillary data sources. Photogramm. Eng. Remote Sens. 67.

Yamada, Y., 2001. Detection of flood-inundated area and relation between the area and micro-geomorphology using SAR and GIS. In: Geoscience and Remote Sensing Symposium, 2001. IGARSS'01. IEEE 2001 International. IEEE, pp. 3282–3284.

Yamazaki, D., O'loughlin, F., Trigg, M.A., Miller, Z.F., Pavelsky, T.M., Bates, P.D., 2014. Development of the global width database for large rivers. Water Resour. Res. 50, 3467–3480.

Zhou, C., Luo, J., Yang, C., Li, B., Wang, S., 2000. Flood monitoring using multi-temporal AVHRR and RADARSAT imagery. Photogramm. Eng. Remote Sens. 66, 633–638.

CHAPTER THIRTEEN

Storm surge and sea level rise: Threat to the coastal areas of Bangladesh

Ali Mohammad Rezaie[1], Celso Moller Ferreira[1] and Mohammad Rezaur Rahman[2]
[1]George Mason University, Fairfax, VA, United States
[2]Bangladesh University of Engineering and Technology, Dhaka, Bangladesh

13.1 Introduction

Storm surge due to hurricanes and tropical cyclones (TCs) is one of the deadliest global natural hazards, killing more than half a million people in the past 60 years (Mohanty et al., 2014), with an estimated average of 13,000 annual deaths around the world (Dilley et al., 2005). In 1970, the "Great Killer" Cyclone took the lives of 300,000 people (Debsarma, 2009; Institute of Water Modeling, 2009) in Bangladesh. Moreover, both satellite altimetry and tidal records show a global rise in sea levels (Church et al., 2004; Church and White, 2006, 2011) and over the next century, sea level is expected to rise at a faster rate than in the last 50 years (Church et al., 2013). Thus, extreme events such as TCs, under a future sea level rise (SLR), pose a major coastal flooding threat to developing countries with low-lying coastal areas (Wong et al., 2013; Barbier 2014).

In this chapter, we describe the most important physical processes related to storm surge and SLR. The chapter also provides insights into flooding in the coastal areas of Bangladesh with an overview on the existing coastal protection measures and discusses potential strategies to build coastal resilience in low-lying developing countries such as Bangladesh.

Extreme Hydroclimatic Events and Multivariate Hazards in a Changing Environment.
DOI: https://doi.org/10.1016/B978-0-12-814899-0.00013-4

13.2 Tropical cyclone-induced storm surge

Storm surge induced by TC is a complex phenomenon, which involves several atmospheric, oceanographic, coastal, and geographic factors. The United States (US) National Oceanic and Atmospheric Administration defines storm surge as an "abnormal" rise in water levels (higher than the projected astronomical tide) generated by severe atmospheric conditions such as a TC, lasting for a period ranging from few minutes to several days. Fig. 13.1 shows the unusual increase in observed water levels at the tidal station in the Barisal River in Bangladesh due to TC Sidr (2007), which made landfall around 1800 UTC on the November 15. TC generally refers to a large-scale closed atmospheric circulation (100 to more than 1000 km from the storm center, Montgomery and Farrell, 1993) triggered by heat transfer from tropical ocean basins (Neumann et al., 1993). These rotating storms are usually generated near the equatorial regions (Palmen, 1948; Gray 1968) during late summer, when the sea surface temperature is at its peak. However, both timing of cyclogenesis and naming of TCs vary with the ocean basin. For example, in the Atlantic Basin, TCs normally form from June to November and are called "Hurricanes" (National Hurricane Center, 2018), while in the Indian Ocean, they are simply called "Cyclones," which usually occur from April to December (Indian Meteorological Department, 2018).

Although the exact physical processes of TC formation are not yet fully understood (Gray, 1968; Gray, 1998; Emanuel, 2003; McTaggart-Cowan et al., 2008; Montgomery, 2016; Muller and Romps, 2018;

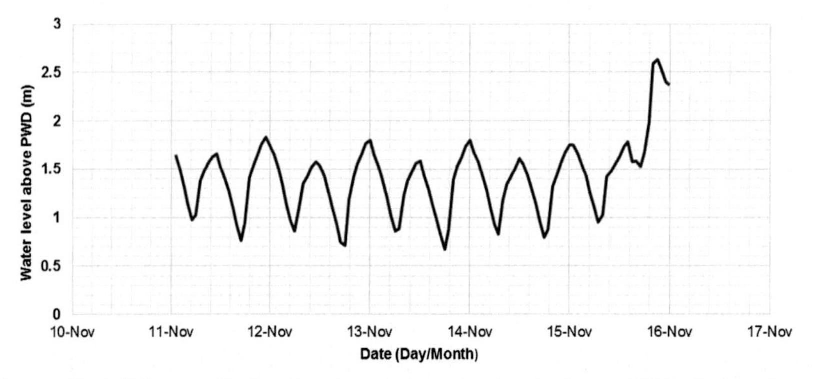

Figure 13.1 "Abnormal" rise in observed water level above PWD during Super Cyclone Sidr (2007) in the Barisal River, Bangladesh. *PWD*, Public water datum.

Montgomery and Farrell, 1993), when sea surface temperature exceeds 26.5°C (Gray, 1968), the near-surface warm and moist air rises upward creating a low pressure zone in the atmosphere. The surrounding air in low altitudes moves towards the low-pressure zone to fill up the pressure deficit in the near-surface atmospheric region. As these processes continue, the rising air containing high moisture content condenses to form thick clouds around 10 km altitude in the troposphere (lowest or near surface atmospheric layer) and creates a cluster of thunderstorms (Gray, 1968). Because of the Earth's rotation and the Coriolis effect, the cluster of thick clouds generates a circular pattern that forms tropical depressions (wind speed 17 ms^{-1}). Additionally, in order for a TC to develop, the spiraling wind disturbance has to be generated at 5° latitude—at least—from the equator (the intertropical Convergence Zone) and with low wind shear (less than 10 m/s) in the mid-troposphere with sufficient vorticity and convergence (Gray 1968). The center of the TC, where the pressure is lowest, is called "Eye," which is approximately 30–60 km in diameter (Weatherford and Gray 1988) and surrounded by an "eyewall," where the wind forcing is the strongest.

After they develop, TCs are driven by the ocean atmospheric circulation and their path depends on the pattern of the air–sea movement of the specific ocean basin (Needham and Keim, 2011). When approaching a coastline, the seawater is driven inland by the combined effect of low pressure and persistent wind over a shallow water body or coastal platform. Thus, the magnitude of the resulting storm surge depends on the storm intensity, size (Irish, Resio and Ratcliff, 2008), forward speed (Rego and Li, 2009), angle of approach, geometry of the basin, continental shelf, regional topography, and local coastal features, such as flood infrastructure and natural features, that impede the intrusion of seawater (Resio and Westerink, 2008; Rego, 2009; Needham and Keim, 2011).

13.3 Climate change and sea level rise

SLR is one of the major consequences of global warming due to the enhanced amount of greenhouse gases in the atmosphere (IPCC 2007; IPCC 2013). Over the past 120 years (1880–2012), the average of the global land and ocean surface temperature has increased by 0.85°C

(0.65°C−1.06°C; Hartmann et al., 2013), and the rate of global mean surface temperature rise in the last 50 years is almost twice the rate over last 100 years (Trenberth and Josey, 2007). Consequently, the melting and loss of ice contents in Greenland and Antarctica (Lemke et al., 2007) are contributing to an increase of global seawater. Thermal expansion of seawaters, as a result of ocean temperature warming (Federation and Lynne, 2013), is also causing the global sea level to rise (Church et al., 2013). Moreover, both sea level records (Church and White, 2006; Jevrejava et al., 2008) and satellite altimetry data (Nerem et al., 2010; Church and White, 2011) show an increase in global seawater levels.

Although sea levels have always changed over geologic timescales (Titus, 1985), the Intergovernmental Panel on Climate Change (IPCC) suggests that a total of 0.19 m rise in global sea level from 1901 to 2010, with a 1.7 (1.5−1.9) mm/year rate of SLR. The rate of SLR during 1993 to 2010 is estimated to be 3.2 (2.8−3.6) mm/year, which is almost double the rate in the past 100 years. In terms of predicting future rises in sea level, there are significant challenges in projecting carbon-cycle feedbacks (Meehl et al., 2007), natural climate variability, rate and extent of glacier losses (Parris et al., 2012; Church et al., 2013), and incorporating complete physical characteristics and mass balance of ice sheet (Church and Gregory, 2001). Thus, depending on different rates of global warming and emission scenarios, the latest end of the century (2100) projection of SLR by the IPCC indicates a 0.26−0.55 m and 0.52−0.98 m SLRs for the lowest (RCP2.6) and the highest (RCP 8.5) greenhouse gas concentration scenarios (Church et al., 2013). In addition, due to different ocean circulation patterns, nonuniform ocean waters warming and salinity variations, the trend and rate of sea level change vary regionally (Cazenave and Llovel 2010; Nicholls and Cazenave 2010; Cazenave and Cozannet 2013; Wong et al., 2013), which is why SLR projections differ from coast to coast.

13.4 Background on the coastal zone of Bangladesh

Bangladesh has historically been vulnerable to storm surge and coastal flooding because of its large exposure to these hazards, its geographic location, demography and socioeconomic conditions in the coastal zone of Bangladesh. From 1877 to 1995, about 154 cyclones occurred on the Bangladeshi coast, which includes 43 severe cyclonic storms, 43 cyclonic storms, 68 tropical depressions (Dasgupta et al., 2010), and at least 70 of them caused major disasters (GoB, 2004a). In order to comprehend Bangladesh's vulnerability to storm surge and coastal flooding, it is important to understand the natural settings and the geomorphology of the coastal zone of the country.

Bangladesh covers an area of about 147,570 km^2 and lies approximately between 20°30′ and 26°40′ N latitude and 88°03′ and 92°40′ E longitude. It is formed by the one of largest tide dominated deltas in the world, the Ganges–Brahmaputra–Meghna (GBM) (Goodbred and Saito, 2010) and all upstream flow from the GBM delta drains into the Bay of Bengal through the coastal areas of Bangladesh (Fig. 13.2). The coastline is approximately 710 km long (MoWR, 2005), with more than 38 million people living in the coastal zone (BBS, 2014). The coastal zone of Bangladesh is delineated (PDO-ICZMP, 2003) based on its vulnerability to: (1) tidal water movements, (2) salinity intrusion, and (3) cyclone and

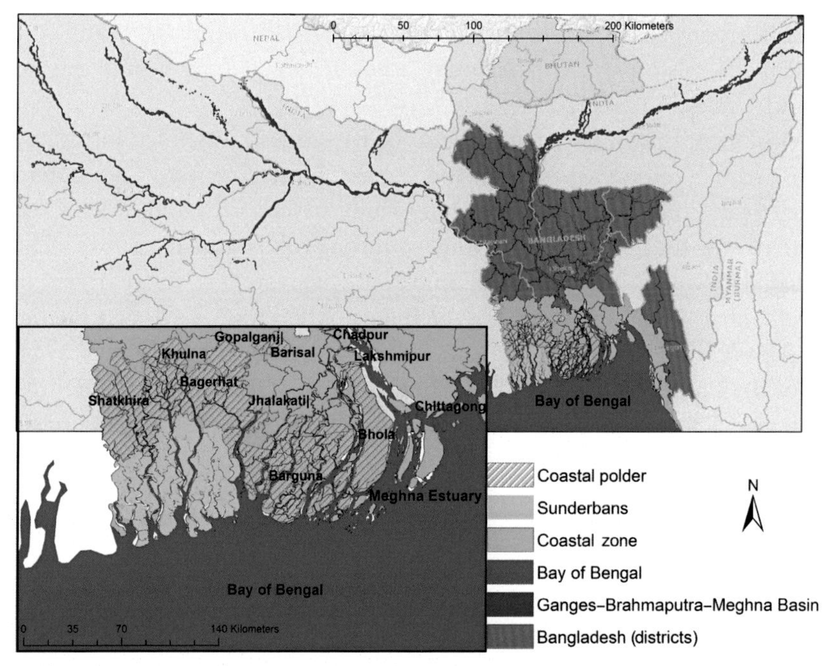

Figure 13.2 Coastal zone of Bangladesh in the Ganges–Brahmaputra–Meghna delta.

storm surge within the administrative boundaries (districts). The coastal region contains 32% of land mass of the country (Islam, 2004) and consists of 19 districts out of 64 districts. The 12 sea facing districts, which are susceptible to all three criteria above, are considered as the "exposed coast," whereas the rest behind is called the "interior coast." The coastal zone of Bangladesh extends up to 370.4 km (200 nautical miles) in the Bay of Bengal, including the "Exclusive Economic Zone" (UNCLOS, 1982; Article 57).

The average population density of the coastal zone is 743 habitants/km^2, where the population density is 482 and 1012 habitants/km^2 in the exposed and interior coast, respectively (PDO-ICZMP, 2003a,b). During the fiscal year of 1999–2000, the per capita gross domestic product for the coastal zone was US$277, which is close to the national average (US $278) (Sarwar, 2005). The coastal population in Bangladesh is mostly dependent on the local natural resources for their livelihood and the major activities are fishing, agriculture, shrimp farming, salt farming, and tourism (PDO-ICZMP, 2003a,b; GoB, 2004b). Due to increased coastal flooding and salinity intrusion, the land use patterns in the coastal zone are changing from traditional crop farming/agriculture to shrimp farming and salt production (Islam 2006). The largest mangrove in the world, the Sundarbans, also plays a significant role for providing livelihoods to around 4 million GoB (2004b).

Moreover, the entire coastal region of Bangladesh is divided into three zones, namely eastern, western, and central coastal region. The eastern coastal zone consists of a 145-km long sandy beach, stretching from Cox' Bazar to the south most tip of the country, Teknaf (Islam, 2001). The Karnafully, Sangu, Matamuhury, and Naf River are the major rivers in this area that drain out to the Bay of Bengal where the Naf River divides Bangladesh from Myanmar (Sarwar, 2005). The eastern zone has an average of 4–5 m elevation with the Chittagong coastal plain surrounded by hill tracts. The central zone is the most active part of coastal Bangladesh due to the continuous processes of accretion and erosion (Ali, 1999). The combined outflow of the GBM river basin drains out to the Bay of Bengal through the Meghna River estuary, and thus, this zone simultaneously receives a large volume of flow and amount of upstream sediment. Finally, the western part is mostly a semi-active delta with low and flat topography. The average elevation of the coastal southwestern part is about 1–2 m (GoB, 2016) and is crisscrossed by a network of interconnected distributaries and estuaries. The region also contains a total of 123 coastal polders (i.e., earthen peripheral embankments) that span for 5107 km to protect coastal communities from saline water intrusion and tidal flooding (more information is provided in the following section). The Sundarbans, named after the predominant "Sundri" trees, is another important feature of the western zone. According to Iftekhar and Islam (2004), the Sundarbans mangrove area lies within a range of 0.9-2.1m above the mean sea level. The mangrove ecosystem not only provide the feeding and breeding grounds for fish and shrimp, but also plays significant role against the impacts of TCs and storm surges (Ali, 1999; Dasgupta et al. 2017).

13.5 Hydromorphological features in the Bangladesh coast

The Bengal is an active tide-dominated delta, influenced by the freshwater inflow and sediment discharge from the GBM river system and the tidal interactions in the coastal regions of Bangladesh (Goodbred and Saito, 2010). On average, the Ganges and Brahmaputra Rivers bring more than 1,037 million tons of sediment load per year to Bangladesh, where 51% of the total sediment is transported to the coastal region, while the rest is deposited in the floodplains and river channels (Islam et al. 1999). However, a recent study by Rahman et al. (2018) suggests that the actual sediment inflow can be around 50% less than previous estimations. The siltation rate in the Ganges is around 2.5 times higher than the rate in the Brahmaputra River basin, although the sediment load is 2.3 times higher in the Brahmaputra River (Islam et al., 1999). The sediment transport increases during the monsoon wet season when the water discharge is higher. Coleman (1969) showed that 87% of the total sediment and 75% of the total discharge occurs during the monsoon season. Thus, it is anticipated that over the geologic timescale, the delta is moving eastwards, towards the central part of the coastal zone, with the majority of upstream flow draining through the Lower Meghna estuary (Allison et al., 2001; Sarker et al., 2013; Brammer, 2014; Rahman et al., 2014). In addition to the natural sedimentation process, human interventions (polders) and regulation of water flow through barrages in the upstream (Akter et al., 2016; Rahman and Rahaman, 2017) have influenced the sedimentation processes in the southwestern part of the coast.

In terms of overall accretion and subsidence, geomorphologic changes are also spatially dynamic in the coastal regions of Bangladesh. Over the last five decades (1943–2008), the Meghna Estuary (Fig. 13.2) has accreted 1700 km^2 of land mass (Sarker et al., 2011) with a deposition rate of about 3.5 mm/year (Darby et al., 2018). While these sedimentation may offset the SLR and coastal erosion in general, the southwestern part of the delta is at risk of tidal flooding and erosion due to deficits in sediment influx (Wilson and Jr, 2015). Several studies also indicate significant changes in the fluvio-tidal regime (Rahman et al., 2014) and geomorphology (Brammer, 2014) in the channels and tributaries of the Meghna Estuary. The estuary, which gains 1880 ha/year of land annually, is highly dynamic in terms of local erosion, accretion, and migration of channels and banks (Meghna Estuary Study, 2001). Furthermore, Brown and Nicholls (2015) estimated an average overall subsidence rate of 5.6 mm/year (with a median of 2.9 mm/year) in the GBM delta based on the measurements from previous studies, suggesting that continued development can increase the rate locally. Brammer (2014) indicates that the rate of subsidence is also higher in the central part of the coastal zone of Bangladesh. Thus, hydromorphological changes in the coast of Bangladesh are spatially dynamic, and consequently, the impacts of coastal flooding are also different over the coastal region.

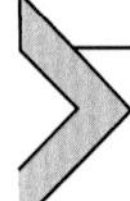

13.6 Coastal flooding and storm surges in Bangladesh

Coastal flooding in Bangladesh is a regular phenomenon that can be attributed to: (1) fluvial, (2) fluvio-tidal, (3) tidal, and (4) TC-induced storm surge floods (Haque and Nicholls, 2018). Due to excessive rainfall-runoff during the monsoon season, the large volume of upstream flow exceeds the conveyance capacity of the rivers, causing fluvial flooding in the coastal areas of the country. The floodplains of the Meghna Estuary are also susceptible to fluvial flooding as it drains the largest portion of the combined GBM flow. Haque and Nicholls (2018) also indicated the importance of fluvio-tidal flooding, which occurs due to seasonal variation of tidal and ebb dominance of estuaries in the coast of Bangladesh. Additionally, the GBM delta has a meso-to macro-tidal range (meso: 2–4 m; macro-tidal >4 m) (Sarker et al., 2011; Bricheno et al., 2016), where tides can encroach up to 100 m (Choudhury and Haque 1990) into the low-lying coast. In the central coastal zone, the mean tidal range varies from 2 to 4 m, with 4–7 m of tidal range in the lower Meghna Estuary and its adjacent islands (Meghna Estuary Study, 2001; BIWTA, 2015). Areas near the Sundarbans in the western zone and Cox's Bazar in the eastern part have a 2–3 and 3.6 m of mean tidal range, respectively (BIWTA, 2015; Haque and Nicholls, 2018). Monsoon wind patterns and velocity can also contribute to tidal flood propagation in the coastal Bangladesh (Haque and Nicholls, 2018).

In terms of TC-induced storm surge flooding, the Bay of Bengal is a potential region for the development of cyclonic storms (Gray, 1968; Mooley and Mohile, 1983; Mohanty, 1994; Ali, 1999), as more than 75% of the total TCs over the globe causing fatalities of 5,000 people—or more–have occurred in the North Indian Ocean (Mohanty et al., 2014). Historical cyclones (e.g. Barisal Cyclone in 1584, Bakrergonj cyclone in 1876, the 1970 cyclone, Urir Char Cyclone in 1985, the April 1991 cyclone, the 2004 cyclones, Sidr in 2007, Aila in 2009) have brought Bangladesh into global attention as a disaster prone country. On average, every 3 years one severe cyclone hits the coast of Bangladesh and five severe cyclones have impacted the country's coast since 1995 (GoB 2009; cited in Dasgupta et al., 2014). Almost every year during pre- (April–May) and post- (October–December) monsoon seasons (Hoque, 1991; Debsarma, 2009; Dasgupta et al., 2010), tropical depressions form near the coast. In 2004, Bangladesh was classified as the world's most vulnerable country to TCs by the (United Nations Development Programme, 2004), as storm surge caused hundreds of thousands of fatalities and billions of dollars in damage in the coastal regions of Bangladesh. According to the estimates by the Government of Bangladesh (GoB, 2004a), approximately 900,000 people have died in cyclonic disasters in the last 35 years (1969–2004) and Chowdhury and Karim (1996) mentioned around 700,000 people since 1960. While mass causalities during TCs has reduced over time (the death toll for the November 12, 1970 and April 29, 1991 TCs were 300,000 and 138,882 people, respectively; Flather, 1994; Debsarma, 2009), the estimated damage and loss due to the

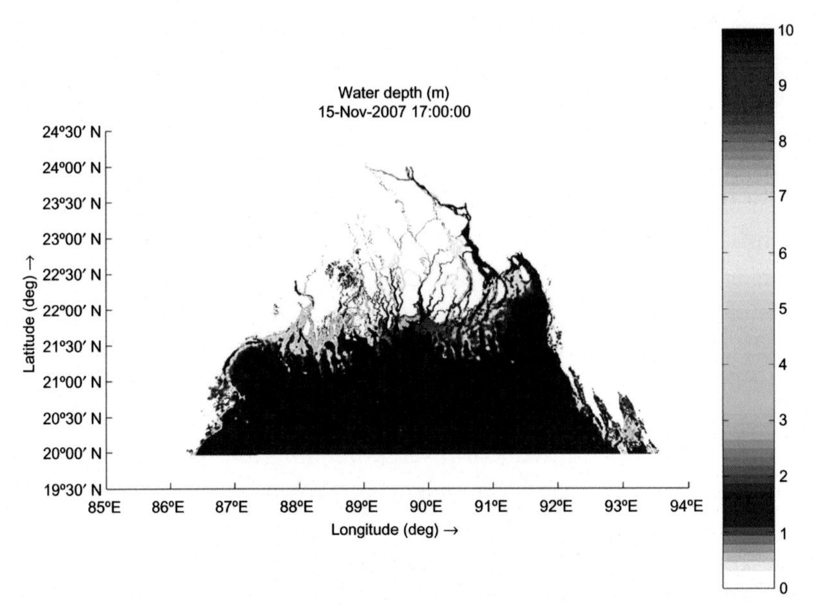

Figure 13.3 Simulated inundation map due to super cyclone Sidr in 2007 (Model Development and simulations were carried out in Rezaie, 2015).

"super cyclone" Sidr was about US$1.67 billion, which affected the livelihood of about 8.9 million people (4234 casualties) (GoB, 2008). Thus, natural disasters associated with cyclones on the coast of Bangladesh are one of the deadliest and costliest in the world.

Moreover, Bangladesh's high vulnerability to TCs is caused by the mortality, inundation, and damage to property due to the densely populated coastal areas, substandard infrastructures, and poor socioeconomic conditions and shallow bathymetry (Murty et al., 1986; Dube et al., 1997; Ali, 1999). The inundation due to storm surge is augmented in the low-lying coast, since the surge amplifies as it approaches the land due to the wide and shallow continental shelf (Murty et al., 1986) and the funneling shape of the coast (Das, 1972). The triangular shape of the coast in the northeast corner of the Meghna Estuary amplifies the surge wave (funneling effect; e.g., Flierl and Robinson, 1972; As-Salek, 1998; Haider et al., 1991; Chowdhury, 1981). High tidal range (SMRC, 2003; Miyan, 2005) and complex tide–surge interaction (Debsarma, 2009) during high tide also intensify the flooding impacts in the coastal regions of Bangladesh. Fig 13.3 shows the simulated inundation in the coastal regions due to 'super cyclone' Sidr in 2007.

13.7 Overview of the coastal protection measures

In order to protect the coast from tidal flooding and saline water intrusion and avoid damage to agricultural production, Bangladesh began the empolderment process in the coastal region during the 1960s under the World Bank funded Coastal Embankment Project (CEP). The project constructed a total of 123 polders (with 49 sea-facing) over the eastern and central part of the coast (Fig. 13.2). Polders are predominantly a Dutch flood protection concept, encircled earthen embankments around natural depressions or low lying areas with gated hydraulic structures in the intersection of the embankments and tidal channels. Although polders initially increased some crop yield by 200%–300% (Nishat, 1988) and provided protection from storm surge, in the 1990s excessive empolderment caused substantial drainage congestion inside the poldered region, severe sedimentation outside the polder, and loss of interconnectivity between major "Beels" (large scale low-lying depressions or topographic lows) and channels in the southwestern region of Bangladesh (Islam, 2006). This consequently led to prolonged waterlogging and inundation of agricultural and household lands, decline in crop production, and increase in unemployment. In 1997, the Khulna–Jessore Drainage Rehabilitation Project was started to solve the water logging problems, which included one-fourth of the CEP area. However, the project was unsuccessful due to conflicts between local stakeholders and the executing government agency, the Bangladesh Water Development Board (BWDB) (ADB, 2007).

Additionally, the disagreement between the two parties regarding the adaptation of nonstructural and indigenous solutions resulted in implementing an ecosystem oriented approach "Tidal River Management (TRM)," which allowed intrusion of tidal flow through breaching of the polder. The TRM allows natural movement of tide water, as, during the flood tide, sea driven sediment are carried and deposited into low-lying basins. On the other hand, the water drains out of the basin during the ebb tide with low sediment load and eroding the riverbeds downstream (Shampa and Pramanik 2012). The implementation of TRM in selected "Beel" areas was also found to improve drainage, navigability between tidal channels and downstream rivers, and induce land reclamation within the low-lying areas (Institute of Water Modeling, 2007). Although the adaptation of TRM is considered economically feasible and socially acceptable, successful implementation of TRM requires overcoming limitations in institutional management, conflicts and disputes among different income group (Rezaie, Islam and Rouf, 2013) with short- and long-term sequential operation and sediment management (Institute of Water Modeling, 2009; 2010).

Furthermore, large-scale afforestation programs have been implemented by the Bangladesh Forest Department during 1966–93 to protect the coast from the impacts of TC. The mangrove afforestation projects planted a total of approximately 120,000 ha of mangroves, which stabilized 60,000 ha of land area and induced economic stimulus through provisioning forest products and employment in the coastal zone of Bangladesh (Saenger, 1993). The damage and impacts of the major TC in 1991 were also limited in the western part of the coast (Saenger, 1993). The Coastal Greenbelt project (1995–2002) implemented forest plantations along 1400 km of coastal roads, improved public awareness related to disasters and nature conservation, and provided alternative income opportunities (ADB, 2005). Akber et al. (2018) show that monetary loss associated with cyclones in the villages sheltered by mangrove and polder is about half compared to the villages not in the shadow of mangrove. Loss incurred per household was highest in villages that are not protected by mangrove and polder.

After the 1991 cyclone, the construction of a Multipurpose Cyclone Shelter and risk zonation (MCSP, 1993) also played a significant role in reducing the impacts of TC and storm surge. During "super cyclone" Sidr, around 2100 cyclone shelters (GoB, 2008) served 1.5 million people impacted by the storm (Paul et al., 2010). However, there is an obvious need for more Multipurpose Cyclone Shelters to provide adequate emergency support to the population (Dasgupta et al., 2010; Ziaul and Ahmed, 2014).

13.8 Modeling, forecasting, and predicting tropical cyclones and storm surge in Bangladesh

In terms of simulating TC-induced storm surge, Das (1972) developed the first notable numerical model for the region, which could predict maximum surge height in the Bay of Bengal using basic cyclone parameters and linearized equations. Further developments in modeling storm surge were made by a series of numerical models developed by Ali (1979) and Johns and Ali (1980), which emphasized the effect of bathymetry and tidal channels in simulating the surge height along the coast. These studies highlighted the importance of lead-time in forecasting TCs and storm surge. However, their models did not incorporate the complex river network in the coastal zone of Bangladesh. A later study by Flather (1994) included coastal flow dynamics through utilizing a combination of 1D and 2D formulations for tidal channels and the open sea, respectively. Although this unified modeling framework was suggested to significantly decrease the lead-time, the prediction of TC was affected by large uncertainties, since it was vastly dependent on atmospheric data related to the genesis and movement of TCs. Furthermore, Dube et al. (1994) presented a real-time surge prediction model, which was applied to forecasting TCs in the east coast of India during 1992–93. The model used a dynamic storm model (Jelesnianski and Taylor, 1973), the same as Flather (1994), to derive the wind field, but neglected the barometric changes due to higher wind stress. While all previous studies advanced the state-of-the-art in simulating hydrodynamic and atmospheric physical processes due to TCs in the Bay of Bengal, none aimed to forecast both storm surge and storm surge induced inundation in the coast of Bangladesh on a real-time basis. Madsen and Jakobsen (2004) developed a modeling system to predict both surge heights and flooding, which consists of three modules: (1) a statistical model for forecasting cyclone track and wind speed; (2) an analytical model for generating wind and pressure fields; and (3) a data assimilation system for incorporating observed wind and pressure intensities in computing cyclonic parameters. While the combined modeling framework was able to track TC to respective inundation in coastal areas, the cumulative uncertainties in the forecast in the three modules were found to be too high for operational purposes.

For operational purposes, the Bangladesh Meteorological Department (BMD) developed a regression and steering-cum-persistence model (STP and STEEPER Model) for tracking cyclones and initially used an empirical formula for storm surge prediction (Khan, 1995; Debsarma 2009). In 2002, BMD implemented the IIT numerical storm surge model (Debsarma, 2009), which covers a large area with several model domains and varying resolution in the North Indian Ocean (low resolution), Bay of Bengal (medium resolution), and shore of the Bay of Bengal and Myanmar (fine resolution). For meteorological inputs, BMD uses the best available data from conventional, radar, and satellite observations and model-derived products, while a Generic Mapping Tool (GMT) is used for mapping, gridding, and visualizing geo-referenced surges on the coast of Bangladesh (Debsarma 2007; 2009). The model was validated for three major cyclones in 1991, 1997, and 1999 in the Bay of Bengal and is currently used for real-time forecasting of surge height. There are still limitations in obtaining wind and tide observations on a real-time basis, and challenges in updating coastal topographic and morphological features in the model. Additionally, real-time predictions of inundation due to storm surge in the coastal areas can support informed decision-making during the emergency management stage, but it requires substantial computational resources that are currently unavailable. Rezaie (2015) proposed a quasi-real-time storm surge inundation prediction system using the coastal hydrodynamic model Delft 3D and an artificial neural network (ANN), which showed that presimulated historical inundation maps, available cyclone data and a proper combination of ANN can provide maps of inundation due to storm surge and TC on a real-time basis. Although the study was carried out for research purposes, a pilot implementation of the prediction system can provide further insights to advance the current prediction system. Moreover, several recent studies have also advanced storm surge inundation modeling in the Bay of Bengal using an unstructured mesh, which greatly improves the discretization of coastal channels, rivers, and polders in the model (Mashriqui et al. 2010; Bhaskaran et al., 2013; Deb and Ferreira 2016).

13.9 Sea level rise in Bangladesh

Bangladesh is considered to be one of the countries most vulnerable to SLR due to the impact on its low-lying coastal areas. In the Bay of Bengal, the magnitude of relative SLR depends on the circulation of the North Indian Ocean, local plate tectonics, the subsidence rate of the GBM delta, and sediment load (Warrick et al., 1996; cited in Karim and Mimura 2008). Alam (1996) suggested that relative SLR and subsidence could negatively impact the coastal regions of the country with severe sedimentation, drainage, and saline water intrusion (both ground and surface water). Several studies (SMRC, 2003, Singh 2002; Sarwar, 2013; GoB, 2016) have indicated an increase in sea level with varying rates in different parts of coastal Bangladesh. Using 22 years (1977–98) of tidal gauge data from three major locations, it is estimated that relative sea level in the western, central, and eastern part of coastal zone is rising at a rate of 4.0, 6.0, and 7.8 mm/year, respectively (SMRC, 2003). Singh (2002) suggested that the higher rate in the eastern coast could be due to the increased precipitation, uplifting of land, and high land subsidence in last decades. However, different rates are found in a study by Sarwar (2013) that collected tidal records from three different sources (BWDB; Bangladesh Inland Water Transport Authority, BIWTA; and Permanent Service for Mean Sea Level) along the coast of Bangladesh. They estimated a higher rate of sea level increase in the central coastal zone (5.73 mm/year) than in the western and eastern zones (3.38 and 1.36 mm/year respectively), indicating a very high vulnerability of the central part near the Meghna Estuary. The SLR assessment by the Climate Change Cell of Bangladesh reported an overall rate of 5.05–7.5 mm/year SLR on the coast of the country (GoB, 2016). It is also indicated that the higher rate of SLR in the central coastal zone is caused by rapid geomorphological changes (recent development of funnel shaped islands), tidal amplification in the Meghna Estuary, and human interventions (embankments, cross-dam implementation) (Pethick and Orford, 2013; GoB, 2016). Additionally, the coastal areas near the Sundarbans mangroves are found to be relatively stable (Auerbach et al., 2015) and SLR estimates are considered more reliable as the area remained free from human or artificial interventions (GoB, 2016).

In terms of future projections, the IPCC postulate a 0.1–0.3 and 0.3–0.6 m rise in SLR in the Bay of Bengal by 2050 and 2100, respectively (Church et al., 2013). One of the early studies by Ali (1999) suggested that a 2°C and 4°C rise in sea surface temperature may cause 0.3 and 1.0 m of SLR in Bangladesh, respectively. Moreover, the Bangladesh National Adaptation Programme of Action recommended a 0.14, 0.32, and 0.88 m of SLR for 2030, 2050, and 2100, respectively (GoB, 2016). Considering SLR corresponding to different Representative Concentration Pathway scenarios from the IPCC's Fifth Assessment Report, Davis et al. (2018) predict that 0.9 million people (by year 2050) to 2.1 million people (by year 2100) could be displaced by direct inundation and that almost all of this movement will occur locally within the southern half of the country.

Regardless of the sea level future rates, SLR will most likely increase the flood depth and flood risk area during storm surges (Karim and Mimura, 2008) with more frequent extreme events in 2050–2100

(Kay et al., 2015). A World Bank assessment by Dasgupta et al. (2010) shows that a 0.27 m rise in SLR and 10% increase in wind speed of a 10-year return period TC could cost an additional US$4.5 billion in damage and loss by 2050. However, the projected increase (approximately 50% by the 2090s) in sediment load in the GBM delta due to anthropogenic climate change (Darby et al., 2018) may offset the negative impacts of SLR in the coastal regions of Bangladesh. Increases in potential monsoon flow due to projected climate scenarios (Whitehead, 2018) can also alter the hydromorphodynamics of rivers, regional subsidence-accretion scenarios, and influence future SLR conditions in coastal areas.

13.10 Way forward to combat, mitigate, and adapt to coastal flooding and sea level rise

The Bangladesh coast is known for its vulnerability to coastal flooding, due to its geographical location, strong tidal influence, and TCs. The geological and geophysical settings of the coastal areas are also highly dynamic in terms of land accretion, subsidence, and human interventions. Additionally, coastal livelihood options and the well-being of coastal populations are highly dependent on local ecosystems and environmental processes. It is expected that climate change and SLR will increase the current flood exposure, damage, and risk in coastal regions (Ali, 1999; Karim and Mimura, 2008; Dasgupta et al., 2010). While the threat of coastal flooding is clearly visible and vulnerability of coastal zones is well known, there is an urgent need for improved coastal monitoring for understanding the physical processes related to TCs, storm surge, and SLR in the area. For example, only four tidal stations provide approximately 30 years of tidal records along the Bangladeshi coast (GoB, 2016). Moreover most tidal gauge data are insufficient to construct time-series analysis for sea level investigation (Sarwar, 2013), which is required for long-term evaluation of sea level changes (Brown and Nicholls, 2015). The Bangladesh Climate Change Cell assessment also emphasized the need to install at least 10 more tidal stations along the coast (GoB, 2016). Due to the rapid and spatially varying rate of erosion and accretion along the coastal zone, it is also essential to obtain fine resolution data of topography, shoreline, and tidal channel networks. The collection of these primary data will not only advance the evaluation of recent changes in the coastal regions of Bangladesh, but also contribute to improved inundation modeling due to storm surge (Rezaie, 2015).

Accurate and reliable cyclone tracks and intensity information is a necessity for an effective forecasting system. At present, BMD uses a combination of track and intensity data from the Indian Meteorological Department and the Joint Typhoon Warning Center for their forecast. These data are available up to one decimal place on the geographic coordinate system (latitude/longitude). More precise information on the genesis and movement of TCs with potential storm surge inundation scenarios could greatly improve the current operational forecasting system. In addition, the Storm Warning Center of BMD provides early storm warnings (24 hours in advance), which has drastically reduced the casualties from TCs over time (The World Bank, 2011). However, during the devastating 1991 cyclone, only very few people in the affected areas responded to the warning and moved to safe sheltered places, but the majority of the population (especially fishermen) could not perceive the severity of the threat. Therefore, an effective, feasible, and easy to understand cyclone warning system, based on an advanced forecasting system, could greatly benefit the current emergency management and reduce the fatalities of TCs and storm surge in coastal Bangladesh (MoWR and FFWC, 2006).

In addition to the challenges in scientific research and data, there is a need for reassessing the effectiveness of coastal polders. It is estimated that with a 0.27 m SLR and 10% increase in cyclone wind speed, 59 polders out of the total 123 will be overtopped (Dasgupta et al., 2010) during storm surge events with a 10-year return period. This can increase the extent of vulnerable areas by 55% with a 2 m rise in inundation depth and displacement of 6–8 million people in coastal areas. Therefore, the design of polder height and strength should be evaluated so it is technically feasible to protect the coast from flooding and storm surge. The long-term impacts of coastal polders should also be taken into consideration as their effects vary within the central and western part of the coastal zone. Currently, the World Bank is implementing phase I of the Coastal Embankment Improvement Project (worth US$400 million) for strengthening flood protection from coastal polders (The World Bank, 2011). During this rehabilitation process it is important to address the root causes of vulnerability so that the recovery after a cyclone is faster. However, Sadik et al. (2018) show that this has not been the case during the recovery process after the Cyclone Aila in 2009. Moreover, there is a deficit in the total number of required cyclone shelters and their distribution over the coastal area is not adequate to provide refuge to the entire coastal community. Construction of more shelters with sufficient spatial distribution, proper maintenance, and income generating initiatives (during the disaster-free time) can contribute to develop coastal resilience in the country (The World Bank, 2011; Mahmood, Dhakal and Keast, 2014; Ziaul and Ahmed, 2014).

Currently, coastal embankments provide the first line of defense against coastal flooding and SLR (GoB, 2004a). Furthermore, the existing coastal ecosystem, such as the Sundarbans mangroves, can reduce the impacts of storm surge acting as buffer against coastal flooding (Sakib et al., 2015; Dasgupta et al., 2017). Coastal afforestation is also found to be technically feasible for stabilizing coastal areas and beneficial to the socio-economy (Saenger, 1993). Therefore, facilitation of natural protection, alongside infrastructure and conventional defense systems, can contribute to ameliorate coastal flood protection in Bangladesh. There is also a need for ecosystem oriented and holistic approaches that will ensure harmonious participation and coordination among different stakeholder groups (government, local agencies, and people). For example, TRM manifests that successful integration of local knowledge, technical expertise, and effective collaboration between government agencies and stakeholders can substantially minimize water-logging issues in the western part of the coast and contribute to long-term coastal zone development. Moreover, the majority of livelihood options in coastal Bangladesh are dependent on natural ecosystems and the services provided by coastal ecosystem are closely linked to the population livelihood and well-being (Adams et al., 2018a,b). Thus, provisioning ecosystem services can play a significant role to avert and alleviate poverty in this region (Adams et al., 2018a,b). The coastal region of Bangladesh is vulnerable to natural hazards and very dynamic in terms of geophysical and socioeconomic conditions. Therefore, the prevention, mitigation, and adaptation initiatives should be dynamic, technically feasible, environmentally friendly, and socially acceptable.

References

ADB, 2005. Project Completion Report Coastal Greenbelt Project (Loan 1353-BAN [SF]) in the People's Republic of Bangladesh.

ADB, 2007. Bangladesh: Khulna–Jessore Drainage Rehabilitation Project. Performance Evaluation Report Project in Bangladesh. Operations Evaluation Department, Asian Development Bank.

Adams, H., et al., 2018a. Characterising associations between poverty and ecosystem services. Ecosystem Services for Well-Being in Deltas. Available from: http://dx.doi.org/10.1007/978-3-319-71093-8.

Adams, H., Adger, W.N., Nicholls, R.J., 2018b. Ecosystem services linked to livelihoods and well-being in the Ganges–Brahmaputra–Meghna delta. Ecosystem Services for Well-Being in Deltas. pp. 29–47. Available from: http://dx.doi.org/10.1007/978-3-319-71093-8.

Akber, M.A., Patwary, M.M., Islam, M.A., Rahman, M.R., 2018. Storm protection service of the Sundarbans mangrove forest, Bangladesh. Nat. Hazards 94 (1), 405–418. Available from: https://doi.org/10.1007/s11069-018-3395-8.

Akter, J., et al., 2016. Evolution of the Bengal delta and its prevailing processes. J. Coastal Res. 321 (5), 1212–1226. Available from: https://doi.org/10.2112/JCOASTRES-D-14-00232.1.

Alam, M., 1996. Subsidence of the Ganges—Brahmaputra Delta of Bangladesh and associated drainage. Sedimentation and Salinity Problems. Springer, Dordrecht, pp. 169–192. Available from: http://dx.doi.org/10.1007/978-94-015-8719-8_9.

Ali, A., 1979. Storm Surges in the Bay of Bengal and Some Related Problems (Ph.D. thesis). University of Reading, England, 227 pp.

Ali, A., 1999. 'Climate change impacts and adaptation assessment in Bangladesh'. Clim. Res. 12 (2–3 Spec. Iss. 6), 109–116. Available from: https://doi.org/10.3354/cr012109.

Allison, M.A., Khan, S.R., Goodbred, S.R., Kuel, S.A., 2001. Stratigraphic evolution of the Late Holocene Ganges–Brahmputra lower delta plan. Sediment. Geol. 155, 317–342.

As-Salek, J.A., 1998. Coastal trapping and funneling effects on storm surges in the Meghna estuary in relation to cyclone hitting Noakhali–Cox's Bazar coast of Bangladesh. J. Phys. Oceanogr. 28 (2), 227–249.

Auerbach, L.W., et al., 2015. 'Flood risk of natural and embanked landscapes on the Ganges–Brahmaputra tidal delta plain'. Nat. Clim. Change 5, 153–157. Available from: https://doi.org/10.1038/NCLIMATE2472.

BBS, 2014. Population & housing census—2011. Bangladesh Bur. Stat. 2, 163–164. Available from: https://doi.org/10.1017/CBO9781107415324.004.

BIWTA, 2015. Bangladesh Tide Tables, 2015. pp. 1–6 and 43–48.

Barbier, E.B., 2014. A global strategy for protecting vulnerable coastal populations. AAAS 345 (6202), 125–1251. Available from: https://doi.org/10.1126/science.1254629.

Bhaskaran, P.K., Nayak, S., Bonthu, S.R., Murty, P.L.N., Sen, D., 2013. Performance and validation of a coupled parallel ADCIRC–SWAN model for THANE cyclone in the Bay of Bengal. Environ. Fluid Mech. 13 (6), 601–623.

Brammer, H., 2014. Bangladesh's dynamic coastal regions and sea-level rise, 1. Climate Risk Management. Elsevier B.V, pp. 51–62. Available from: http://dx.doi.org/10.1016/j.crm.2013.10.001.

Bricheno, L.M., Wolf, J., Islam, S., 2016. Tidal intrusion within a mega delta: an unstructured grid modelling approach. Estuarine Coastal Shelf Sci. 182, 12–26. Available from: https://doi.org/10.1016/j.ecss.2016.09.014.

Brown, S., Nicholls, R.J., 2015. Subsidence and human influences in mega deltas: the case of the Ganges–Brahmaputra–Meghna, Sci. Total Environ. 527–528. pp. 362–374. Available from: http://dx.doi.org/10.1016/j.scitotenv.2015.04.124.

Cazenave, A., Cozannet, G. Le, 2013. Sea level rise and its coastal impacts. Earth Future 2, 15–34. Available from: https://doi.org/10.1002/2013EF000188. Received.

Cazenave, A., Llovel, W., 2010. Contemporary sea level rise. Annu. Rev. Mar. Sci. 2 (1), 145–173. Available from: https://doi.org/10.1146/annurev-marine-120308-081105.

Choudhury, J.U., Haque, A., 1990. Permissible water withdrawal based upon prediction of salt-water intrusion in the Meghna delta. In: The Hydrological Basis for Water Resources Management Proceedings of the Beijing Symposium. Publication No. 197. International Association of Hydrological Sciences (IAHS), Wallingford.

Chowdhury, J.U., Karim, M.F., 1996. A risk-based zoning of storm surge prone area of the Ganges tidal plain. J. Civil Eng. (Institution of Engineering, Bangladesh) 24, 221–233.

Chowdhury, A.M., 1981. Cyclone in Bangladesh. A Report. Bangladesh Space Research and Remote Sensing Organization, Dhaka.

Church, J.A. Gregory, J.M., 2001. Changes in sea level. In: Climate Change 2001: The Scientific Basis. 639. Available from: <http://www.grida.no/climate/ipcc_tar/wg1/pdf/TAR-11.pdf>.

Church, J.A., White, N.J., 2006. A 20th century acceleration in global sea-level rise. Geophys. Res. Lett. 33 (1), 94–97. Available from: https://doi.org/10.1029/2005GL024826.

Church, J.A., White, N.J., 2011. Sea-level rise from the late 19th to the early 21st century. Surv. Geophys. 32 (4–5), 585–602. Available from: https://doi.org/10.1007/s10712-011-9119-1.

Church, J.A., et al., 2004. Estimates of the regional distribution of sea level rise over the 1950-2000 period. J. Clim. 17 (13), 2609–2625. Available from: http://dx.doi.org/10.1175/1520-0442(2004)017<2609:EOTRDO>2.0.CO;2.

Church, J.A., et al., 2013. Sea level change. In: Climate Change 2013: The Physical Science Basis. Contribution of Working Group I to the Fifth Assessment Report of the Intergovernmental Panel on Climate Change. pp. 1137–1216. https://doi.org/10.1017/CB09781107415315.026.

Coleman, J.M., 1969. Brahmaputra River: channel processes and sedimentation. Sediment. Geol. 3, 129–239.

Darby, S.E., Nicholls, R.J., Rahman, M.M., Brown, S., Karim, R., 2018. A sustainable future supply of fluvial sediment for the Ganges–Brahmaputra Delta. In: Nicholls, R., Hutton, C., Adger, W., Hanson, S., Rahman, M., Salehin, M. (Eds.), Ecosystem Services for Well-Being in Deltas. Palgrave Macmillan, Cham, pp. 277–291. Available from: http://dx.doi.org/10.1007/978-3-319-71093-8.

Das, P.K., 1972. A prediction model for storm surges in the Bay of Bengal. Nature 239, 211–213.

Dasgupta, S., et al., 2010. Vulnerability of Bangladesh to cyclones in a changing climate: potential damages and adaptation cost. World 16, 54. Available from: https://doi.org/10.1111/j.1467-7717.1992.tb00400.x.

Dasgupta, S., et al., 2014. Cyclones in a changing climate: the case of Bangladesh. Clim. Dev 2014 (6), 96–110.

Dasgupta, S. et al., 2017. Mangroves as Protection from Storm Surges in Bangladesh. p. 33. Available from: <http://documents.worldbank.org/curated/en/798551511274945303/pdf/WPS8251.pdf>.

Davis, K.F., et al., 2018. A universal model for predicting human migration under climate change: Examining future sea level rise in Bangladesh. Environ. Res. Lett. 13 (6).

Deb, M., Ferreira, C.M., 2016. Simulation of cyclone-induced storm surges in the low-lying delta of Bangladesh using coupled hydrodynamic and wave model (SWAN + ADCIRC). J. Flood Risk Manage. Available from: https://doi.org/10.1111/jfr3.12254.

Debsarma, S.K., 2007. Numerical simulations of storm surges in the Bay of Bengal. In: First JCOMM Scientific and Technical Symposium on Storm Surges. Seoul, Republic of Korea, 2–6 October 2007.

Debsarma, S.K., 2009. Simulations of storm surges in the Bay of Bengal. Mar. Geodesy J. 32 (2), 178–198.

Dilley, M., et al., 2005. Natural Disaster Hotspots. Available from: https://doi.org/10.1596/0-8213-5930-4.
Dube, S.K., Rao, A.D., Sinha, P.C., Chittibabu, P., 1994. A real time storm surge prediction system: An Application to east coast of India. Proc. Indian natn. Sci. Acad. 60, 157–170.
Dube, S.K., Rao, A.D., Sinha, P.C., Murty, T.S., Bahulayan, N., 1997. Storm surges in the Bay of Bengal and Arabian Sea: the problem and its prediction. Mausam 48 (2), 283–304.
Emanuel, K., 2003. Tropical cyclones. Annu. Rev. Earth Planet. Sci. 31 (1), 75–104. Available from: https://doi.org/10.1146/annurev.earth.31.100901.141259.
Federation, K.R., Lynne, D., 2013. Observations: ocean pages. In: Climate Change 2013 - The Physical Science Basis. 255–316. https://doi.org/10.1017/CBO9781107415324.010.
Flather, R.A., 1994. A storm surge prediction model for the northern Bay of Bengal with application to the cyclone disaster in April 1991. J. Phys. Oceanogr. 24, 172–190.
Flierl, G.R., Robinson, A.R., 1972. Deadly surges in the Bay of Bengal: dynamics and storm tide tables. Nature 239, 213–215.
GoB, 2004a. Living in the Coast: Problems, Opportunities and Challenges, Development. Program Development Office for Integrated Coastal Zone Management Plan Project, Water Resources Planning Organization, Dhaka, Bangladesh.
GoB, 2004b. Living in the Coast—People and Livelihoods. Program Development Office for Integrated Coastal Zone Management Plan Project, Water Resources Planning Organization, Dhaka, Bangladesh.
GoB, 2008. Cyclone Sidr in Bangladesh Damage, Loss and Needs Assessment for Disaster Recovery and Reconstruction.
GoB, 2016. Assessment of Sea Level Rise on Bangladesh Coast Through Trend Analysis. Climate Change Cell (CCC), Department of Environment, Ministry of Environment and Forests, Bangladesh.
Government of the People's Republic of Bangladesh (GoB), 2009. Bangladesh climate change strategy and action plan 2009. Retrieved from https://moef.gov.bd/site/page/97b0ae61-b74e-421b-9cae-f119f3913b5b/BCCSAP-2009.
Goodbred, S.L., Saito, Y., 2010. Tide-dominated deltas. Principles of Tidal Sedimentology. pp. 1–621. Available from: http://dx.doi.org/10.1007/978-94-007-0123-6.
Gray, W.M., 1968. Global view of the origin of tropical disturbances and storms. Mon. Weather Rev. 96 (10), 669–700. Available from: http://dx.doi.org/10.1175/1520-0493(1968)096 < 0669:GVOTOO > 2.0.CO;2.
Gray, W.M., 1998. Meteorology, and atmospheric physics the formation of tropical cyclones. Meteorol. Atmos. Phys. 67, 37–69. Available from: https://doi.org/10.1007/BF01277501.
Haider, R., Rahman, A., Huq, S., 1991. Cyclone'91: An Environmental and Perceptual Study. Centre for Advanced Studies (BACS), Bangladesh.
Haque, A., Nicholls, R.J., 2018. Floods and the Ganges–Brahmaputra–Meghna Delta. 147–159. https://doi.org/10.1007/978-3-319-71093-8.
Hartmann, D.L., et al., 2013. Observations: atmosphere and surface. In: Climate Change 2013 the Physical Science Basis: Working Group I Contribution to the Fifth Assessment Report of the Intergovernmental Panel on Climate Change, 9781107057, pp. 159–254. https://doi.org/10.1017/CBO9781107415324.008.
Hoque, M.M., 1991. Field Study and Investigation on the Damage Caused by Cyclones in Bangladesh: A Report on the April 1991 Cyclone; Cyclone damage in Bangladesh, Report on Field Study and Investigations on the Damage Caused by the Cyclone in Bangladesh in 29–30 April 1991, 1991. United Nations Centre for Regional Development, Nagoya, Japan, p. 75.

IPCC, 2007. Climate Change 2007 Synthesis Report. Intergovernmental Panel on Climate Change [Core Writing Team IPCC]. https://doi.org/10.1256/004316502320517344.

IPCC, 2013. Climate change 2013: the physical science basis. Contribution of working group I to the fifth assessment report of the Intergovernmental Panel on Climate Change. Intergovernmental Panel on Climate Change, Working Group I Contribution to the IPCC Fifth Assessment Report (AR5). Cambridge Univ. Press, New York, p. 1535. Available from: http://dx.doi.org/10.1029/2000JD000115.

Institute of Water Modeling, 2007. Monitoring the Effects of Beel Khukshia TRM Basin and Dredging of Hari River for Drainage Improvement of Bhabodah Area. Final Report. IWM, Dhaka.

Institute of Water Modeling, 2009. Mathematical Modeling for Planning & Design of Beel Kapalia Tidal Basin for Tidal River Management (TRM) & Sustainable Drainage Improvement. Final Report. Volume I: Main Report. IWM, Dhaka.

Institute of Water Modeling, 2010. Feasibility Study and Detailed Engineering Design for Long Term Solution of Drainage Problems in the Bhabodaho Area. Draft Final Report. IWM, Dhaka.

Iftekhar, M.S., Islam, M.R., 2004. Managing mangroves in Bangladesh: a strategy analysis. J. Coastal Conserv. 10, 139–146.

Indian Meteorological Department, Genesis, 2018. Available from: <http://www.rsmcnewdelhi.imd.gov.in/index.php?option = com_content&view = article&id = 51&Itemid = 199&lang = en> (accessed 30.08.18).

Irish, J.L., Resio, D.T., Ratcliff, J.J., 2008. The influence of storm size on hurricane surge. J. Phys. Oceanogr. 38 (9), 2003–2013. Available from: https://doi.org/10.1175/2008JPO3727.1.

Islam, M.R. (Ed.), 2004. Where Land Meets the Sea: A Profile of the Coastal Zone of Bangladesh. The University Press Limited, Dhaka.

Islam, M.R. 2006. Managing diverse land uses in coastal Bangladesh: institutional approaches. In: Environment and Livelihoods in Tropical Coastal Zones. 237–248. https://doi.org/10.1079/9781845931070.0237.

Islam, M.R., et al., 1999. The Ganges and Brahmaputra rivers in Bangladesh: basin denudation and sedimentation. Hydrol. Processes 13 (17), 2907–2923. Available from: http://dx.doi.org/10.1002/(SICI)1099-1085(19991215)13:17 < 2907::AID-HYP906 > 3.0.CO;2-E.

Islam, M.S., 2001. Sea-Level Changes in Bangladesh: The Last Ten Thousand Years. Asiatic Society of Bangladesh, Dhaka.

Jelesnianski, C.P., Taylor, A.D., 1973. NOAA Technical Memorandum. ERL, WMPO-3, 33 pp.

Jevrejava, S., et al., 2008. Recent global sea level acceleration started over 200 years ago? Geophys. Res. Lett. 35 (8), 8–11. Available from: https://doi.org/10.1029/2008GL033611.

Johns, B., Ali, A., 1980. The numerical modelling of storm surges in the Bay of Bengal. Q. J. R. Meteorol. Soc. 106, 1–8. Available from: https://doi.org/10.1002/qj.49710644702.

Karim, M.F., Mimura, N., 2008. Impacts of climate change and sea-level rise on cyclonic storm surge floods in Bangladesh. Global Environ. Change 18 (3), 490–500. Available from: https://doi.org/10.1016/j.gloenvcha.2008.05.002.

Kay, S., et al., 2015. Environmental Science Processes & Impacts Modelling the Increased Frequency of Extreme Sea Levels in the Ganges–Brahmaputra–Meghna Delta due to sea Level Rise and Other Effects of Climate Change. Royal Society of Chemistry, pp. 1311–1322. Available from: http://dx.doi.org/10.1039/c4em00683f.

Khan, S.R., 1995. Geomorphic Characterization of Cyclone Hazards Along the Coast of Bangladesh, MSc Thesis, ITC, Enschede, The Netherlands.

Lemke, P., et al., 2007. Observations: changes in snow, ice and frozen ground. Climate Change 2007: The Physical Science Basis. Contribution of Working Group I to the Fourth Assessment Report of the Intergovernmental Panel on Climate Change. pp. 337–383. Available from: http://dx.doi.org/10.1016/j.jmb.2004.10.032.

MCSP, 1993. Multipurpose Cyclone Shelter Programme. Planning and Implementation Issues, Part 4 Volume IV. Prepared for Planning Commission, Government of Bangladesh and United Nations Development Programme/World Bank, July.

Madsen, H., Jakobsen, F., 2004. Cyclone induced storm surge and flood forecasting in the northern Bay of Bengal. Coastal Eng. 51 (4), 277–296. Available from: https://doi.org/10.1016/j.coastaleng.2004.03.001 (2004).

Mahmood, M., Dhakal, S.P., Keast, R., 2014. The State of Multi-Purpose Cyclone Shelters in Bangladesh. https://doi.org/10.1108/F-10-2012-0082.

Mashriqui, H., Kemp, G., Heerden, I., Binselam, A., Yang, Y., Streva, K., et al., 2010. Experimental storm surge forecasting in the Bay of Bengal. Wind Storm Storm Surge Mitigation. pp. 15–25.

McTaggart-Cowan, R., et al., 2008. 'Climatology of tropical cyclogenesis in the North Atlantic (1948–2004)'. Mon. Weather Rev. 136 (4), 1284–1304. Available from: https://doi.org/10.1175/2007MWR2245.1.

Meehl, G.A., et al., 2007. 2007: Global climate projections. In: Climate Change 2007: Contribution of Working Group I to the Fourth Assessment Report of the Intergovernmental Panel on Climate Change. pp. 747–846. https://doi.org/10.1080/07341510601092191.

Meghna Estuary Study, 2001. Hydro-morphological dynamics of the Meghna estuary. In: Meghna Estuary Study (MES) Project. Bangladesh Water Development Board, Dhaka.

Ministry of Water Resources (MoWR), 2005. Coastal Zone Policy, Government of the People's Republic of Bangladesh. Dhaka, Bangladesh.

Ministry of Water Resources (MoWR) and Bangladesh Water Development Board Flood Forecasting and Warning Centre (FFWC), 2006. Consolidation and Strengthening of Flood Forecasting and Warning Services. Final Report Volume V – Project Extension Report.

Miyan, M.A., 2005. Cyclone disaster mitigation in Bangladesh. In: Proceedings of the Second Regional Technical Conference on Tropical Cyclones, Storm Surges and Floods. World Meteorological Organization Technical Document (WMO/TD-No 1272), Tropical Cyclone Programme, Report No. TCP-51. Published by Secretariat of the World Meteorological Organization, Geneva, Switzerland.

Mohanty, U.C., 1994. Tropical cyclone in the Bay of Bengal and deterministic method for prediction of their trajectories, Sadhana. Acad. Proc. Eng. Sci. 19, 567–582.

Mohanty, U.C., Mohapatra, M., Singh, O.P., Bandyopadhyay, B.K., Rathore, L.S. (Eds.), 2014. Monitoring and Prediction of Tropical Cyclones in the Indian Ocean and Climate Change. Springer, Dordrecht. Available from: https://doi.org/10.1007/978-94-007-7720-0.

Montgomery, M., Farrell, B., 1993. 'Tropical Cyclone Formation'. Am. Meteorol. Soc. 50 (2), 286.

Montgomery, M.T., 2016. Recent advances in tropical cyclogenesis, Advanced Numerical Modeling and Data Assimilation Techniques for Tropical Cyclone Predictions, 1. pp. 561–587. Available from: http://dx.doi.org/10.5822/978-94-024-0896-6_22.

Mooley, D.A., Mohile, C.M., 1983. A study of cyclonic storms incident in the different sections on the coast around Bay of Bengal. Mausam. 34, 139–152.

Muller, C.J., Romps, D.M., 2018. Acceleration of tropical cyclogenesis by self-aggregation feedbacks. Proc. Natl. Acad. Sci. U.S.A. 115 (12), 2930–2935. Available from: https://doi.org/10.1073/pnas.1719967115.

Murty, T.S., Flather, R.A., Henry, R.F., 1986. The storm surge problem in the Bay of Bengal. Prog. Oceanogr. 16, 195–233.

National Hurricane Center, Tropical Cyclone Climatology, 2018. Available from: <https://www.nhc.noaa.gov/climo/> (accessed 30.08.18).

Needham, H.F., Keim, B.D. 2011. Storm surge: Physical processes and an impact scale, in Recent Hurricane Research—Climate, Dynamics, and Societal Impacts In: Lupo, E. (Ed.), Intech Open Access, Croatia.

Nerem, R.S., et al., 2010. Estimating mean sea level change from the TOPEX and Jason altimeter missions. Mar. Geodesy 33 (sup1), 435–446. <https://doi.org/10.1080/01490419.2010.491031>.

Neumann, C.J., et al., 1993. Tropical Cyclones of the North Atlantic Ocean, 1871–1986. Washington, D.C. Available from: <http://www.nws.noaa.gov/oh/hdsc/Technical_papers/TP55.pdf>.

Nicholls, R.J., Cazenave, A., 2010. Sea-level rise and its impact on coastal zones. Science 328 (5985), 1517–1520. Available from: https://doi.org/10.1126/science.1185782.

Nishat, A., 1988. Review of present activities and state of art of the coastal areas of Bangladesh. In: Coastal Area Resource Development and Management. Part II. Coastal Area Resource Development and Management Association (CARDMA), Dhaka, Bangladesh, pp. 23–35.

Palmen, E.H., 1948. On the formation and structure of tropical cyclones. Geophys. Univ. Helsinki 3 (1948), 26–38.

Parris, A., et al., 2012. Global sea level rise scenarios for the United States National Climate Assessment. Climate Program Office, Springfield, MD, p. 2012.

Paul, B.K., et al., 2010. Cyclone evacuation in Bangladesh: tropical cyclones Gorky (1991) vs. Sidr (2007). Environ. Hazards 9 (1), 89–101. Available from: https://doi.org/10.3763/ehaz.2010.SI04.

Pethick, J., Orford, J.D., 2013. Rapid rise in effective sea-level in southwest Bangladesh: its causes and contemporary rates, Global Planet. Change, 111, pp. 237–245. <https://doi.org/10.1016/j.gloplacha.2013.09.019>.

Program Development Office for Integrated Coastal Zone Management Plan (PDO-ICZMP), 2003a. Coastal Livelihoods; Situations and Contexts, Working Paper wp015. Water Resources Planning Organization, Ministry of Water Resources, Bangladesh. Dhaka.

Program Development Office for Integrated Coastal Zone Management Plan (PDO-ICZMP), 2003b. Delineation of the Coastal Zone, Working Paper wp005. Water Resources Planning Organization, Ministry of Water Resources, Bangladesh. Dhaka.

Rahman, M.M. et al., 2014. A Preliminary Assessment of the Impact of Fluvio-Tidal Regime on Ganges-Brahmaputra-Meghna Delta and Its Impact on the Ecosystem Resources, Proceedings of the International Conference on Climate Change Impact and Adaptation, Gazipur, Bangladesh, November 15-17, 2013.

Rahman, M., et al., 2018. Recent Sediment Flux to the Ganges–Brahmaputra–Meghna Delta System, 643. Science of the Total Environment. Elsevier B.V, pp. 1054–1064, <https://doi.org/10.1016/j.scitotenv.2018.06.147>.

Rahman, M.M., Rahaman, M.M., 2017. Impacts of Farakka barrage on hydrological flow of Ganges river and environment in Bangladesh. Sustain. Water Resour. Manage. 1–14. <https://doi.org/10.1007/s40899-017-0163-y>.

Rego, J.L., 2009. Storm Surge Dynamics Over Wide Continental Shelves: Numerical Experiments Using the Finite-Volume Coastal Ocean Model, C (August). Available from: <http://etd.lsu.edu/docs/available/etd-06302009-161650/unrestricted/rego-diss.pdf>.

Rego, J.L., Li, C., 2009. On the importance of the forward speed of hurricanes in storm surge forecasting: a numerical study. Geophys. Res. Lett. 36 (7), 1–5. Available from: https://doi.org/10.1029/2008GL036953.

Resio, D.T., Westerink, J.J., 2008. Modeling the physics of storm surges. Phys. Today 61 (9), 33–38. Available from: https://doi.org/10.1063/1.2982120.

Rezaie, A.M., 2015. Quasi-Real Time Prediction of Storm Surge Inundation for the Coastal Region of Bangladesh (MSc. thesis). Bangladesh University of Engineering and Technology, Dhaka, Bangladesh.

Rezaie, A.M., Islam, T., Rouf, T., 2013. Limitations of institutional management and socio-economic barriers of Tidal River Management, a semi-natural process to save bhabodaho from water-logging problem. In: Fukuoka, S., Nakagawa, H., Sumi, T., Zhang, H. (Eds.), Advances in River Sediment Research. CRC Press, Taylor & Francis Group, London, pp. 2173–2181.

SMRC, 2003. The Vulnerability Assessment of the SAARC Coastal Region due to Sea Level Rise: Bangladesh Case Study. SAARC Meteorological Research Center, Dhaka.

Sadik, M.S., et al., 2018. A study on cyclone Aila recovery in Koyra, Bangladesh: evaluating the inclusiveness of recovery with respect to predisaster vulnerability reduction. Int. J. Disaster Risk Sci. 9 (1), 28–43.

Saenger, P., 1993. Land from the sea: the mangrove afforestation program of Bangladesh. Ocean Coastal Manage. 20, 23–39.

Sakib, M., Nihal, F., Haque, A., Rahman, M., Ali, M., 2015. Sundarban as a buffer against storm surge flooding. World J. Eng. Technol. 3, 59–64. Available from: https://doi.org/10.4236/wjet.2015.33C009.

Sarker, M.H., Akter, J., Ferdous, M.K., Noor, F., 2011. Sediment dispersal processes and management in coping with climate change in the Meghna Estuary, Bangladesh. Sediment Problems and Sediment Management in Asian River Basins. pp. 203–217.

Sarker, M.H., Akter, J., Rahman, M.M., 2013. Century-scale dynamics of the Bengal delta and future development. In: Fourth International Conference on Water & Flood Management (ICWFM-2013), August 2013, pp. 91–104.

Sarwar, M.G.M., 2005. Impacts of sea level rise on the coastal zone of Bangladesh. MSc Dissertation. Programme in environmental science, Lund University, Sweden.

Sarwar, G.M., 2013. Sea-level rise along the coast of Bangladesh. Disaster Risk Reduction Approaches in Bangladesh. Springer. Available from: http://dx.doi.org/10.1007/978-4-431-54252-0.

Shampa, Pramanik, I.M., 2012. Tidal river management (TRM) for selected coastal area of Bangladesh to mitigate drainage congestion. Int. J. Sci. Technol. Res. 1 (5), 1–6.

Singh, O.P., 2002. Spatial variation of sea level trend along the Bangladesh coast. Mar. Geodesy 25 (3), 205–212. Available from: https://doi.org/10.1080/01490410290051536.

The World Bank, 2011. The Cost of Adapting to Extreme Weather Events in a Changing Climate. 1–64. Available from: <http://siteresources.worldbank.org/INTBANGLADESH/Resources/BDS28ClimateChange.pdf>.

Titus, J.G., 1985. Sea level rise and wetland loss: an overview. In: Titus, J.G. (Ed.), Greenhouse Effect, Sea Level Rise and Coastal Wetlands. U.S. Environmental Protection Agency, Washington, DC, pp. 1–35.

Trenberth, K.E., Josey, S.A., 2007. Observations: surface and atmospheric climate change. Changes 164 (236–336), 235–336. Available from: https://doi.org/10.5194/cp-6-379-2010.

UNCLOS, 1982. United Nations Convention on the Law of the Sea, Montego Bay, 10 December 1982.

United Nations Development Programme (UNDP), 2004. A Global Report: Reducing Disaster Risk: A Challenge for Development. UNDP Bureau for Crisis Prevention and Recovery, New York. Available from: www.undp.org/bcpr.

Warrick, R.A., Bhuiya, A.H., Mitchell, W.M., Murty, T.S., Rasheed, K.B.S., 1996. Sea-level changes in the Bay of Bengal. In: Warrick, R.A., Ahmad, Q.K. (Eds.), The Implications of Climate and Sea-level Change for Bangladesh. Kluwer Academic Publishers, Dordrecht.

Weatherford, C., Gray, W.M., 1988. Typhoon structure as revealed by aircraft reconnaissance. Part II: Structural variability. Mon. Weather Rev. 116, 1044–1056.

Whitehead, P.G., 2018. Biophysical modelling of the Ganges, Brahmaputra, and Meghna catchment. In: Nicholls, R., Hutton, C., Adger, W., Hanson, S., Rahman, M., Salehin, M. (Eds.), Ecosystem Services for Well-Being in Deltas. Palgrave Macmillan, Cham.

Wilson, C.A., Jr, S.L.G., 2015. Construction and maintenance of the Ganges–Brahmaputra–Meghna delta: linking process, morphology, and stratigraphy. Annu. Rev. Available from: https://doi.org/10.1146/annurev-marine-010213-135032.

Wong, P.P., Losada, I.J., Gattuso, J.-P., Hinkel, J., Khattabi, A., McInnes, K.L., et al., 2013. Chapter 5: Coastal systems and low-lying areas. Climate Change 2014: Impacts, Adaptation, and Vulnerability. Part A: Global and Sectoral Aspects. Contribution of Working Group II to the Fifth Assessment Report of the Intergovernmental Panel on Climate Change. Cambridge University Press, pp. 1–85. Available from: http://dx.doi.org/10.1017/CBO9781107415379.010.

Ziaul, M., Ahmed, F., 2014. Multipurpose uses of cyclone shelters: quest for shelter sustainability and community development. Int. J. Disaster Risk Reduction 9, 1–11. <https://doi.org/10.1016/j.ijdrr.2014.03.007>.

Further reading

The World Bank, 2018. The World Bank Projects: Coastal Embankment Improvement Project—Phase I (CEIP-I). The World Bank. Available from: <http://projects.worldbank.org/P128276/coastal-embankment-improvement-project-phase-1ceip-1?lang = en&tab = overview> (accessed 08.09.018).

CHAPTER FOURTEEN

Hazard assessment and forecasting of landslides and debris flows: A case study in Northern Italy

Marco Borga
Department of Land, Environment, Agriculture and Forestry, University of Padova Agripolis, Legnaro, Italy

14.1 Introduction

A landslide is a type of mass wasting process that results in the downward and outward movement of slope-forming materials including rock, soil, artificial fill, or a combination of these, under the influence of gravity (Hungr et al., 2001, 2014). Landslides involve flowing, sliding, toppling, falling, or spreading, and many landslides exhibit a combination of different types of movements, at the same time or during the lifetime of the landslide. The various types of landslides can be differentiated by the kinds of material involved and the mode of movement (Varnes, 1978).

Landslides in which the sliding surface is located within the soil mantle or weathered bedrock (typically to a depth from few decimeters to some meters) are called shallow landslides (Fig. 14.1). The movement mechanism can be classified as a slide, not involving significant internal distortion of the moving mass. The material involved in shallow landslides includes both earth (material smaller than 2 mm) and debris (material larger than 2 mm). Precipitation-induced shallow landslides are triggered during rainstorms or periods of rapid snowmelt when shear strength is reduced because of an increase in pore-water pressure.

Shallow landslides frequently mobilize to form debris flows (termed DFs hereinafter) (Iverson, 1990). These are a form of rapid mass movement in which masses of poorly sorted sediment, agitated and saturated

Extreme Hydroclimatic Events and Multivariate Hazards in a Changing Environment.
DOI: https://doi.org/10.1016/B978-0-12-814899-0.00014-6

Figure 14.1 Shallow landslides triggered by a heavy storm.

with water, surge down slopes in response to gravitational attraction. Both solid and fluid forces vitally influence the motion, distinguishing DFs from related phenomena such as rock avalanches and sediment-laden water floods. Another common DF initiation mechanism is associated with channel bed erosion and mobilization by surface runoff (Coe et al., 2008a,b; Kean et al., 2013): a DF occurs when a critical surface discharge, rather than a critical groundwater level, is reached (Berti and Simoni, 2005; Larsen et al., 2006; Godt and Coe, 2007).

Landslides are present on all continents, and play an important role in the evolution of landscapes. Landslides are also a ubiquitous hazard in terrestrial environments with slopes, incurring human fatalities in urban settlements, along transport corridors, and at sites of rural industry. Recently, global landslide databases have shown the extent to which landslides impact on society and identified areas most at risk (Petley, 2012, Froude and Petley, 2018). Global data from Froude and Petley (2018) related to fatal nonseismic landslides, covering the period from January 2004 to December 2016, shows that in total 55,997 people were killed in 4862 distinct landslide events. Examination of a catalogue of landslides and DFs compiled by Salvati et al. (2010) for Italy revealed that in the 59-year period 1950–2008 most of the 2204 landslides, which have resulted in at least 4103 fatalities in Italy, were rainfall-induced shallow landslides or DFs.

The following sections provide a concise summary on selected key areas relevant for rainfall-induced shallow landsliding and DFs risk management. We expand upon these themes by illustrating: (1) the triggering mechanisms; (2) methods for hazard assessment; and (3) methods for shallow landsliding and DFs forecasting.

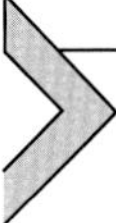

14.2 Shallow landslides and the initiation process

14.2.1 Background: the infinite-slope stability model

Limit equilibrium theory is often used to analyze the stability of natural slopes. A number of methods and procedures based on limit equilibrium principles have been developed for this purpose. Regardless of the specific procedures, the following principles (Morgenstern and Sangrey, 1978) are common to all methods of limit equilibrium analysis:

- A failure surface or mechanism is postulated.
- The shearing resistance required to equilibrate the failure mass is calculated by means of statics. The potential failure mass is assumed to be in a state of "limit equilibrium," and the shear strength of the soil or rock in the failure mass is mobilized everywhere along the slip surface.
- The calculated shearing resistance required for equilibrium is compared with the available shear strength. This comparison is made in terms of the factor of safety, which is defined as the factor by which the shear strength parameter must be reduced in order to bring the slope into a state of limiting equilibrium along a given slip surface.
- The mechanism or slip surface with the lowest factor of safety is found by iteration.

Planar infinite slope analysis has been widely applied to the determination of natural slope stability, particularly where the thickness of the soil mantle is small compared with the slope length and where landslides are due to the failure of a soil mantle that overlies a sloping drainage barrier. The drainage barrier may be bedrock or a denser soil mass. In this case, soil depth is obviously the depth to the drainage barrier. However, a translational failure plane may develop at any hydraulic conductivity contrast where positive pore water pressure can develop. Therefore, the depth to the failure plane may be much less than the depth to competent bedrock. Examples of such conditions include loose near-surface soil in thick glacial or slope failure deposit, loose volcanics overlying denser soil layers, and loose colluvial soil overlying decomposed residual soil common in granitic terrain.

The infinite slope model relies on several simplifying assumptions. These are reviewed in detail by Hammond et al. (1992). It is important to recognize that all forces in infinite slope analysis are assumed to vary in only the direction normal to the ground surface. The static determinacy and mathematical simplicity that results from this assumption make infinite slope analysis uniquely well suited for drawing unambiguous conclusions about the effects of ground water flow on slope stability (Iverson, 1990). The principal disadvantage of infinite-slope analysis is that mechanical one-dimensionality precludes accurate assessments of slopes in which ground water flow or topography produce forces that vary in directions other than the slope-normal direction (Iverson and Reid, 1992). Actually, it is found (Reid and Iverson, 1992) that an order of magnitude difference between the soil and subsurface stratum can cause groundwater to flow nearly parallel to the drainage barrier. However, deviations from slope parallel flow can occur, especially in deeper soils, and the flow vector will affect the calculation of shear strength (Iverson and Major, 1986, 1987).

The role played by vegetation in improving slope stability is well recognized, and comprehensive reviews may be found in the literature (Gray and Sotir, 1996; Cohen and Schwarz, 2017). The most obvious way in which woody vegetation enhances slope stability is via basal and lateral root reinforcement. Deeply penetrating vertical taproots and sinker roots provide the main contribution to the stability of slopes vis-à-vis resistance to shallow landsliding. Mechanical restraint against sliding only extends as far as the depth of root penetration. In addition, the root must penetrate across the failure surface to have a significant effect. The most effective restraint is provided where roots penetrate across the soil mantle into fractures or fissures in the underlying bedrock or where roots penetrate into a residual soil or transition zone whose density and shear strength increase with depth. Several researchers have used the tensile strength of individual roots in mathematical models to estimate the root resistance per unit area (Waldron, 1977; Wu et al., 1988; Gray and Sotir, 1996; Arnone et al., 2016; Cislaghi et al., 2017). These models are all similar in that they resolve the tensile force that develops in the root during shear into a tangential component that directly resists shear, and a normal component that increases the confining stress on the shear plan, thereby increasing the frictional component of soil shear strength.

Practical application of the findings obtained from these models is difficult, also because it has been shown that not all roots mobilize their maximum strength at the same time. Indeed, methods for the quantification of different types of root reinforcement mechanisms have been through a succession of models in the last few decades, starting with the assumption of the simultaneous breakage of all roots (Wu et al., 1979; Waldron and Dakessian, 1981) to the application of fiber bundle models that consider the progressive failures of roots (Schwarz et al., 2010; Cohen et al., 2011).

An in-depth analysis of root strength and root area ratio values of forest species in Northern Italy is reported by Bischetti et al. (2005), while Giadrossich et al. (2017) provides a comprehensive review of methods measure the mechanical behavior of tree roots. Literature review provided by Hammond et al. (1992, p. 58) shows that root strength may occasionally exceed 20 kPa with a mean value around 5 kPa over 11 studies.

14.2.2 Role of the different parameters

The shear strength S (kPa) of the soil at the failure plane can be expressed by:

$$S = C + (\sigma - u)\tan\varphi \tag{14.1}$$

where C is the effective soil cohesion (kPa), σ is the total normal stress on the failure plane (kPa), u is the pore water pressure on the failure plane (kPa), and ϕ is the internal friction angle of the soil (degrees).

The infinite-slope stability model provides a one-dimensional model for failure of shallow soils that neglects arching and lateral root reinforcement. Because of the geometry of an infinite slope, overall stability can be determined by analyzing the stability of a single, vertical element in the slope. End effects in the sliding mass can be neglected, and so too can lateral forces on either side of the vertical element, which are assumed to be opposite and equal. Under these assumptions the factor of safety (FS) for a dry, nonvegetated hillslope with cohesive soils is given by:

$$FS = \frac{C_s + \cos^2\theta \rho_d gD \tan\varphi}{D\rho_d g \sin\theta \cos\theta} \tag{14.2}$$

where C_s is soil cohesion (kPa), θ is slope angle (degrees), ρ_d is dry soil density (kg m^{-3}), g is gravitational acceleration, and D the vertical soil depth (m). Eq. (14.2) shows that for cohesive soils there is a fairly large decrease in FS with increasing D, with FS approaching the value ($\tan\phi/\tan\theta$) for large values of D. When estimating values for soil shear strength, one should keep in mind that the soil at the failure plane may not have the same properties as the bulk of the overlying material. Examples of this situation include thin clay seams at the failure plane, or a frictional resistance between soil and schist or phyllite bedrock that is less than within soil mass itself. Therefore, sampling or testing the upper soil material may give inappropriate values.

For vegetated sites, root systems contribute to soil strength by providing an artificial cohesion that can be added to effective soil cohesion in the Mohr–Coulomb equation. Then, the effects of rooting strength and vegetation surcharge can be incorporated into Eq. (14.2) as follows:

$$\mathrm{FS} = \frac{C_r + C_s + (\cos^2\theta \rho_d gD + W\cos\theta)\tan\varphi}{D\rho_d g \sin\theta \cos\theta + W\sin\theta} \tag{14.3}$$

where C_r is root cohesion (kPa) and W is vegetation surcharge (kPa). The effect of tree surcharge on FS depends on D. FS increases (decreases) with increasing W for values of D larger (smaller) than $(C_s/C_r)(W/\rho_d\, g\cos\theta)$. It should be noted that the threshold depth increases with increasing the slope angle.

Eq. (14.3) applies at the lithic contact when one assumes that roots intersect this plane (Sidle, 1992). If roots act to strengthen the soil only around the perimeter of the failure (e.g., deeper soils or shallow rooted vegetation) and not at the basal plane, then values of root cohesion must be reduced. Hammond et al. (1992) proposed a method for estimating an apparent value of C_r for such cases by comparing C_r values that give the same factor of safety for the infinite slope equation as for a three-dimensional bloc model (Burroughs, 1984). These authors suggested that for soil depths less than 1.5 m and block widths between 6 and 12 m, C_r values should be multiplied by 0.15 to 0.34.

The factor of safety for a nonvegetated slope with slope-parallel seepage, simplified for wet and dry soil density the same, is given by

$$\mathrm{FS} = \frac{C_s + \cos^2\theta\left[\rho_s g(D - D_w) + (\rho_s g - \rho_w g)D_w\right]\tan\varphi}{D\rho_s g \sin\theta \cos\theta} \tag{14.4}$$

where ρ_s is wet soil density (kg/m^3), ρ_w is the density of water (kg/m^3), and D_w is the vertical height of the water table within the soil layer (m).

Since steady-state flow conditions are assumed, effective stress parameters (effective friction angle and effective cohesion) must be used in the analysis. For vegetated slopes, FS is computed as follows

$$FS = \frac{C_r + C_s + \left[\rho_s g(D - D_w)\cos^2\theta + (\rho_s g - \rho_w g)D_w \cos^2\theta + W\cos\theta\right]\tan\varphi}{D\rho_s g \sin\theta \cos\theta + W\sin\theta} \quad (14.5)$$

Fig. 14.2 illustrates the relative importance of each variable in Eq. (14.5). This sensitivity analysis plot shows the relative value of FS as a function of the relative value of the input variables for a selected set of central values. It is obvious from this figure that increasing soil and root strength will increase FS, and increasing slope, soil depth, groundwater-soil depth ratio, and tree surcharge will decrease the FS. Generally, the FS is more sensitive to slope, soil internal friction, soil depth, and groundwater-soil depth ratio, and less to tree surcharge. The relative sensitivity of the FS to the other variables will change depending on the central value selected. This is illustrated by Fig. 14.3, in which only the central value of soil depth has been changed from 1.00 to 0.40 m. It is noted that FS becomes more sensitive to soil and root cohesion and less sensitive to groundwater–soil depth ratio and soil internal friction when the central value for soil depth is decreased. These trends should be

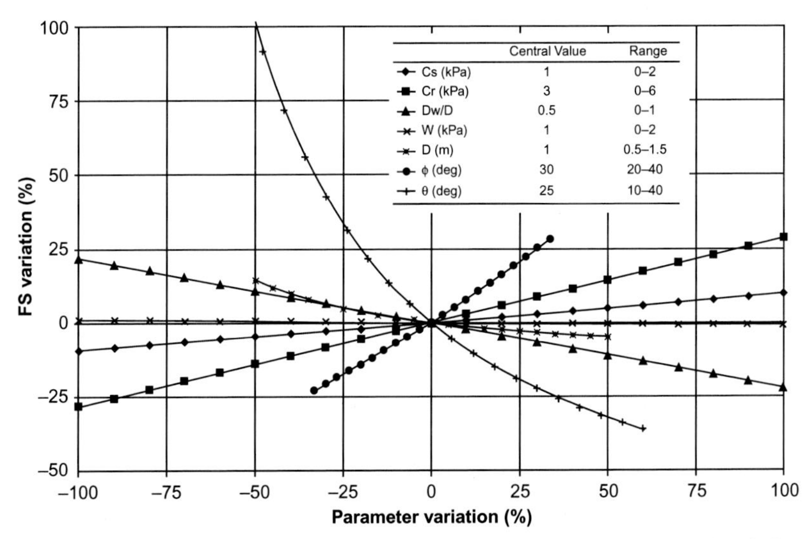

Figure 14.2 Sensitivity plot for the infinite slope equation with central soil depth value equal to 1.0 m.

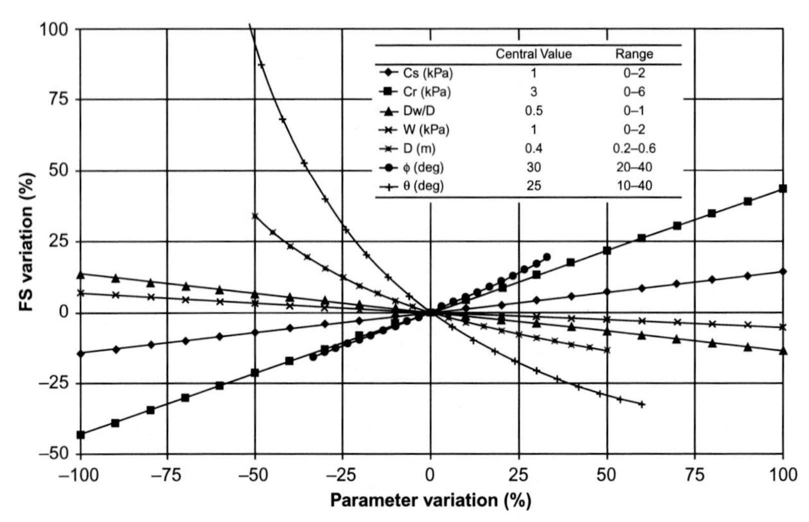

Figure 14.3 Sensitivity plot for the infinite slope equation with central soil depth value equal to 0.4 m.

expected, because frictional strength is more important in conditions of high normal stress, and cohesive stress is more important in conditions of low normal stress.

14.3 Debris flows initiation processes

DFs can be triggered by a variety of mechanisms. Commonly, DFs occur when landslides transform into rapidly flowing masses. Many DFs in undisturbed landscapes begin as discrete landslides on steep (greater than about 15°) hillslopes as a result of a relatively rapid influx of large amounts of water (e.g., Iverson et al., 1997). However, in headwater catchments with active sediment supplies and steep channels, many DFs initiate by mobilization of sediments that accumulate in channels (Takahashi, 1991; Marchi et al., 2002; Hürlimann et al., 2003). Material stored in channels can be mobilized by rainfall runoff surges and act as the source sediment for DFs. In these cases, DFs occur when flood surges erode and entrain channel sediment, i.e., when a critical surface discharge, rather than a critical groundwater level, is reached (Berti and Simoni, 2005; Godt and Coe, 2007; Kean et al., 2013).

The initiation of DFs in natural channels is related to the instability of the coarse debris that fills the channel bottom. The concentration of runoff in the channel bottom produces a water stream that causes erosion of the debris surface layer that extends to the layers below with whole or partial mobilization of the bed material (Gregoretti et al., 2016). The inclusion of bed material in the water stream generates the DF. The triggering areas are located where the morphology of the channel allows both the deposit of debris and the concentration of runoff to produce a water stream.

A number of conceptual and physically based models for channel bed mobilization initiation processes has been developed based on results from sediment transport experiments in steep flumes (e.g., Tognacca et al., 2000; Gregoretti, 2000; Armanini et al., 2005). In some of these experiments, a critical discharge of water is observed to create a DF surge by eroding the sediment by hydrodynamic forces from the top down. Models for the incipient motion of particles on steep slopes provide a framework to predict DF thresholds from channel bed mobilization in real situations, as shown by Berti and Simoni (2005) and by Gregoretti and Dalla Fontana (2008).

The incipient discharge model for predicting DF threshold, developed by Tognacca et al. (2000) and Gregoretti (2000), and based on the approach of Schoktlitsch (1943) and Bathurst et al. (1987), relates the threshold unit-width peak discharge $q_{w,t}$ (m^2/s), required to mobilize channel bed materials and therefore trigger the DF in steep channel beds, to the local channel bed slope θ and to the mean grain size d_M (m) as follows:

$$q_{w,t} = \beta \cdot \frac{{d_M}^{1.5}}{\tan(\theta)^b} \tag{14.6}$$

where β and b are empirical constants. Eq. (14.6) shows that the smaller the debris material grain size or the higher the bed slope, the lower the unit-width discharge required to mobilize the channel deposits.

Gregoretti (2000) examined critical hydrologic and geomorphic conditions for channel bed stability from experimental tests in laboratory flumes filled with a layer of three gravel sizes with approximately 12°–20° slope angles. The experimental results provided the parameters $\beta = 0.78$ and $b = 1.27$ for the critical incipient discharge equation for DF initiation.

Attempts have been made (e.g., Berti and Simoni, 2005; Gregoretti and Dalla Fontana, 2008) to extend predictions of DF thresholds to real situations, generally assuming the availability of information on the grain size characteristics of the debris material of the initiation zone (Gregoretti and Dalla Fontana, 2008). However, these characteristics are difficult to determine in practice due to the unsorted nature of the granular material usually involved (grain-size distribution often ranging often over four orders of magnitude) which introduces uncertainty and subjectivity in the grain-size distribution determination. Moreover, the systematic assessment of particle size distribution at potential debris-flow initiation sites is hardly possible, especially if a number of channels has to be investigated.

14.4 Shallow landslides and debris flows hazard assessment

Landslide hazard assessment is a key step towards landslide risk management. Hazard assessment for rainfall-induced shallow landslide and DFs should include the following steps: mapping of landslide/DFs susceptible areas; mapping of landslide/DFs magnitude and frequency; and, finally, methods for defining when landslides and DFs are likely to occur.

14.4.1 Mapping of landslide/debris flows susceptible areas

A variety of approaches has been proposed in the last four decades to map landslide susceptible areas: (1) multivariate analysis of factors characterizing observed sites of slope instability (Carrara et al., 1991; Reichenbach et al., 2018); (2) stability ranking based on criteria such as slope, lithology, land form, or geologic structure (Hollingsworth and Kovacs, 1981; Montgomery et al., 1991); (3) process-based approaches using deterministic analyses to delineate landslide potential (Hammond et al., 1992; Montgomery et al., 1998; Canli et al., 2018). This last approach emerged in the last 20 years and is undergoing rapid evolution, driven in part by new observational technology. With this approach, slope-instability theory is commonly coupled to soil-water models to calculate grid-cell-scale factors of safety (e.g., Montgomery and Dietrich, 1994; Borga et al., 1998; Montgomery et al., 1998; Iverson, 2000; Simoni et al., 2008; Baum et al., 2010). For example, in recent years, methods have advanced from the

Figure 14.4 Debris flows deposition area.

necessity of evaluating saturated soil thicknesses to models that evaluate time-dependent rainfall infiltration into unsaturated soils (Iverson, 2000; Savage et al., 2003; Baum et al., 2010; Canli et al., 2018). These more advanced models can potentially be used to provide spatially and temporally explicit information on hillslope instability.

To be exhaustive, the mapping of landslide susceptible areas has to include both the potentially unstable slopes as well as propagation areas (Fig. 14.4). The last decades have witnessed the development of continuum-mechanics numerical models to investigate DFs propagation on a complex three-dimensional topography (e.g., O'Brien et al., 1993; Armanini, 2013; Rosatti et al., 2015; von Boetticher et al., 2017). On the basis of the number of phases considered, these approaches can be distinguished in one-phase and two-phase models, generally neglecting the air in the interstices and assuming a completely saturated material.

14.4.2 Characterization of landslide/debris flows magnitude and frequency

Once the susceptible areas have been defined, magnitude of the landslide phenomena is a key parameter in landslide hazard assessment. Hungr (1997) defined landslide intensity as a set of spatially distributed parameters describing the destructiveness of the landslide. These parameters include the maximum movement velocity, the depth of moving mass, and the potential volume of material that can be mobilized from hillslopes and channels and thus deposited in the inundation zone. Rickenmann (1999) recommends geomorphic assessments of potential sediment volumes stored in channels as the most reliable method of characterizing debris-flow volume. Marchi and D'Agostino (2004) described such an approach for basins in the Dolomites in Italy. In a different approach, Cannon et al. (2004) used data from recently burned basins throughout the Western United States to develop a multivariate statistical model that can be used to estimate the potential peak discharges of a DF issuing from a basin mouth as a function of the extent of the burn, basin gradient, material properties, and the triggering rainfall.

14.5 Methods for landslides/debris flows forecasting: rainfall intensity-duration thresholds

A common goal of landslide research is to improve the ability to forecast the occurrence of the events in space and time. Susceptibility maps and models are typically used to estimate where landslides and DFs are likely to occur (Borga et al., 1998; Guzzetti et al., 2005; Baum et al., 2010), while rainfall thresholds are determined to define the meteorological conditions that, when reached or exceeded, are likely to result in DFs (e.g., Caine, 1980; Guzzetti et al., 2008; Chen et al., 2011; Jakob et al., 2012). Rainfall thresholds can be defined by adopting physical (process-based, conceptual) or empirical (historical, statistical) approaches (Aleotti 2004; Guzzetti et al., 2007, 2008; Schneuwly-Bollschweiler and Stoffel, 2012; Borga et al., 2014; Peruccacci et al., 2017 and references therein), and most commonly take the form of a simple relationship linking rainfall duration to other measures of rainfall, including rainfall intensity or event-cumulated rainfall (Aleotti 2004; Guzzetti et al., 2007). A widely used rainfall threshold relationship consists of a power law model that links rainfall depth R to the rainfall duration D (Reichenbach et al., 1998; Aleotti, 2004; Guzzetti et al., 2007):

$$R = \alpha D^{\delta} \tag{14.7}$$

where α and δ are parameters that adapt the power law model to the observations. Such a relationship is termed rainfall-duration (*RD*) threshold hereinafter. Brunetti et al. (2010) and Peruccacci et al. (2012, 2017) proposed the frequentist method to identify the power law model from the empirical data. With this procedure, the exponent parameter δ is identified from the linear regression of the log transformation of the R, D pairs, while the multiplicative parameter α is set to represent a desired probability of exceedance.

The general acceptance of rainfall thresholds for DF forecasting and warning emerges from the considerable literature proposing rainfall thresholds (e.g., Jibson, 1989; Wilson and Wieczorek, 1995; Deganutti et al., 2000; Jakob and Weatherly, 2003; Aleotti 2004; Cannon and Gartner, 2005; Guzzetti et al., 2007, 2008; Chen et al., 2011; Jakob et al., 2012; Borga et al., 2014 and references therein). Despite the popularity, the definition and exploitation of rainfall thresholds for forecasting of the possible occurrence of DFs is hampered by uncertainties, the sources of which remain largely undetermined. A common problem with the operational use of rainfall thresholds for DFs is the occurrence of false positives (i.e., rainfall conditions that should have resulted in landslides that were not reported) and false negatives (i.e., rainfall conditions that should have not resulted in landslides that were reported) (Staley et al., 2013). A high False Alarm Rate (i.e., the fraction of forecasted events that did not occur) may reduce the public's willingness to respond to the warnings (Barnes et al., 2007; Staley et al., 2013), but it is unavoidable given the large uncertainties inherent in the threshold-based forecasting of DF occurrence. In part, this uncertainty is associated with the fact that rainfall is not the only causative factor for DF initiation (Aleotti and Chowdhury, 1999). Other factors play a role, including hydrologic, geologic, and climatologic conditions, introducing variability in the thresholds. Investigators have recognized the issue, and recent efforts were made to quantify the dependence of rainfall thresholds on hydrologic and geologic factors (Wieczorek and Glade, 2005; Frattini et al., 2009; Peruccacci et al., 2012; Berti et al., 2012a,b).

A recognized component of the uncertainty associated with the estimation and use of rainfall thresholds for landslides forecasting is related to the large spatial and temporal variability of the precipitation. Due to the lack of rain gauges at the exact location where the slope failures occur, the rainfall that triggers single or multiple DFs is generally unknown. The amount of rainfall resulting in DFs is most often estimated based on data from the neighboring raingauges. Based on data from the Eastern Italian Alps, Nikolopoulos et al. (2014) have shown that the rainfall estimation uncertainty translates into a systematic underestimation of the rainfall thresholds. These uncertainties have consequences in the operational use of the thresholds, leading to a step degradation of the performances of the rainfall threshold for identification of DFs occurrence. Surprisingly, although rainfall estimation uncertainty was long recognized as an important factor in definition of the thresholds (Caine, 1980), few studies so far have explicitly investigated its effect on the estimation of the thresholds. The recent availability of accurate radar rainfall estimates for DF triggering storms offers unprecedented access to the space-time distribution of rainfall in the DF initiation region and opens new perspectives in the examination of the impact of rainfall-sampling errors on *RD* thresholds identification and use (Marra et al., 2016). Some results are discussed in the next section.

14.5.1 Rainfall sampling error and its effect on the identification of the *RD* threshold for the debris flows case

A number of studies (Nikolopoulos et al., 2014, 2015; Marra et al., 2014, 2016; Destro et al., 2017) on the effects of rainfall sampling uncertainties on the identification of the RD threshold have been carried out in the Upper Adige River basin in Northern Italy (Fig. 14.5), close to Trento (9700 km^2).

This is a mountainous region with more than 64% of the area located above 1500 m a.s.l. Rainfall over the region is monitored by a network of 120 raingauges, which corresponds to an average spatial density of about 1/80 km^{-2}. A C-band, Doppler weather radar, located at 1860 m a.s.l. on the top of Mt. Macaion in a central position on the region (Fig. 14.5), provides quantitative rainfall estimates every 5 minutes, at a spatial resolution of 1 km. Detailed information about the instrument are reported in Marra et al. (2014). A database reporting location and date of occurrence of DFs is available for the area and includes more than 400 events during the period 2000–12. These data have been obtained by systematic post-event surveys executed after each event. Ten DF triggering rainfall events that occurred in the study area between 2005 and 2010 have been examined in detail. This period is selected based on the availability of high resolution, quality controlled radar rainfall estimates. The 10 selected events are among the most severe storms that hit the region during this period and triggered a total of 82 DF that caused casualties and significant damage to infrastructures. Radar data used in these studies have been carefully processed and corrected for error sources associated with (1) signal attenuation in heavy rain, (2) wet radome attenuation, (3) beam blockage, and

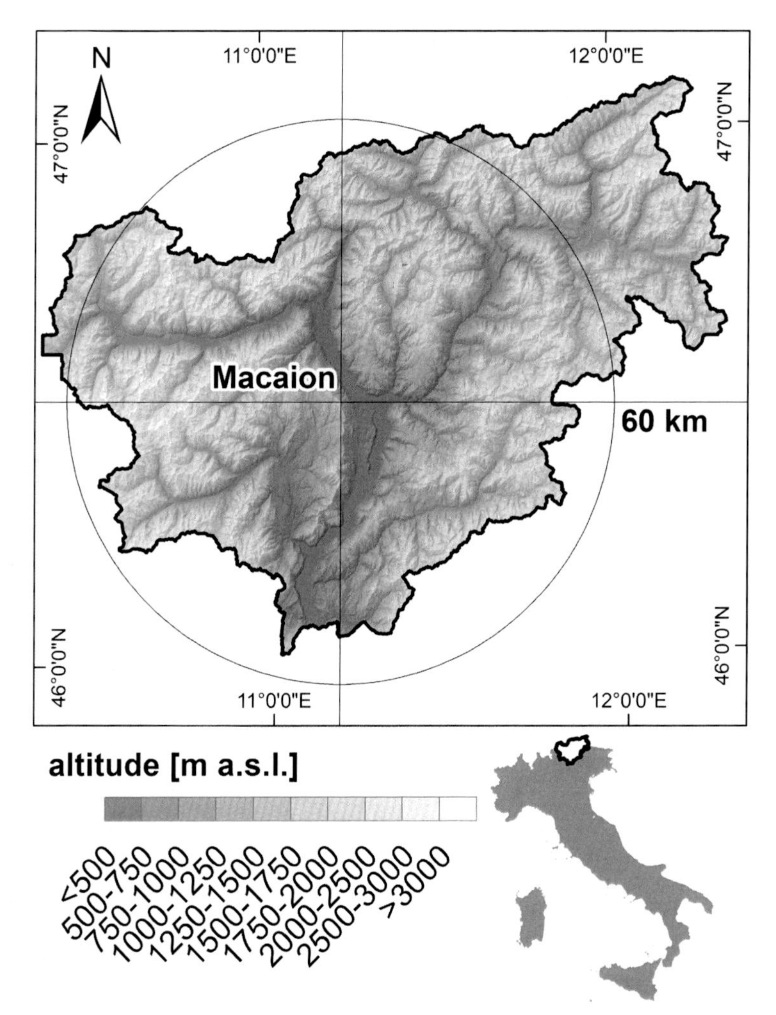

Figure 14.5 Map of the Upper Adige river basin closed ad Trento, with terrain elevation. Location of the Macaion weather radar with the range circle at 60 km is also shown.

(4) vertical profile of reflectivity, according to the procedures described in detail by Marra et al. (2014). In addition, the corrected radar-rainfall fields are further adjusted according to the mean field bias derived, on an event-basis, comparing radar observations with rain gauge data.

The durations of the rainfall events range from 1.5 hours (August 1, 2005) to 26 hours (July 16–17, 2009), while the rainfall depths at DF locations range from 14.6 (July 30, 2009) to 180.8 mm (August 14–15, 2010).

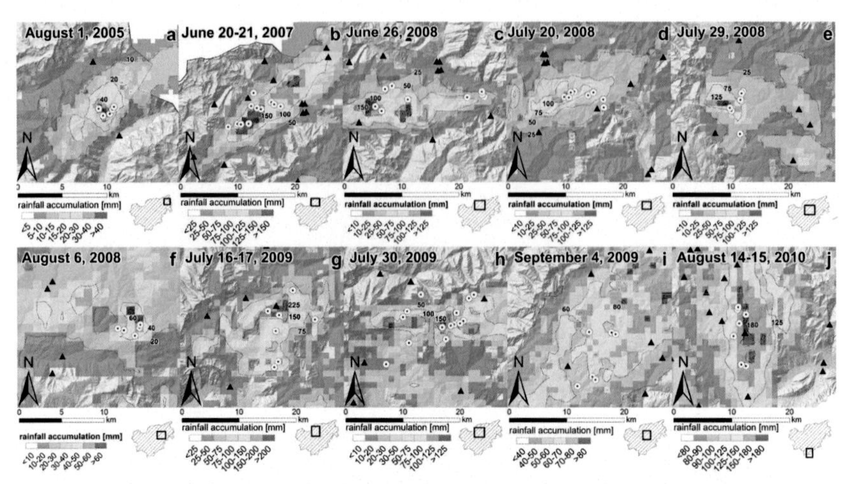

Figure 14.6 Maps of event rainfall depth for the 10 rainfall events (A–J). Dotted circles represent the triggered debris flows, black triangles the raingauges. Events are presented in chronological order: (A) August 1, 2005; (B) June 20–21, 2007; (C) June 26, 2008; (D) July 20, 2008; (E) July 29, 2008; (F) August 6, 2008; (G) July 16–17, 2009; (H) July 30, 2009; (I) September 4, 2009; (J) August 14–15, 2010. Location of each rainfall map relative to the Upper Adige river basin is shown using a black rectangular box superimposed on a mini-map of the basin.

All events occurred during the summer months and thus are considered to be representative of the DF seasonal regime and of the climatological characteristics of the triggering storms (Nikolopoulos et al., 2015). Fig. 14.6 shows the maps of radar rainfall depths for the 10 storm events, together with the location of raingauges and of the triggered DF. The figure shows that in many cases DF initiation points correspond spatially to the main rainfall depth peaks. However, other cases are located in areas with relatively low rain depths. These observations underline both the impact that topography, geology, vegetation, and sediment supply have on DF triggering and the uncertainty that may still affect the radar rainfall estimates used in this work. Fig. 14.6 also provides an appreciation of the large rainfall spatial variability characterizing these events and of the corresponding rain gauge sampling problem. In all cases, except July 16–17, 2009 and August 14–15, 2010, the rainfall core area characterized by the highest rain depth is not sampled by raingauges.

The rainfall spatial organization is examined by using a Lagrangian approach centered on the DF and focusing on a 21×21 km^2 squared area around each initiation point. Each rainfall field is normalized, dividing it by the value of the rainfall depth over the DF triggering point in order to

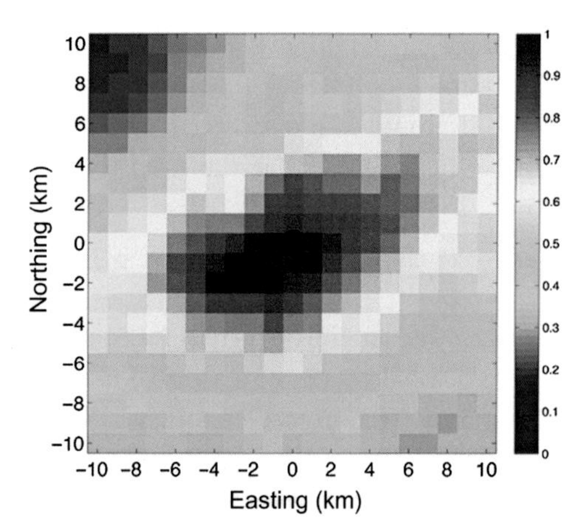

Figure 14.7 Mean normalized rain depth pattern around DF locations, obtained averaging the 82 rainfall depth patterns normalized over the rainfall over each DF location. The small circle identifies the DF location. *DF*, Debris flow.

make each rainfall field comparable to the others. The mean rainfall pattern, obtained by averaging the 82 normalized maps, is reported in Fig. 14.7. This figure shows a clear peak in the mean normalized fields with the peak location corresponding to the DF initiation point. On average, the precipitation decreases with increasing the distance from the DF location. The pattern is far from being symmetric: the highest values of the normalized rainfall depths are arranged along a SW−NE direction. Interestingly, this direction is consistent with the displacement direction of the convective cells observed during strong convective precipitation over the study area.

Marra et al. (2016) used a simulation approach to show how the space-time organization of rainfall depth around the DF controls the bias in the identification of the *RD* rainfall threshold relationship, when the relationship is identified by using data from a raingauge located at a distance *h* from the DF location based on the nearest neighbor approach. The focus on this specific estimation technique is motivated by findings by Nikolopoulos et al. (2015), who has shown that, on average, this approach provides less underestimated estimations of DF triggering rainfall with respect to more elaborated interpolation methods. With the simulation approach, Marra et al. (2016) assumed that the high-resolution radar rainfall data represents the space-time patterns of true precipitation over the DF window. Raingauge measurements were simulated by sampling these rainfall fields by assuming that the subgrid rainfall spatial variability is negligible.

By randomly sampling radar-rainfall fields at distance h from each DF, Marra et al. (2016) obtained a set of N_{DF} point measurements mimicking raingauges. Following a Monte Carlo exercise, the process was repeated to generate an ensemble of N_{DF} sets that was used to derive the following power law regression:

$$\hat{R}_l(h|D) = \pi_l(h)D^{\rho_l(h)} \tag{14.8}$$

where $\pi_l(h)$ and $\rho_l(h)$ are the empirical regression coefficients identified for the lth element of the ensemble. Marra et al. (2016) used 1000 Monte Carlo replications, a number that was found to be enough to reach stable results.

Fig. 14.8 shows the resulting RD regressions for three distances ($h = 1$, 5, and 10 km) from the DF location. Shaded areas represent the 5th−95th

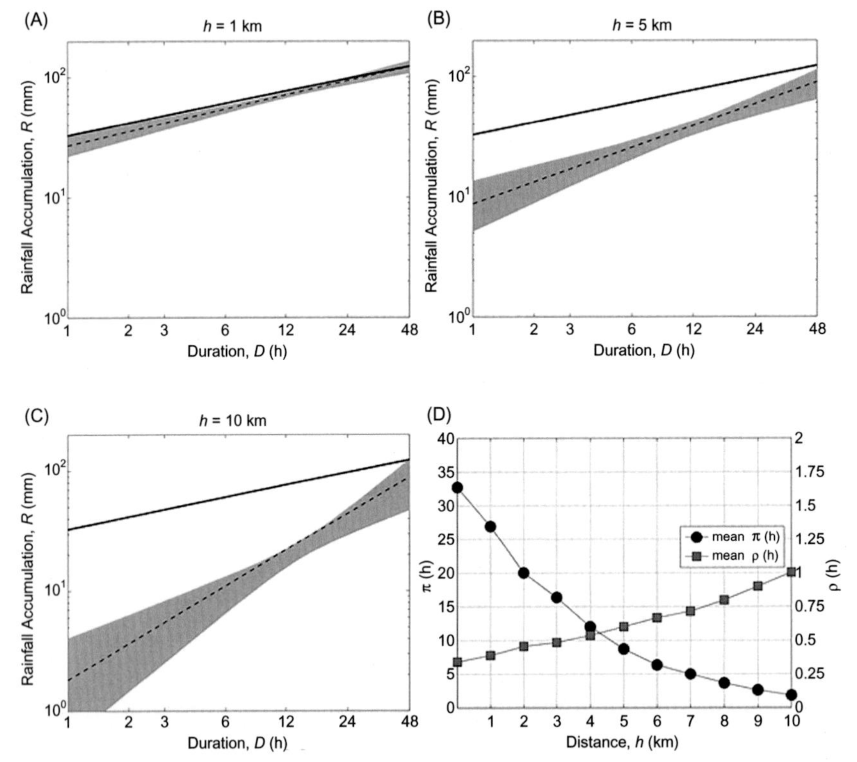

Figure 14.8 Ensembles of RD regressions derived from the Monte Carlo (MC) sampling at distances of 1 (A), 5 (B) and 10 km (C). The gray areas correspond to the 5th−95th quantile of the distribution of the ensembles, with the dashed line corresponding to the average relationship. Black line corresponds to the reference regression (obtained by using the radar data). Panel (D) shows the regression coefficients for distances between 0 and 10 km.

quantile ensembles of the regression lines; dashed lines are the relationships obtained using the mean values of the multiplicative $\pi(h)$ and exponent $\rho(h)$ parameters. The figure clearly shows the impact of rainfall sampling at varying distances. At 5 km distance the information about the reference regression line is completely missed. The figure shows also the behavior of the mean values of the multiplicative ($\pi_{\mathrm{mean}}(h)$) and exponent ($\rho_{\mathrm{mean}}(h)$) parameters, for distances from 0 to 10 km. As expected, the mean of the multiplicative parameter decreases monotonically down to 8.7 (1.8) at 5 km (10 km) distance. Conversely the mean of the exponent increases up to 0.60 (1.0) at 5 km (10 km) distance.

These results show the need of alternative rainfall estimation methods for DFs forecasting. The vast advancements in satellite-based rainfall estimation over the last couple of decades have led to a number of global rainfall products with various spatiotemporal resolutions. Although several evaluation studies have shown that these datasets can be associated with significant systematic and random errors (Maggioni et al., 2016), they still provide the only source of rainfall information over many areas around the globe and thus offer a unique opportunity for landslide hazard analysis and forecasting in those areas (Brunetti et al., 2018). Nikolopoulos et al. (2017) presented an intercomparison of several satellite-based products in the context of rainfall thresholds for DF prediction. These researchers showed that the parameters of the satellite-based thresholds differ less than 30% from the corresponding rain gauge-based parameters. Results from these authors further suggest that the adjustment of satellite-based estimates (either gauge-based or by applying an error model) together with spatial resolution has an important impact on the estimation of the accumulation—duration threshold.

14.6 Conclusions

This chapter provides an overview on hazards related to shallow landslides and DFs by illustrating: (1) the triggering mechanisms; (2) methods for hazard assessment; and (3) methods for shallow landsliding and DFs forecasting. For this last topic, the chapter provides a survey of recent literature on the role that the spatial organization of shallow landslides and DFs-triggering rainfall events may have on the identification and use of rainfall thresholds for landslide forecasts. This summary shows that the rainfall depth field systematically exhibits a peak corresponding to or close to the DF triggering location with rain depth decreasing with the distance. For the case studies considered here, the mean rainfall depth observed at 5 km (10 km) distance is around 70% (40%) of the rainfall depth measured over the DF initiation point. This implies that, measuring rainfall away from the DF initiation points, the rainfall estimate will be underestimated and that this underestimation will increase with distance. It is shown also that the features of the space-time organization of the DF triggering rainfall explains the biases in the identification of the rainfall thresholds when these thresholds are identified by using raingauges at a given distance from the DF initiation point. These results show the need for integrating remote sensing rainfall information in early warning systems for DFs and landslides. Results summarized in this chapter show that satellite rainfall estimates might be an important additional data source for developing continental or global landslide warning systems.

References

Aleotti, P., 2004. A warning system for rainfall-induced shallow failures. Eng. Geol. 73, 247−265.

Aleotti, P., Chowdhury, R., 1999. Landslide hazard assessment: summary review and new perspectives. Bull. Eng. Geol. Environ. 58, 21−44−44.

Armanini, A., 2013. Granular flows driven by gravity. J. Hydraul. Res. 51 (2), 111−120.

Armanini, A., Carpart, H., Fraccarollo, L., Larcher, M., 2005. Rheological stratification in experimental free-surface flows of granular−liquid mixtures. J. Fluid Mech. 532, 269−319. Available from: https://doi.org/10.1017/S0022112005004283.

Arnone, E., Caracciolo, D., Noto, L.V., Preti, F., Bras, R.L., 2016. Modeling the hydrological and mechanical effect of roots on shallow landslides. Water Resour. Res. 52 (11), 8590−8612.

Barnes, L.R., Gruntfest, E.C., Hayden, M.H., Schultz, D.M., Benight, C., 2007. False alarms and close calls: a conceptual model of warning accuracy. Weather Forecast 22, 1140−1147.

Bathurst, J.C., Graf, W.H., Cao, H.H., 1987. Bed load discharge equations for steep mountain rivers. In: Thorne, Bathurst, Hey (Ed.), Sediment Transport in Gravel Bed Rivers. John Wiley and Sons, pp. 453−477.

Baum, R.L., Godt, J.W., Savage, W.Z., 2010. Estimating the timing and location of shallow rainfall-induced landslides using a model for transient, unsaturated infiltration. J. Geophys. Res. Earth Surf. 115 (3), F03013.

Berti, M., Simoni, A., 2005. Experimental evidences and numerical modelling of debris flow initiated by channel runoff. Landslides 2 (3), 171−182.

Berti, M., Martina, M., Franceschini, S., Pignone, S., Simoni, A., Pizziolo, M., 2012a. Probabilistic rainfall thresholds for landslide occurrence using a Bayesian approach. J. Geophys. Res. Earth Surf. (2003−2012) 117 447−458.

Berti, M., Crucil, G., Degetto, M., De Vido, G., Gregoretti, C., Martina, M., et al., 2012b. Hydrologic response in the initiation area of the Dimai debris flow (Dolomites, Italian Alps). Rend. Online Soc. Geol. Ital. 21, 564−566.

Bischetti, G.B., Chiaradia, E.A., Simonato, T., Speziali, B., Vitali, B., Vullo, P., et al., 2005. Root strength and root area ratio of forest species in Lombardy (Northern Italy). Plant Soil 278 (1-2), 11–22.

Borga, M., Dalla Fontana, G., Da Ros, D., Marchi, L., 1998. Shallow landslide hazard assessment using a physically based model and digital elevation data. Environ. Geol. 35, 81–88.

Borga, M., Stoffel, M., Marchi, L., Marra, F., Jakob, M., 2014. Hydrogeomorphic response to extreme rainfall in headwater systems: flash floods and debris-flows. J. Hydrol. 518, 194–205.

Brunetti, M.T., Peruccacci, S., Rossi, M., Luciani, S., Valigi, D., Guzzetti, F., 2018. Rainfall thresholds for the possible occurrence of landslides in Italy. Nat. Hazards Earth Syst. Sci. 10, 447–458.

Brunetti, M.T., Melillo, M., Peruccacci, S., Ciabatta, L., Brocca, L., 2018. How far are we from the use of satellite rainfall products in landslide forecasting? Remote Sens. Environ. 210, 65–75.

Burroughs, E.R., 1984. A landslide hazard rating for portions of the Oregon coast range. In: O'Loughlin, C.L., Pearce, A.J. (Eds.), Symposium on Effects of Forest Land Use on Erosion and Slope Stability, May 7–11, 1984. University of Hawaii, Honolulu, pp. 265–274.

Caine, N., 1980. The rainfall intensity-duration control of shallow landslides and debris flows. Geografiska Annaler 62A, 23–27.

Canli, E., Mergili, M., Thiebes, B., Glade, T., 2018. Probabilistic landslide ensemble prediction systems: Lessons to be learned from hydrology. Nat. Hazards Earth Syst. Sci. 18 (8), 2183–2202.

Cannon, S., Gartner, J., 2005. Wildfire-related debris flow from a hazards perspective. In: Jakob, M., Hungr, O. (Eds.), Debris Flow Hazards and Related Phenomena. Springer, Berlin Heidelberg, pp. 363–385.

Cannon, S.H., Gartner, J.E., Rupert, M.G., Michael, J.A., 2004. Emergency assessment of debris-flow hazards from basins burned by the Cedar and Paradise fires of 2003, southern California. In: U.S. Geological Survey Open-File Report 2004-1011.

Carrara, A., Cardinali, M., Detti, R., Guzzetti, F., Pasqui, V., Reichenbach, P., 1991. GIS techniques and statistical models in evaluating landslide hazard. Earth Surf. Processes Landforms 16 (5), 427–445.

Chen, J.C., Huang, W.S., Jan, C.D., Tsai, Y.F., 2011. Rainfall conditions for the initiation of debris flows during Typhoon Morakot in the Chen-Yu-Lan watershed in central Taiwan. In: Genevois, R., Hamilton, D.L., Prestininzi, A. (Eds.), Fifth International Conference on Debris flow Hazards Mitigation: Mechanics, Prediction and Assessment, Casa Editrice Università La Sapienza, Roma, pp. 31–36.

Cislaghi, A., Chiaradia, E.A., Bischetti, G.B., 2017. Including root reinforcement variability in a probabilistic 3D stability model. Earth Surf. Processes Landforms 42 (12), 1789–1806.

Coe, J.A., Kinner, D.A., Godt, J.W., 2008a. Initiation conditions for debris flows generated by runoff at Chalk Cliffs, central Colorado. Geomorphology 96 (3–4), 270–297. Available from: https://doi.org/10.1016/j.geomorph.2007.03.017.

Coe, J.A., Cannon, S.H., Santi, P.M., 2008b. Introduction to the special issue on debris flows initiated by runoff, erosion, and sediment entrainment in western North America. Geomorphology 96 (3–4), 247–249. Available from: https://doi.org/10.1016/j.geomorph.2007.05.001.

Cohen, D., Schwarz, M., 2017. Tree-root control of shallow landslides. Earth Surf. Dyn. 5 (3), 451–477.

Cohen, D., Schwarz, M., Or, D., 2011. An analytical fiber bundle model for pullout mechanics of root bundles. J. Geophys. Res. Earth Surf. 116 (3), F03010.

Deganutti, A.M., Marchi, L., Arattano, M., 2000. Rainfall and debris flow occurrence in the Moscardo basin (Italian Alps). In: Wieczorek, G.F., Naeser, N.D. (Eds.), Debris flow hazards mitigation—Mechanics, Prediction, and Assessment. Balkema, Rotterdam, pp. 67–72.

Destro, E., Marra, F., Nikolopoulos, E.I., Zoccatelli, D., Creutin, J.D., Borga, M., 2017. Spatial estimation of debris flows-triggering rainfall and its dependence on rainfall return period. Geomorphology 278, 269–279.

Frattini, P., Crosta, G., Sosio, R., 2009. Approaches for defining thresholds and return periods for rainfall-triggered shallow landslides. Hydrol. Processes 23 (10), 1444–1460.

Froude, M.J., Petley, D.N., 2018. Global fatal landslide occurrence from 2004 to 2016. Nat. Hazards Earth Syst. Sci. 18 (8), 2161–2181.

Giadrossich, F., Schwarz, M., Cohen, D., Cislaghi, A., Vergani, C., Hubble, T., et al., 2017. Methods to measure the mechanical behaviour of tree roots: a review. Ecol. Eng. 109, 256–271.

Godt, J.W., Coe, J.A., 2007. Alpine debris flows triggered by a 28 July 1999 thunderstorm in the central Front Range, Colorado. Geomorphology 84, 80–97.

Gray, D.H., Sotir, R.B., 1996. Biotechnical and Soil Bioengineering Slope Stabilization. John Wiley and Sons, New York, p. 378.

Gregoretti, C., 2000. The initiation of debris flow at high slopes: experimental results. J. Hydraul. Res. 38 (2), 83–88.

Gregoretti, C., Dalla Fontana, G., 2008. The triggering of debris flow due to channel-bed failure in some alpine headwater basins of the Dolomites: analyses of critical runoff. Hydrol. Process. 22, 2248–2263.

Gregoretti, C., Degetto, M., Bernard, M., Crucil, G., Pimazzoni, A., De Vido, G., et al., 2016. Runoff of small rocky headwater catchments: field observations and hydrological modelling. Water Resour. Res. 52 (10), 8138–8158.

Guzzetti, F., Reichenbach, P., Cardinali, M., Galli, M., Ardizzone, F., 2005. Probabilistic landslide hazard assessment at the basin scale. Geomorphology 72 (1-4), 272–299.

Guzzetti, F., Peruccacci, S., Rossi, M., Stark, C.P., 2007. Rainfall thresholds for the initiation of landslides in central and southern Europe. Meteorol. Atmos. Phys. 98, 239–267. Available from: https://doi.org/10.1007/s00703-007-0262-7.

Guzzetti, F., Peruccacci, S., Rossi, M., Stark, C.P., 2008. The rainfall intensity-duration control of shallow landslides and debris flows: an update. Landslides 5, 3–17. Available from: https://doi.org/10.1007/s10346-007-0112-1.

Hammond, C., Hall, D., Miller, S., Swetik, P., 1992. Level 1 Stability Analysis (LISA) Documentation for Version 2.0. Gen. Tech. Rep. INT-285. U.S. Dep. of Agric., For. Serv., Intermountain Res. Stn, Ogden, Utah, p. 190.

Hollingsworth, R., Kovacs, G.S., 1981. Soil slumps and debris flows: prediction and protection. Bull. Assoc. Eng. Geol. 18, 17–28.

Hungr, O., 1997. Some methods of landslide hazard intensity mapping. In: Cruden, D. M., Fell, R. (Eds.), Landslide Risk Assessment. Proceedings International Workshop on Landslide Risk Assessment, Honolulu, February 19–21, 1997, Balkema, Rotterdam, pp. 215–226.

Hungr, O., Evans, S.G., Bovis, M.J., Hutchinson, J.N., 2001. A review of the classification of landslides of the flow type. Environ. Eng. Geosci. 7 (3), 221–238.

Hungr, O., Leroueil, S., Picarelli, L., 2014. The Varnes classification of landslide types, an update. Landslides 11 (2), 167–194.

Hürlimann, M., Rickenmann, D., Graf, C., 2003. Field and monitoring data of debris-flow events in the Swiss Alps. Can. Geotech. J. 40, 161–175.

Iverson, R.M., 1990. Groundwater flow field in infinite slopes. Géotechnique 40, 139–143.

Iverson, R.M., 2000. Landslide triggering by rain infiltration. Water Resour. Res. 36, 1897–1910. Available from: https://doi.org/10.1029/2000WR900090.

Iverson, R.M., Major, J.J., 1986. Groundwater seepage vectors and the potential for hillslope failure and debris flow mobilization. Wat. Resour. Res. 22, 1543–1548.

Iverson, R.M., Major, J.J., 1987. Rainfall, ground water flow, and seasonal movement at minor creek landslide, northwestern California: physical interpretation of empirical relations. Geol. Soc. Am. Bull. 99, 579–594.

Iverson, R.M., Reid, M.E., 1992. Gravity-driven groundwater flow and slope failure potential: 1 Elastic effective-stress model. Wat. Resour. Res. 28, 925–938.

Iverson, R.M., Reid, M.E., Lahusen, R.G., 1997. Debris-flow mobilization from landslides. Annu. Rev. Earth Planet. Sci. 25, 85–136.

Jakob, M., Weatherly, H., 2003. A hydroclimatic threshold for landslide initiation on the North Shore Mountains of Vancouver, British Columbia. Geomorphology 54, 137–156.

Jakob, M., Owen, T., Simpson, T., 2012. A regional real-time debris flow warning system for the District of North Vancouver, Canada. Landslides 9, 165–178.

Jibson, R., 1989. Debris flows in southern Puerto Rico. In: Geological Society of America Special Papers 236. pp. 29–56.

Kean, J.W., McCoy, S.W., Tucker, G.E., Staley, D.M., Coe, J.A., 2013. Runoff-generated debris flows: observations and modeling of surge initiation, magnitude, and frequency. J. Geophys. Res. Earth Surf. 118, 2190–2207. Available from: https://doi.org/10.1002/jgrf.20148.

Larsen, I.J., Pederson, J.L., John, C., Schmidt, J.C., 2006. Geologic versus wildfire controls on hillslope processes and debris flow initiation in the Green River canyons of Dinosaur National Monument. Geomorphology 81, 114–127.

Maggioni, V., Meyers, P.C., Robinson, M.D., 2016. A review of merged high resolution satellite precipitation product accuracy during the Tropical Rainfall Measuring Mission (TRMM) era. J. Hydrometeor. 17, 1101–1117. Available from: https://doi.org/10.1175/JHM-D-15-0190.1.

Marchi, L., D'Agostino, V., 2004. Estimation of debris-flow magnitude in the Eastern Italian Alps. Earth Surf. Processes Landforms 29 (2), 207–220.

Marchi, L., Arattano, M., Deganutti, A.M., 2002. Ten years of debris-flow monitoring in the Morcardo Torrent (Italian Alps). Geomorphology 46, 1–17.

Marra, F., Nikolopoulos, E.I., Creutin, J.D., Borga, M., 2014. Radar rainfall estimation for the identification of debris-flow occurrence thresholds. J. Hydrol. 519 (2014), 1607–1619.

Marra, F., Nikolopoulos, E.I., Creutin, J.D., Borga, M., 2016. Space–time organization of debris flows-triggering rainfall and its effect on the identification of the rainfall threshold relationship. J. Hydrol. 541, 246–255.

Montgomery, D.R., Dietrich, W.E., 1994. A physically-based model for the topographic control on shallow landsliding. Wat. Resour. Res. 30, 1153–1171.

Montgomery, D.R., Wright, R.H., Booth, T., 1991. Debris flow hazard migration for colluvium-filled swales. Bull. Assoc. Eng. Geol. 28, 299–319.

Montgomery, D.R., Sullivan, K., Greenberg, H.R., 1998. Regional test of a model for shallow landsliding. Hydrol. Processes 12, 943–955.

Morgenstern, N.R., Sangrey, D.A., 1978. Methods of stability analysis. In: Schuster, R.L., Krizek, R.J. (Eds.), Landslides: Analysis and Control, Transp. Res. Board, Sp. Rep., 176. Nat Acad of Sci, pp. 155–171.

Nikolopoulos, E.I., Crema, S., Marchi, L., Marra, F., Guzzetti, F., Borga, M., 2014. Impact of uncertainty in rainfall estimation on the identification of rainfall thresholds for debris flow occurrence. Geomorphology 221, 286–297.

Nikolopoulos, E.I., Marra, F., Creutin, J.-D., Borga, M., 2015. Estimation of debris flow triggering rainfall: influence of rain gauge density and interpolation methods. Geomorphology 243, 40–50.

Nikolopoulos, E.I., Destro, E., Maggioni, V., Marra, F., Borga, M., 2017. Satellite rainfall estimates for debris flow prediction: an evaluation based on rainfall accumulation–duration thresholds. J. Hydrometeorol. 18 (8), 2207–2214.

O'Brien, J.S., Julien, P.Y., Fullerton, W.T., 1993. Two-dimensional water flood and mudflow simulation. J. Hydraul. Res. 119 (2), 244–261.

Peruccacci, S., Brunetti, M.T., Luciani, S., Vennari, C., Guzzetti, F., 2012. Lithological and seasonal control on rainfall thresholds for the possible initiation of landslides in central Italy. Geomorphology 139-140, 79–90.

Peruccacci, S., Brunetti, M.T., Gariano, S.L., Melillo, M., Rossi, M., Guzzetti, F., 2017. Rainfall thresholds for possible landslide occurrence in Italy. Geomorphology 290, 39–57.

Petley, D., 2012. Global patterns of loss of life from landslides. Geology 40 (10), 927–930.

Reichenbach, P., Cardinali, M., De Vita, P., Guzzetti, F., 1998. Regional hydrological thresholds for landslides and floods in the Tiber River Basin (central Italy). Environ. Geol. 35 (2-3), 146–159.

Reichenbach, P., Rossi, M., Malamud, B.D., Mihir, M., Guzzetti, F., 2018. A review of statistically-based landslide susceptibility models. Earth-Sci. Rev. 180, 60–91.

Reid, M.E., Iverson, R.M., 1992. Gravity-driven groudwater flow and slope failure potential, 2, Effects of slope morphology, material properties, and hydraulic heterogeneity. Wat. Resour. Res. 28, 939–950.

Rickenmann, D., 1999. Empirical relationships for debris flows. Nat. Hazards 19, 47–77.

Rosatti, G., Zorzi, N., Begnudelli, L., Armanini, A., 2015. Evaluation of the Trent2D model capabilities to reproduce and forecast debris-flow deposition patterns through a back analysis of a real event. Engineering Geology for Society and Territory—Volume 2: Landslide Processes 1629–1633.

Salvati, P., Bianchi, C., Rossi, M., Guzzetti, F., 2010. Societal landslide and flood risk in Italy. Nat. Hazards Earth Syst. Sci. 10 (3), 465–483.

Savage, W.Z., Godt, J.W., Baum, R.L., 2003. A model for spatially and temporally distributed shallow landslide initiation by rainfall infiltration. In: Rickenmann, D., Chen, C. (Eds.), Debris-Flow Hazards Mitigation—Mechanics, Prediction, and Assessment—Proceedings of the Third International Conference on Debris Flow Hazards, Davos, Switzerland, September 10–13, 2003. Millpress, Rotterdam.

Schneuwly-Bollschweiler, M., Stoffel, M., 2012. Hydrometeorological triggers of periglacial debris flows – a reconstruction dating back to 1864. J. Geophys. Res. Earth Surf. 117, F02033.

Schoktlitsch, A., 1943. Berechnung der Geschiebefracht. Wasser und Energiewirtshaft, N. 1 (in German).

Schwarz, M., Cohen, D., Or, D., 2010. Root-soil mechanical interactions during pullout and failure of root bundles. J. Geophys. Res. Earth Surf. 115, F04035. Available from: https://doi.org/10.1029/2009JF001603.

Sidle, R.C., 1992. Theoretical model of the effects of timber harvesting on slope stability. Water Resour. Res. 28, 1897–1910.

Simoni, S., Zanotti, F., Bertoldi, G., Rigon, R., 2008. Modelling the probability of occurrence of shallow landslides and channelized debris flows using GEOtop-FS. Hydrol. Process. 22, 532–545. Available from: https://doi.org/10.1002/hyp.6886.

Staley, D., Kean, J., Cannon, S., Schmidt, K., Laber, J., 2013. Objective definition of rainfall intensity–duration thresholds for the initiation of post-fire debris flows in southern California. Landslides 10, 547–562.

Takahashi, T., 1991. Debris flow. In: IAHR Monograph. A.A. Balkema, Rotterdam, The Netherlands.
Tognacca, C., Bezzola, G.R., Minor, H.E., 2000. Threshold criterion for debris-flow initiation due to channel bed failure. In: Wieczorek, G.F., Naeser, N.D. (Eds.), Proceedings of the Second International Conference on Debris Flow Hazards Mitigation, Taipei, Taiwan. A.A. Balkema, Rotterdam, pp. 89–97.
Varnes, D.J., 1978. Slope movement types and processes. In: Schuster, R.L., Krizek, R.J. (Eds.), Landslides, Analysis and Control, Special Report 176: Transportation Research Board. National Academy of Sciences, Washington, DC, pp. 11–33.
von Boetticher, A., Turowski, J.M., McArdell, B.W., Rickenmann, D., Hürlimann, M., Scheidl, C., et al., 2017. DebrisInterMixing-2.3: a finite volume solver for three-dimensional debris-flow simulations with two calibration parameters—Part 2: Model validation with experiments. Geosci. Model Dev. 10 (11), 3963–3978.
Waldron, I.J., 1977. The shear resistance of root-permeated homogeneous and stratified soil. J. Soil Sci. Soc. Am. 41, 843–849.
Waldron, L.J., Dakessian, S., 1981. Soil reinforcement by roots: calculation of increased soil shear resistance form root properties. Soil Sci. 132, 427–435.
Wieczorek, G.F., Glade, T., 2005. Climatic factors influencing occurrence of debris flows. In: Jakob, M., Hungr, O. (Eds.), Debris Flow Hazards and Related Phenomena. Springer, Berlin, pp. 325–362.
Wilson, R.C., Wieczorek, G.F., 1995. Rainfall thresholds for the initiation of debris flows at La Honda, California. Environ. Eng. Geosci. 1, 11–27.
Wu, T.H., McKinnell III, W.P., Swanston, D.N., 1979. Strength of tree roots and landslides on Prince of Wales Island, Alaska. Can. Geotech. J. 16, 19–33. Available from: https://doi.org/10.1139/t79-003.
Wu, T.H., McOmber, R.M., Erb, R.T., Beal, B.E., 1988. A study of soil root interaction. J. Geotech. Eng., ASCE 114 (12), 1351–1375.

Further reading

Blijenberg, H.M., 1995. In-situ strength test of coarse, cohesionless debris on scree slopes. Eng. Geol. 39, 137–146.
Carrara, A., 1983. Multivariate models for landslide hazard evaluation. Math. Geol. 15, 402–426.
D'Agostino, V., Bertoldi, G., 2014. On the assessment of the management priority of sediment source areas in a debris-flow catchment. Earth Surf. Processes Landforms 39 (5), 656–668.
Guzzetti, F., Mondini, A.C., Cardinali, M., Fiorucci, F., Santangelo, M., Chang, K.T., 2012. Landslide inventory maps: new tools for an old problem. Earth-Sci. Rev. 112 (1), 42–66.
Kirschbaum, D.B., Adler, R., Hong, Y., Kumar, S., Peters-Lidard, C., Lerner-Lam, A., 2012. Advances in landslide nowcasting: evaluation of a global and regional modeling approach. Environ. Earth Sci 66, 1683–1696. Available from: https://doi.org/10.1007/s12665-011-0990-3.
Palladino, M.R., Viero, A., Turconi, L., Brunetti, M.T., Peruccacci, S., Melillo, M., et al., 2018. Rainfall thresholds for the activation of shallow landslides in the Italian Alps: the role of environmental conditioning factors. Geomorphology 303, 53–67.
Wieczorek, G.F., 1996. Landslide triggering mechanisms. In: Turner, A.K., Schuster, R.L. (Eds.), Landslides: Investigation and Mitigation. Transportation Research Board, National Research Council, Washington DC, pp. 76–90. , Special Report.

CHAPTER FIFTEEN

Snow avalanches

Sven Fuchs[1], Margreth Keiler[2] and Sergey Sokratov[3]
[1]Institute of Mountain Risk Engineering, University of Natural Resources and Life Sciences, Vienna, Austria
[2]Institute of Geography, University of Bern, Bern, Switzerland
[3]Faculty of Geography, M.V. Lomonosov Moscow State University, Moscow, Russian Federation

After seasonally frozen ground, the seasonal snow cover has the second largest extent of any component of the cryosphere. With a mean annual area of approximately 26 million km^2 worldwide (Barry and Thian, 2011), snow cover affects a large part of the populated areas and provides a considerable share of hydroclimatic hazards, above all in mountain regions (Fuchs et al., 2015a). In parallel, snow cover plays an important role as an economic factor in mountain areas (e.g., hydropower, agriculture, tourism, etc.; Callaghan et al., 2011). Correspondingly, snow cover is a determinant of snow avalanches and other hazards in these regions (McClung and Schaerer, 2006; Keiler et al., 2010) as a consequence of its seasonal dynamics and variation. Even though in comparison to other hydroclimatic hazards, such as heavy precipitation and flooding, snow avalanches are relatively limited in their spatial and temporal extent, they cause the majority of winter fatalities both in settlements (e.g., Fuchs et al., 2013, 2017b; Kazakova et al., 2017) and in the tourism and leisure industry (e.g., Badoux et al., 2016; Höller, 2017; Spencer and Ashley, 2011; Walcher et al., in press; Jekich et al., 2016), as well as significant infrastructure loss worldwide (Schweizer, 2008). Therefore, the understanding of temporal and spatial issues of avalanche formation relative to hydroclimate and snowpack development is crucial (McClung, 2005; Schweizer, 2008; Schweizer et al., 2003), in particular for regional avalanche forecasting. Temporal and spatial variability of snow cover and snow pack is strongly related to local and regional temperature regimes and precipitation patterns, hence the hydroclimate (Grünewald et al., 2013). The necessity of involving new and/or innovative methods using remote sensing techniques in identifying areas endangered by snow avalanches is obvious; however, it is not the same for highly-explored regions, such as the Central European mountains or the American

Extreme Hydroclimatic Events and Multivariate Hazards in a Changing Environment.
DOI: https://doi.org/10.1016/B978-0-12-814899-0.00015-8

Rocky Mountains, and relatively unexplored regions, such as mountain areas in artic regions of the Russian Federation. While for the former detailed and extensive documentation of snow avalanches as well as release conditions and triggers can be used to validate remote sensing data and to further develop data processing models (Eckerstorfer et al., 2017), for the latter remote sensing may be used for avalanche detection and forecast only with limited possibilities of validation due to an overall data-scarceness (Fuchs et al., 2017b).

15.1 Hazard characteristics

Snow avalanches are a well-known hazard type and are defined as a sudden release of snow masses and ice on slopes, sometimes containing portion of rocks, soil, and vegetation; and by definition the downhill trajectory exceeds 50 m (Wilhelm, 1975). Avalanche observations are reliable indicators for snow instability, and a relationship between high avalanche risk and high avalanche activity exists (Schweizer et al., 2003). According to the speed of the moving snow, avalanches can be distinguished from creeping and gliding movements of snow.

A number of classifications of snow avalanches exists (Kuroda, 1967; De Quervain et al., 1981; Dzyuba and Laptev, 1984). An international classification used by the majority of scientists and practitioners in the field has become accepted worldwide and classifies avalanches according to their release type, the shape of the trajectory, and the type of movement (De Quervain et al., 1981), see Table 15.1. Various conditions result in a release of avalanches, spanning from heavy snowfall to sudden temperature increase, but the prediction of individual avalanche formation is extremely challenging due to the high spatial variability and transient nature of the snowpack (Schweizer et al., 2003).

Generally, snow avalanches start from terrain that is steeper than about 30°–45° and favors snow accumulation (Wilhelm, 1975). On terrain less than about 15° snow avalanches start to decelerate and finally stop. Snow avalanche formation differs according to different volumes, repeatability and dynamic characteristics (McClung and Schaerer, 2006). While loose snow avalanches are released from a more or less definable point in a comparatively cohesionless surface layer of either wet or dry snow, slab avalanches involve the release of a cohesive slab over an extended plane

Table 15.1 International snow avalanche classification.

Zone	Criterion	Characteristic and denomination	
Origin	Manner of starting	From a point	From a line
		Loose snow avalanche	**Slab avalanche**
	Position of failure layer	Within the snowpack	On the ground
		Surface-layer avalanche	**Full-depth avalanche**
	Liquid water in snow	Absent	Present
		Dry-snow avalanche	**Wet-snow avalanche**
Transition	Form of path	Open slope	Gully or channel
		Unconfined avalanche	**Channeled avalanche**
	Form of movement	Snowdust cloud	Flowing along ground
		Powder snow avalanche	**Flowing snow avalanche**
Deposition	Surface roughness of deposit	Coarse	Fine
		Coarse deposit	**Fine deposit**
	Liquid water in deposit	Absent	Present
		Dry deposit	**Wet deposit**
	Contamination of deposit	No apparent contamination	Rock debris, soil, branches, trees
		Clean deposit	**Contaminated deposit**

Source: After De Quervain, M.R., De Crécy, L., Lachapelle, E.R., Lossev, K., Shoda, M., Nakamura, T., 1981. Avalanche Atlas. Illustrated International Avalanche Classification. UNESCO, Paris; Fuchs, S., Keiler, M., Sokratov, S. 2015a. Snow and avalanches. In: Huggel, C., Carey, M., Clague, J.J., Kääb, A. (Eds.), The High-Mountain Cryosphere: Environmental Changes and Human Risks. Cambridge University Press, Cambridge.

of weakness. Slab avalanche activity is highest soon after snow storms because of the additional load on the existing snow layers (Schweizer et al., 2003). The existence of a weak layer below a cohesive slab layer is a prerequisite for the development of dry slab avalanches. This weak layer is either a result of the metamorphism inside the snow pack or a buried surface hoar. Crystals formed by kinetic grain growth such as surface hoar or depth hoar (Fruehauf et al., 2009) together with changes in response to temperature and water vapor gradients variability can also be accompanied by the formation of a solid and icy layer on top of the snow pack, restricting the connection of new-fallen snow with the older snow below the solid layer, and often forms the horizon at which the snow masses start to move downhill. Differently to the causes of snow avalanche release, the mechanism of avalanche movement and corresponding distances and forces are rather well described (Fuchs et al., 2015a).

Flow velocities of snow avalanches vary between 50 and 200 km/h for large dry snow avalanches, whereas wet avalanches are considerably denser and slower (20–100 km/h, McClung and Schaerer, 2006). If the avalanche path is steep, dry snow avalanches may generate a powder cloud. Depending on the type of avalanche the moved amount of snow is variable, and in combination with the high velocities the induced damage may vary significantly (Fuchs et al., 2013).

Besides natural triggering by overloading or internal weakening of the snowpack, snow avalanches can also be triggered artificially—unlike most other rapid mass movements—through localized, rapid, near-surface loading by, for example, people (usually unintentionally) or intentionally by explosives used as part of avalanche control programs or industrial activities (Mokrov et al., 2000). Occasionally, snow avalanches have also been triggered by large earthquakes (Stethem et al., 2003). While naturally released avalanches mainly threaten buildings and infrastructure, human-triggered avalanches are the main threat to recreationists in mountain areas.

15.2 Remote sensing of snow avalanches

In operational avalanche forecasting, avalanche formation is usually assessed heuristically by weighting avalanche-contributing factors such as terrain, precipitation, wind, temperature, and snow stratification (Bründl et al., 2010), that is, the complex interaction between terrain, snowpack, and meteorological conditions (Schweizer et al., 2003). These data are generally measured at specific accessible but isolated points such as weather stations, snow profiles, or avalanche activity observations (Bühler, 2012). On-site testing of snow pack properties in the field, terrain-based reconnaissance of avalanche activity and dynamics, and modeling of both are used to assess avalanche formation and run-out. However, for defining the release conditions, the monitoring of snow accumulation and the modeling of both snowpack properties and stability still are major areas of research (Bründl et al., 2010) because snow pack characteristics vary substantially within close adjacency to a measurement location (Schweizer et al., 2008). Moreover, field-based approaches require weather conditions to allow measurement campaigns and snow stability to assess the often remote avalanche starting zones and to study snow parameters in situ. Therefore, temporal and spatial data gaps result, leading to some uncertainties and biased statistical analysis in snow avalanche hazard and risk assessment (Eckerstorfer et al., 2016; Keiler et al., 2006).

To overcome these challenges, remote sensing techniques spanning from ground-based devices over airborne systems to spaceborne imagery are increasingly used. Given repeated measurements, remote sensing of snow avalanches allows for a comprehensive and unbiased monitoring in a temporarily consistent and spatially continuous way. While remote sensing of snow cover already has a long history (Scherer et al., 2013), remote sensing of snow avalanches is only recently becoming mature (Eckerstorfer and Bühler, 2015). Nevertheless, in contrast to other hydroclimatic hazards such as river flooding, the spatial extent of areas affected by snow avalanches is relatively small (Fuchs et al., 2015b; Jongman et al., 2014), which limits the some remote sensing methods to those that only provide a medium-to-small resolution.

Remote sensors can either be passive or active. Passive optical sensors make use of the reflected sunlight either in the visible electromagnetic spectrum, or in the near infrared and shortwave infrared part. In contrast, LiDAR (light detection and ranging) and radar (radio detection and ranging) sensors are actively emitting radiation and measure the reflection from the surface. LiDAR is based on the visible and near infrared and radar on the microwave region of the spectrum, the latter having the advantage of acquiring data also during darkness (e.g., polar night, nighttime) and bad weather conditions (Eckerstorfer et al., 2016). According to this differentiation the applicability and informative value of remotely-sensed data on snow avalanches differs.

15.2.1 Passive sensors

Optical remote sensing using passive sensors is based on measurement of the reflective properties of snow and is therefore dependent on solar radiation as a primary energy source. As a result, the use of passive sensors is restricted by daylight, weather conditions, and cloud cover. The size and spatial extent of snow avalanches can be detected by using contrast differences between avalanche deposits and surrounding (undisturbed) snowpack, due to snow properties such as increased snow depth, snow density, and surface roughness causing cast shadows. The assessment of these snow properties, however, is influenced by sensor illumination and observation angles as well as different atmospheric conditions caused by altitude variations, limiting avalanche detection using optical sensors (Scherer et al., 2013; Eckerstorfer et al., 2016). Moreover, coarse resolution sensors have the disadvantage of recording many overlaying signals in one pixel (e.g., different snow types, snow-free patches, rocks, cast shadow) particularly in alpine areas with complex terrain (Bühler, 2012).

15.2.1.1 Ground-based passive sensors

Time-lapse photography may be used in order to obtain dynamic information on avalanche activity, as shown by Christiansen (2001) for the development of the snow cover in avalanche-starting zones or Van Herwijnen and Simenhois (2012) and Hendrikx et al. (2012) for snow gliding. Results may be linked to meteorological data in order to better understand underlying process behavior, and can further be used to improve our understanding of avalanche dynamics, as recently shown with respect to the release time and size of wet snow avalanches by Van Herwijnen et al. (2013).

Techniques of photogrammetry may also be used to assess spatiotemporal snow distribution, as shown by Wirz et al. (2011) and Gauthier et al. (2014), for a discussion see Westoby et al. (2012). Most of the studies report the reconstruction of high-resolution topography using commercially available digital single-lens reflex (SLR) cameras in combination with low-cost 3D modeling software, which makes the entire procedure highly affordable in comparison to the use of airborne or even most of the spaceborne sensors.

The use of ground-based passive sensors provides nearly real-time data; however, the spatial coverage is relatively limited. According to Eckerstorfer et al. (2016), application ranges vary between a few meters to several hundreds of meters with a temporal resolution between minutes and days (mostly depending on the power input available). A further shortcoming is the often missing spatial orientation and scale, which demands postprocessing and georeferencing.

15.2.1.2 Airborne passive sensors

Airborne aerial imagery has been used for decades in geosciences for documentation and monitoring, as well as for topographic mapping of snow avalanches (Akif'eva, 1980). Two facts reduce the applicability with respect to snow avalanches: (1) the relatively high costs for obtaining suitable georeferenced images through aerial surveys; and (2) the fact that most of the aerial imagery is recorded during summertime when the visibility and the flight conditions are better compared to wintertime. As such, operational use of spatially inclusive and comprehensive images is limited. Nevertheless, after large avalanche events such as those that occurred in some regions in the European Alps in 1999 surveys are commissioned, sometimes executed by the respective national air force, and the resulting images can at least be used to map the outline of snow avalanche release and run-out of specific events (Fuchs and McAlpin, 2005). Moreover, Korzeniowska et al. (2017) recently reported on the use of near-infrared aerial imagery for regional snow avalanche detection, based on object-based image analysis.

To overcome the shortcomings of commercial aerial photography, the recent technical progress with respect to remotely piloted aircraft systems ("drones") may be promising. Eckerstorfer et al. (2015) reported on creating a high-resolution orthophoto mosaic to compute both avalanche debris outline and avalanche debris volume for a test site in Norway. More frequently, drones have been used to gather accurate snow depth data, which can be used in the case of available high-resolution digital elevation models to compute the spatial distribution of snow depth. This information can be further used in combination with expert knowledge for avalanche forecast (Bühler et al., 2016), but until now not as a stand-alone-method for an assessment of an avalanche hazard and risk. Nevertheless, due to restrictions by air traffic control in many countries, the potential of such systems may not be fully used. As such, even if the aerial survey can be preprogrammed and in principle the vehicle could undertake missions under fully autonomous flight mode, regulatory authorities regularly require contact flight which limits the operating distances considerably.

At the same time, however, the number, versatility, and reliability of alternative platforms to be used in large-scale aerial photography have been increasing (Aber et al., 2010). Unmanned aerial vehicles including platforms such as ultralite vehicles, kites, blimps, or remote-controlled model aircrafts are nowadays also used to carry small commercial cameras or other inexpensive sensors, often sufficient to provide quickly needed information in emergency situations (Kerle, 2013)—which is information on the event size and a rough estimation of affected persons or buildings, without any standardized georeferencing. Thus, a standard for application on snow avalanche hazard and risk is outstanding.

15.2.1.3 Spaceborne passive sensors

Satellite remote sensing is widespread in natural hazard management, in particular because of the short temporal orbiting and spatial coverage, ranging from low to very high resolution (Bello and Aina, 2014). Nevertheless, so far only a few studies on snow avalanches are available, while for other hazard types affecting larger areas the use of spaceborne sensors is largely adopted (Kerle, 2013).

Snow avalanches are visually recognized in satellite images as rough surface characterized by a line-shaped pattern oriented in the aspect direction of the terrain (Larsen et al., 2013). Two recent studies showed the potential for automated avalanche detection using such texture features. The main challenges were related to overillumination and underexposure in parts of the imagery, caused by mountain topography in combination with a low solar angle in high latitudes leading to a lack of visual contrast between the avalanche debris and the surrounding snowpack (Lato et al., 2012; Larsen et al., 2013). Lato et al. (2012), as well as more recently Eckerstorfer et al. (2016), provided an overview on possible platforms to be used in spaceborne avalanche research, concluding that so far, most of the high-resolution data available is relatively expensive in comparison to the expected outcomes. The exception is the recently launched Sentinel satellites of the European Earth Observation Program “Copernicus” which should allow for analyzing snow avalanches using the visible as well as near and shortwave infrared spectrum, with data being free-of-charge.

15.2.2 Active sensors

Active sensors are based on the emission of radiation and the reflection measurement from the illuminated surface. Platforms use ground-based and airborne LiDAR and ground-based, air-, and spaceborne radio detection and ranging (radar) sensors.

15.2.2.1 Ground-based active sensors

Prokop et al. (2008, 2015) report on possibilities of snow avalanche assessment using ground-based LiDAR based on earlier studies on spatial snow depth measurements (Prokop, 2008). The principle of the applied LiDAR sensor was to collect 3D data from temporally changing landscape surfaces and to measure the height differences in order to achieve information on extent and volume of the snow masses. The study had clearly shown the potential for assessing both mass loss in the avalanche starting zone and mass gain in the run-out area with a horizontal resolution of decimeters given a measurement distance of several hundred meters. Due to the relative fast development of LiDAR sensors and a parallel technological software development, recent studies expect an increase in ground-based LiDAR applications with respect to millimeter-scale range accuracy in natural hazard management (Deems et al., 2013), such as the study of Deems et al. (2015) reporting on the mapping of snow distribution and snow depth in avalanche starting zones.

A similar principle in measuring height differences allows ground-based radar detection for snow avalanches (Martinez-Vazquez and Fortuny-Guasch, 2008). As reported in Eckerstorfer et al. (2016), snow avalanches represent a significant physical change in the backscattering, leading to temporal decorrelation in the data, and can therefore be spatially assessed. Nevertheless, physical changes in the snowpack may also lead to similar decorrelation, which makes a distinction challenging. Fast application times and acquisition intervals below 30 seconds together with a horizontal resolution of meters at a distance of 1 km, in contrast, are reported as added values of this method, given good calibration, co-registration filtering, and processing of images (Caduff et al., 2015). Moreover, radar allows for measurement independent of weather and light conditions with measurement ranges between 50 m and around 18 km (Eckerstorfer et al., 2016). As for ground-based LiDAR, these techniques require image acquisition before and after an avalanche event, so that they are usually only applied in experimental settings where the measurement data is available.

15.2.2.2 Airborne active sensors

According to Eckerstorfer et al. (2016) there are few studies only showing the successful use of LiDAR airborne detection of snow avalanches, providing only little information on LiDAR performance. Vallet et al. (2000) reported on snow volume measurements within an accuracy of

20–30 cm, while Chrustek and Wezyk (2009) concluded that the method has some advantages compared to traditional ground-based devices for avalanche mapping in smaller mountain regions where the avalanche frequency is closely related to changes in the local terrain.

Airborne radar remote sensing of snow avalanches seems to be so far restricted to rescue operations for avalanche victims and an associated sensor development (e.g., Instanes et al., 2004; Fruehauf et al., 2009; Jaedicke, 2003).

15.2.2.3 Spaceborne active sensors

Spaceborne remote sensing using active sensors using the radar technology traces back already two decades, when Wiesmann et al. (2001) reported on the detection of avalanche debris on synthetic aperture radar images, taking some of the avalanches during the 1999 winter in Switzerland as an example. Further applications include those of Bühler et al. (2009) who applied change detection on the backscattering information of data acquired on different dates to show the potential of spaceborne avalanche detection. However, as also shown by Malnes et al. (2013), the validation of results needs comprehensive fieldwork and/or optical airborne data due to a high percentage of false classification with respect to the avalanche outlines. The challenge of temporal availability, together with high acquisition costs have largely been solved since the Sentinel 1 satellite of the European "Copernicus" program is operational (Eckerstorfer et al., 2016), nevertheless, the ground resolution of the radar sensor may still be too low. Other higher-resolution satellites include the TerraSAR-X (Germany) and the Cosmo-SkyMed (Italy), both offer relatively moderately-priced or free-of-charge data in the case of scientific use.

15.3 Snow avalanche risk

The general advantages of remotely sensed data on natural hazards (e.g., a safe data acquisition with high temporal and spatial resolution) are opposed to the needs for a reliable assessment of risk, that is, spatially explicit data on hazard, exposure, and vulnerability. The concept of risk, defined in its broadest way as hazard times consequences, has been introduced in disaster management since experiences suggested that elements at risk (such as buildings, infrastructure lines) and vulnerability should be increasingly considered within the framework of hazard management to reduce losses. Starting in the 1990s as the United Nations International Decade for Natural Disaster Reduction (United Nations General Assembly, 1989), the primary focus was shifted from hazards and their consequences to the physical and socioeconomic dimensions of risk and toward a wider understanding, assessment, and management of natural hazards. This highlighted the integration of approaches to risk reduction into a broader context between sciences and humanities (Fuchs et al., 2011). If snow avalanche risk is considered by the potential exposure of a system, resulting from the convolution of hazard and consequences during a certain period of time at a certain location, it becomes obvious that risk is not static (Fuchs et al., 2015a, 2013). The main challenge of risk assessment is rooted in the system dynamics driven by both geophysical and social forces, stressing the need for an integrative management approach (Bründl et al., 2010).

Exposure and vulnerability to snow avalanches, however, are characteristics of risk that have—to our knowledge—not been successfully assessed quantitatively using remote sensing techniques because of two reasons: temporal and spatial scale and scale-dependency. These issues have currently been discussed for other hazard types (such as floods, where usually larger areas are affected, which makes a general assessment of exposure and vulnerability possible). However, due to the discussed limited spatial extent of snow avalanches, together with the fact that they often occur during a period of limited availability of remote sensing data, exposure and vulnerability can only barely be assessed so far at the necessary resolution.

15.3.1 Exposure

The temporal variability and spatial extent of exposure has an important influence on the assessment of avalanche risk (Fuchs et al., 2015b), even if this risk is comparatively low compared to other hydroclimatic hazards (Fraefel et al., 2004; Fuchs et al., 2015b, 2017b). In many avalanche-prone regions, socioeconomic developments in the human-made environment have led to an asset concentration and a shift in urban and suburban population. Long-term changes in building numbers and values exposed to snow avalanches show a significant increase in numbers and values in many regions (Campbell et al., 2007; Gardner and

Dekens, 2007; Keiler, 2004; Shnyparkov et al., 2012). These can be so far only roughly assessed using multitemporal remotely sensed data, whereas classical (governmental) object-based and georeferenced data turned out to be very accurate in terms of information content needed in risk assessment (Fuchs et al., 2015b).

To give an example, while in many rural and urban settlements of the European Alps the total number of buildings exposed to snow avalanches had almost tripled since the 1950s, the proportional increase in the number of buildings was significantly lower than the proportional increase in the building value (Fuchs et al., 2017a). Buildings inside hazard-prone areas showed a lower average value than buildings outside those areas (Fuchs et al., 2015b). Nevertheless, the total amount of buildings exposed to snow avalanches is only minor in the Eastern European Alps compared to other mountain hazards (only approximately 4% of exposed buildings are exposed to snow avalanches, including multihazard risk, see Fig. 15.1), an issue that was also recently addressed for the Western Alps (Röthlisberger et al., 2017).

Another example of exposure of settlements to snow avalanches also illustrates the discussed dilemma: Grab and Linde (2014) used daily Moderate Resolution Imaging Spectroradiometer snow cover images for a 7-year period to establish the frequency and spatial extent of snowfalls across a test site in Southern Africa. In addition, a digital shape file containing the location, name, and district attributes of villages affected was used to assist in the construction of a village exposure to snow index. Hence, in this example, exposure to avalanches was not directly measured using remote sensing, but this information was gathered by merging the

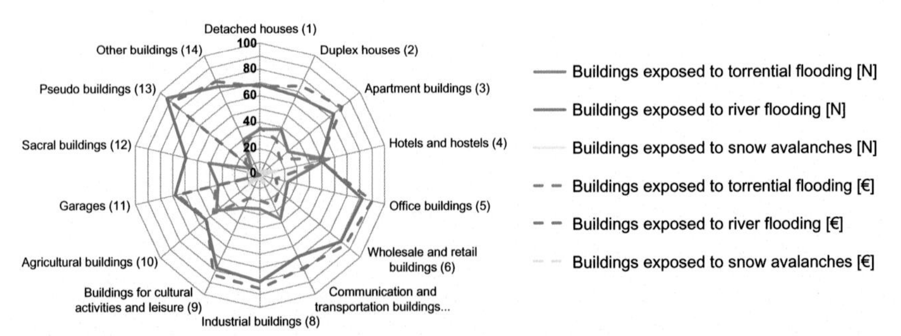

Figure 15.1 Radar chart of exposed buildings in the Eastern European Alps (Republic of Austria) to different types of mountain hazards. *Courtesy of Fuchs, S., Keiler, M., Zischg, A., 2015b. A spatiotemporal multi-hazard exposure assessment based on property data. Nat. Hazard. Earth Syst. Sci. 15, 2131, with permission.*

respective information on snowfall adjacent to villages and roads with ground-based geographic information system (GIS) data.

Short-term fluctuations in exposure supplemented the underlying long-term dynamics, in particular with respect to temporary variations of population movements or commuting citizens in settlements and of vehicles on the infrastructure network (Zischg et al., 2005; Keiler et al., 2005) as well as with respect to different management strategies (Fuchs et al., 2007; Holub and Fuchs, 2008). By implementing a quantifying fluctuation model it was proven by Keiler et al. (2005) how strong variations affect avalanche risk in mountain resorts. Similarly, Margreth et al. (2003), as well as Hendrikx and Owens (2008), were pointing to the dynamics in avalanche exposure on traffic infrastructure, and Fuchs et al. (2013) addressed the multiple temporal scales of exposure on high alpine traffic routes. Such information, however, is only barely available when only using remote-sensing data.

15.3.2 Vulnerability

Vulnerability, broadly defined, is the potential for loss (Fuchs and Thaler, 2018a), and includes elements of exposure (people, places, infrastructure at risk from a hazard), sensitivity (the degree to which the people, places, or infrastructure are harmed), and coping (the skills, resources, and opportunities of people and places to survive, absorb the impacts, and manage the adverse outcomes). Despite the growing amount of studies recently published (see for an overview, e.g., Fuchs and Thaler, 2018b), current approaches are still driven by a divide between natural and social sciences, leading to different research methods, concepts, and results. Whereas social scientists tend to view vulnerability and resilience as representing the set of socioeconomic factors that determine people's ability to cope with stress or changes (e.g., Cutter et al., 2003; Fekete and Hufschmidt, 2014), natural scientists and engineers often view both terms focusing on the likelihood of occurrence of specific hazards, and associated impacts on the built environment (e.g., Papathoma-Köhle et al., 2017). The assessment of physical, social, economic, and institutional vulnerability requires social and economic data of certain spatial and temporal resolution, not necessarily corresponding to the available resolution of geophysical and climatic data used in snow avalanche hazard assessment. With respect to snow avalanche risk, population density and land use are maybe the most prominent direct drivers for vulnerability in mountain regions. Apart from the overall population number, it is also the population distribution and composition, such as the level of urbanization and household size, as well as the increasing effects of counterurbanization (Kaltenborn et al., 2009; Löffler and Steinicke, 2006) which define the level of vulnerability to mountain hazards (Fuchs, 2009; Papathoma-Köhle and Thaler, 2018). However, as noted by Papathoma-Köhle et al. (2011), studies focusing on the vulnerability assessment of communities and buildings to snow avalanches are significantly fewer in number than similar studies regarding other disaster types, probably due to lack of sufficient data on avalanche damages to exposed elements, and to our knowledge none of the available studies is based on remotely sensed data.

15.4 Conclusions

It has been shown in the previous sections that a variety of remote sensing platforms exist that are potentially suitable for the detection of snow avalanches. Depending on the purpose and availability of resources, these include passive and active, as well as optical, LiDAR, and radar sensors. So far, however, the detection of snow avalanches can hardly be accomplished by only using such data, mainly because of the lack of reliability of ground-based proof, and the overall challenges in professional avalanche forecast (Bründl et al., 2010; Schweizer et al., 2003). Instead, remotely sensed data may be used in addition to traditional data sources in the evaluation of snow avalanche hazard and risk, and provide some advantages that have to be further assessed in future research. Overall, remote sensing techniques are relatively safe in comparison to in situ field measurements, in particular during periods of high avalanche danger or during bad-weather conditions. Remotely sensed information may therefore support the ongoing improvements in operational avalanche forecasting, but will so far not replace ground-based fieldwork and data collection, statistical data analysis, and modeling.

It has been reported by Eckerstorfer et al. (2016) that a minimum spatial resolution of 30 m is needed to successfully detect avalanches using remote sensing platforms with only minor omission and commission errors. Nevertheless, a spatial resolution of 30 m is not fine enough to assess exposure or even elements of vulnerability in a reliable manner.

A minimum temporal resolution of a few days would be optimal to eliminate detection uncertainties, which may be due to heavy snowfall in the target region or changes of physical snow properties leading to changes in the reflection behavior of the snow pack.

There has been certain development towards detection of snowpack properties using remote sensing. As such, remote sensing may support the detection of snow avalanche hazard and risk as some basic information needed for further computing hazard and risk may be gathered (Veitinger and Sovilla, 2016). Depending on the scale and accessibility of the area, the geographical location, the latitude, and altitude, remotely-sensed information can be fed into the traditional ground-based assessment, and, given accurate validation techniques of snow avalanche detection using available sensors, may then be further developed toward a semi-automated and automated hazard classification. The highest potential is offered by the recently launched Sentinel satellites, but also by a further application of terrestrial LiDAR and LiDAR scans using remotely piloted aircraft systems. These options, however, require a snow-free high-resolution digital elevation model so that snow masses can be computed precisely. Moreover, in data-scarce and remote areas when data are available with high temporal resolution, remote sensing of snowpack and snow avalanches may also be used to build up an avalanche cadaster based on hindcasting (Frauenfelder et al., 2010). The combination of remotely-assessed relief parameters together with climatic parameters has shown to be promising for small-scale snow avalanche hazard estimation in remote regions and was used to compile small-scale maps of snow avalanche activity, repeatability, and the factors of the snow avalanches formation (Kotlyakov, 1997).

To conclude, the currently available remote sensing data of high resolution required for snow avalanche assessment are still expensive and rarely have sufficient temporal resolution. The algorithms for automated detection of either positions or the timing of snow avalanches are still underdeveloped and require expert control, comparable to the workload needed during traditional field surveys. However, the appearance of new platforms such as remotely piloted aircraft systems or new satellites such as the Sentinel generation, together with a gradual increase in available snow avalanches data from different regions, can eventually fill the existing gaps between the requirements of operational snow avalanches forecasts and the data availability.

Acknowledgment

Sergey Sokratov received funding from the Russian Science Foundation, grant number 16-17-00104 "Snow avalanches and debris flows risk at the territory of Russia: estimation, forecast and mitigation measures".

References

Aber, J., Marzolff, I., Ries, J., 2010. Small-Format Aerial Photography. Elsevier, Amsterdam.

Akif'eva, K.B., 1980. Procedural Manual on Interpretation of Aerial Photographs in Snow Avalanches Studies [in Russian]. Gidrometeoizdat, Moscow.

Badoux, A., Andres, N., Techel, F., Hegg, C., 2016. Natural hazard fatalities in Switzerland from 1946 to 2015. Nat. Hazard. Earth Syst. Sci. 16, 2747–2768.

Barry, R., Thian, Y.G., 2011. The Global Cryosphere. Cambridge University Press, Cambridge.

Bello, O.M., Aina, Y.A., 2014. Satellite remote sensing as a tool in disaster management and sustainable development: towards a synergistic approach. Procedia Soc. Behav. Sci. 120, 365–373.

Bründl, M., Bartelt, P., Schweizer, J., Keiler, M., Glade, T., 2010. Review and future challenges in snow avalanche risk analysis. In: Alcántara-Ayala, I., Goudie, A. (Eds.), Geomorphological Hazards and Disaster Prevention. Cambridge University Press, Cambridge, pp. 49–61.

Bühler, Y., 2012. Remote sensing tools for snow and avalanche research. Proceedings, 2012 International Snow Science Workshop, Anchorage, Alaska, pp. 264–268.

Bühler, Y., Hüni, A., Christen, M., Meister, R., Kellenberger, T., 2009. Automated detection and mapping of avalanche deposits using airborne optical remote sensing data. Cold Reg. Sci. Technol. 57, 99–106.

Bühler, Y., Adams, M.S., Bösch, R., Stoffel, A., 2016. Mapping snow depth in alpine terrain with unmanned aerial systems (UASs): potential and limitations. Cryosphere 10, 1075–1088.

Caduff, R., Wiesmann, A., Bühler, Y., Pielmeier, C., 2015. Continuous monitoring of snowpack displacement at high spatial and temporal resolution with terrestrial radar interferometry. Geophys. Res. Lett. 42, 813–820.

Callaghan, T.V., Johansson, M., Brown, R.D., Groisman, P.Y., Labba, N., Radionov, V., et al., 2011. Changing snow cover and its impacts. In: Amap (Ed.), Snow, Water, Ice and Permafrost in the Arctic (SWIPA): Climate Change and the Cryosphere. Arctic Monitoring and Assessment Programme, Oslo, pp. 4.1–4.58.

Campbell, C., Bakermans, L., Jamieson, B., Stethem, C. (Eds.), 2007. Current and Future Snow Avalanche Threats and Mitigation Measures in Canada. o.O.: Public Safety Canada.

Christiansen, H.H., 2001. Snow-cover depth, distribution and duration data from northeast Greenland obtained by continuous automatic digital camera. Ann. Glaciol. 32, 102–108.

Chrustek, P., Wezyk, P., 2009. Using high resolution LiDAR data to estimate potential avalanche release areas on the example of Polish mountain regions. Proceedings, 2009 International Snow Science Workshop. Davos, Switzerland, pp. 495–499.

Cutter, S., Boruff, B., Shirley, W., 2003. Social vulnerability to environmental hazards. Soc. Sci. Q. 84, 242–261.

De Quervain, M.R., De Crécy, L., Lachapelle, E.R., Lossev, K., Shoda, M., Nakamura, T., 1981. Avalanche Atlas. Illustrated *International Avalanche Classification*. UNESCO, Paris.

Deems, J.S., Painter, T.H., Finnegan, D.C., 2013. Lidar measurement of snow depth: a review. J. Glaciol. 59, 467–479.
Deems, J.S., Gadomski, P.J., Vellone, D., Evanczykc, R., Lewinter, A., Birkeland, K.W., et al., 2015. Mapping starting zone snow depth with a ground-based lidar to assist avalanche control and forecasting. Cold Reg. Sci. Technol. 120, 197–204.
Dzyuba, V.V., Laptev, M.N., 1984. Geneticheskaya klassifikatsiya i diagnosticheskie priznaki snezhnykh lavin [Genetic classification and diagnostic features of snow avalanches, in Russian]. Materialy glyatsiologicheskikh issledovanii [Data Glaciol. Stud.] 50, 97–104.
Eckerstorfer, M., Bühler, Y., 2015. Remote sensing of snow avalanches: potential and limitation for operational use. Avalanche Rev. 33, 14–15.
Eckerstorfer, M., Solbø, S.A., Malnes, E., 2015. Using "structure-from-motion" photogrammetry in mapping snow avalanche debris. In: Kriz, K. (Ed.), Wiener Schriften zur Geographie und Kartographie. University of Vienna, Vienna, pp. 171–187.
Eckerstorfer, M., Bühler, Y., Frauenfelder, R., Malnes, E., 2016. Remote sensing of snow avalanches: recent advances, potential, and limitations. Cold Reg. Sci. Technol. 121, 126–140.
Eckerstorfer, M., Malnes, E., Müller, K., 2017. A complete snow avalanche activity record from a Norwegian forecasting region using Sentinel-1 satellite-radar data. Cold Reg. Sci. Technol. 144, 39–51.
Fekete, A., Hufschmidt, G., 2014. From application to evaluation: addressing the usefulness of resilience and vulnerability. Int. J. Disaster Risk Sci. 5, 1–2.
Fraefel, M., Schmid, F., Frick, E., Hegg, C., 2004. 31 Jahre Unwettererfassung in der Schweiz. In: Mikoš, M., Gutknecht, D. (Eds.), Internationales Symposion Interpraevent, May 24–27, 2004, Riva del Garda. I/45–56.
Frauenfelder, R., Kronholm, K., Solberg, R., Larsen, S.-Ø., Salberg, A.-B., Larsen, J.O., et al., 2010. DUE AvalRS: remote-sensing derived avalanche inventory data for decision support and hind-cast after avalanche events. In: Lacoste-Francis, H. (Ed.), Proceedings of ESA Living Planet Symposium, Held on June 28–July 2, 2010 at Bergen in Norway. Noordwijk: ESA Communications.
Fruehauf, F., Heilig, A., Schneebeli, M., Fellin, W., Scherzer, O., 2009. Experiments and algorithms to detect snow avalanche victims using airborne ground-penetrating radar. IEEE Trans. Geosci. Remote Sens. 47, 2240–2251.
Fuchs, S., 2009. Susceptibility versus resilience to mountain hazards in Austria—paradigms of vulnerability revisited. Nat. Hazard. Earth Syst. Sci. 9, 337–352.
Fuchs, S., McAlpin, M.C., 2005. The net benefit of public expenditures on avalanche defence structures in the municipality of Davos, Switzerland. Nat. Hazard. Earth Syst. Sci. 5, 319–330.
Fuchs, S., Thaler, T., 2018a. Introduction. In: Fuchs, S., Thaler, T. (Eds.), Vulnerability and Resilience to Natural Hazards. Cambridge University Press, Cambridge, pp. 1–13.
Fuchs, S., Thaler, T. (Eds.), 2018b. Vulnerability and Resilience to Natural Hazards. Cambridge University Press, Cambridge.
Fuchs, S., Thöni, M., McAlpin, M.C., Gruber, U., Bründl, M., 2007. Avalanche hazard mitigation strategies assessed by cost effectiveness analyses and cost benefit analyses—evidence from Davos, Switzerland. Nat. Hazard. 41, 113–129.
Fuchs, S., Kuhlicke, C., Meyer, V., 2011. Editorial for the special issue: vulnerability to natural hazards—the challenge of integration. Nat. Hazard. 58, 609–619.
Fuchs, S., Keiler, M., Sokratov, S.A., Shnyparkov, A., 2013. Spatiotemporal dynamics: the need for an innovative approach in mountain hazard risk management. Nat. Hazard. 68, 1217–1241.

Fuchs, S., Keiler, M., Sokratov, S., 2015a. Snow and avalanches. In: Huggel, C., Carey, M., Clague, J.J., Kääb, A. (Eds.), The High-Mountain Cryosphere: Environmental Changes and Human Risks. Cambridge University Press, Cambridge, pp. 50–70.

Fuchs, S., Keiler, M., Zischg, A., 2015b. A spatiotemporal multi-hazard exposure assessment based on property data. Nat. Hazard. Earth Syst. Sci. 15, 2131.

Fuchs, S., Röthlisberger, V., Thaler, T., Zischg, A., Keiler, M., 2017a. Natural hazard management from a coevolutionary perspective: exposure and policy response in the European Alps. Ann. Am. Assoc. Geographers 107, 382–392.

Fuchs, S., Shnyparkov, A., Jomelli, V., Kazakov, N.A., Sokratov, S., 2017b. Editorial to the special issue on natural hazards and risk research in Russia. Nat. Hazard. 88, 1–16.

Gardner, J., Dekens, J., 2007. Mountain hazards and the resilience of social–ecological systems: lessons learned in India and Canada. Nat. Hazard. 41, 317–336.

Gauthier, D., Conlan, M., Jamieson, B., 2014. Photogrammetry of fracture lines and avalanche terrain: potential applications to research and hazard mitigation projects. Proceedings, 2014 International Snow Science Workshop, Banff, Canada, pp. 109–115.

Grab, S.W., Linde, J.H., 2014. Mapping exposure to snow in a developing African context: implications for human and livestock vulnerability in Lesotho. Nat. Hazard. 71, 1537–1560.

Grünewald, T., Stötter, J., Pomeroy, J.W., Dadic, R., Moreno Baños, I., Marturià, J., et al., 2013. Statistical modelling of the snow depth distribution in open alpine terrain. Hydrol. Earth Syst. Sci. 17, 3005–3021.

Hendrikx, J., Owens, I., 2008. Modified avalanche risk equations to account for waiting traffic on avalanche prone roads. Cold Reg. Sci. Technol. 51, 214–218.

Hendrikx, J., Peitzsch, H.E., Fagre, D.B., 2012. Monitoring glide avalanches using time-lapse photography. Proceedings, 2012 International Snow Science Workshop, Anchorage, Alaska, pp. 872–877.

Höller, P., 2017. Avalanche accidents and fatalities in Austria since 1946/47 with special regard to tourist avalanches in the period 1981/82 to 2015/16. Cold Reg. Sci. Technol. 144, 89–95.

Holub, M., Fuchs, S., 2008. Benefits of local structural protection to mitigate torrent-related hazards. In: Brebbia, C., Beriatos, E. (Eds.), Risk Analysis VI WIT, Southampton, pp. 401–411.

Instanes, A., Lønne, I., Sandaker, K., 2004. Location of avalanche victims with ground-penetrating radar. Cold Reg. Sci. Technol. 38, 55–61.

Jaedicke, C., 2003. Snow mass quantification and avalanche victim search by ground penetrating radar. Surv. Geophys. 24, 431–445.

Jekich, B.M., Drake, B.D., Nacht, J.Y., Nichols, A., Ginde, A.A., Davis, C.B., 2016. Avalanche fatalities in the United States: a change in demographics. Wilderness Environ. Med. 27, 46–52.

Jongman, B., Koks, E.E., Husby, T.G., Ward, P.J., 2014. Increasing flood exposure in the Netherlands: implications for risk financing. Nat. Hazard. Earth Syst. Sci. 14, 1245–1255.

Kaltenborn, B.P., Andersen, O., Nellemann, C., 2009. Amenity development in the Norwegian mountains: effects of second home owner environmental attitudes on preferences for alternative development options. Landscape Urban Plann. 91, 195–201.

Kazakova, E., Lobkina, V., Gensiorovskiy, Y., Zhiruev, S., 2017. Large-scale assessment of avalanche and debris flow hazards in the Sakhalin region, Russian Federation. Nat. Hazard. 88, 237–251.

Keiler, M., 2004. Development of the damage potential resulting from avalanche risk in the period 1950–2000, case study Galtür. Nat. Hazard. Earth Syst. Sci. 4, 249–256.

Keiler, M., Zischg, A., Fuchs, S., Hama, M., Stötter, J., 2005. Avalanche related damage potential—changes of persons and mobile values since the mid-twentieth century, case study Galtür. Nat. Hazard. Earth Syst. Sci. 5, 49–58.

Keiler, M., Sailer, R., Jörg, P., Weber, C., Fuchs, S., Zischg, A., et al., 2006. Avalanche risk assessment—a multi-temporal approach, results from Galtür, Austria. Nat. Hazard. Earth Syst. Sci. 6, 637–651.

Keiler, M., Knight, J., Harrison, S., 2010. Climate change and geomorphological hazards in the eastern European Alps. Philos. Trans. A Math. Phys. Eng. Sci. 368, 2461–2479.

Kerle, N., 2013. Remote sensing of natural hazards and disasters. In: Bobrowski, P. (Ed.), Encyclopedia of Natural Hazards. Springer, Dordrecht, pp. 837–847.

Korzeniowska, K., Bühler, Y., Marty, M., Korup, O., 2017. Regional snow-avalanche detection using object-based image analysis of near-infrared aerial imagery. Nat. Hazard. Earth Syst. Sci. 17, 1823–1836.

Kotlyakov, V.M. (Ed.), 1997. World Atlas of Snow and Ice Resources. Russian Academy of Sciences, Moscow.

Kuroda, M., 1967. Classification of snow avalanches. Physics of Snow and Ice. Proceedings (International Conference on Low Temperature Science, Sapporo, August 14–19, 1966). Institute of Low Temperature Science, Hokkaido University, Sapporo.

Larsen, S.Ø., Salberg, A.-B., Solberg, R., 2013. Automatic avalanche mapping using texture classification of optical satellite imagery. In: Lasaponara, R., Masini, N., Biscione, M. (Eds.), Proceedings of the 33rd EARSeL Symposium 2013, "Towards Horizon 2020", pp. 399–409.

Lato, M.J., Frauenfelder, R., Bühler, Y., 2012. Automated detection of snow avalanche deposits: segmentation and classification of optical remote sensing imagery. Nat. Hazard. Earth Syst. Sci. 12, 2893–2906.

Löffler, R., Steinicke, E., 2006. Counterurbanization and its socioeconomic effects in high mountain areas of the Sierra Nevada (California/Nevada). Mt. Res. Dev. 26, 64–71.

Malnes, E., Eckerstorfer, M., Larsen, Y., Frauenfelder, R., Jonsson, A., Jaedicke, C., et al., 2013. Remote sensing of avalanches in northern Norway using Synthetic Aperture Radar. Proceedings, 2013 International Snow Science Workshop. Grenoble, France, pp. 955–959.

Margreth, S., Stoffel, L., Wilhelm, C., 2003. Winter opening of high alpine pass roads—analysis and case studies from the Swiss Alps. Cold Reg. Sci. Technol. 37, 467–482.

Martinez-Vazquez, A., Fortuny-Guasch, J., 2008. A GB-SAR processor for snow avalanche identification. IEEE Trans. Geosci. Remote Sens. 46, 3948–3956.

McClung, D., 2005. Risk-based definition of zones for land-use planning in snow avalanche terrain. Can. Geotech. J. 42, 1030–1038.

McClung, D., Schaerer, P., 2006. The Avalanche Handbook. The Mountaineers, Seattle, WA.

Mokrov, E., Chernouss, P., Fedorenko, Y., Husebye, E., 2000. The influence of seismic effect on avalanche release. In: American Avalanche Association, (Ed.), Proceedings of the 2000 International Snow Science Workshop, October 1–6, 2000, Big Sky, Montana. American Avalanche Association, Bozeman, MT, pp. 338–341.

Papathoma-Köhle, M., Thaler, T., 2018. Institutional vulnerability. In: Fuchs, S., Thaler, T. (Eds.), Vulnerability and Resilience to Natural Hazards. Cambridge University Press, Cambridge, pp. 98–123.

Papathoma-Köhle, M., Kappes, M., Keiler, M., Glade, T., 2011. Physical vulnerability assessment for alpine hazards: state of the art and future needs. Nat. Hazard. 58, 645–680.

Papathoma-Köhle, M., Gems, B., Sturm, M., Fuchs, S., 2017. Matrices, curves and indicators: a review of approaches to assess physical vulnerability to debris flows. Earth Sci. Rev. 171, 272–288.

Prokop, A., 2008. Assessing the applicability of terrestrial laser scanning for spatial snow depth measurements. Cold Reg. Sci. Technol. 54, 155–163.

Prokop, A., Schirmer, M., Rub, M., Stocker, M., 2008. A comparison of measurements methods: terrestrial laser scanning, tachymetry and snow probing for the determination of the spatial snow-depth distribution on slopes. Ann. Glaciol. 49, 210–216.

Prokop, A., Schön, P., Singer, F., Pulfer, G., Naaim, M., Thibert, E., et al., 2015. Merging terrestrial laser scanning technology with photogrammetric and total station data for the determination of avalanche modeling parameters. Cold Reg. Sci. Technol. 110, 223–230.

Röthlisberger, V., Zischg, A., Keiler, M., 2017. Identifying spatial clusters of flood exposure to support decision making in risk management. Sci. Total Environ. 598, 593–603.

Scherer, D., Hall, D.K., Hochschild, V., König, M., Winther, J.-G., Duguay, C.R., et al., 2013. Remote sensing of snow cover. In: Duguay, C.R., Pietroniro, A. (Eds.), Remote Sensing in Northern Hydrology: Measuring Environmental Change. American Geophysical Union, Washington, DC, pp. 7–38.

Schweizer, J., 2008. On the predictability of snow avalanches. In: Campbell, C., Conger, S., Haegeli, P. (Eds.), Proceedings of ISSW 2008. Whistler, Canada, pp. 21–27, pp. 688–692.

Schweizer, J., Jamieson, B., Schneebeli, M., 2003. Snow avalanche formation. Rev. Geophys. 41, 1016.

Schweizer, J., Kronholm, K., Jamieson, B., Birkeland, K., 2008. Review of spatial variability of snowpack properties and its importance for avalanche formation. Cold Reg. Sci. Technol. 51, 253–272.

Shnyparkov, A.L., Fuchs, S., Sokratov, S.A., Koltermann, K.P., Seliverstov, Y.G., Vikulina, M.A., 2012. Theory and practice of individual snow avalanche risk assessment in the Russian arctic. Geogr. Environ. Sustain. 5 (3), 64–81.

Spencer, J., Ashley, W., 2011. Avalanche fatalities in the western United States: a comparison of three databases. Nat. Hazard. 58, 31–44.

Stethem, C., Jamieson, B., Schaerer, P., Liverman, D., Germain, D., Walker, S., 2003. Snow avalanche hazard in Canada—a review. Nat. Hazard. 28, 487–515.

United Nations General Assembly, 1989. International decade for natural disaster reduction. United Nations General Assembly Resolution 236 Session 44 of 22 December 1989. A-RES-44-236.

Vallet, J., Skaloud, J., Koelbl, O., Merminod, B., 2000. Development of a helicopter-based integrated system for avalanche mapping and hazard management. Int. Arch. Photogramm. Remote Sens. 33, 565–572.

Van Herwijnen, A., Simenhois, R., 2012. Monitoring glide avalanches using time-lapse photography. Proceedings, 2012 International Snow Science Workshop, Anchorage, Alaska, pp. 899–903.

Van Herwijnen, A., Berthod, N., Simenhois, R., Mitterer, C., 2013. Using time-lapse photography in avalanche research. Proceedings, 2013 International Snow Science Workshop. Grenoble, France, pp. 950–954.

Veitinger, J., Sovilla, B., 2016. Linking snow depth to avalanche release area size: measurements from the Vallée de la Sionne field site. Nat. Hazard. Earth Syst. Sci. 16, 1953–1965.

Walcher, M., Haegeli, P., Fuchs, S., in press. Risk of death and major injury from natural hazards in mechanized backcountry skiing in Canada. Wilderness Environ. Med.

Westoby, M.J., Brasington, J., Glasser, N.F., Hambrey, M.J., Reynolds, J.M., 2012. "Structure-from-Motion" photogrammetry: a low-cost, effective tool for geoscience applications. Geomorphology 179, 300–314.

Wiesmann, A., Wegmueller, U., Honikel, M., Strozzi, T., Werner, C.L., 2001. Potential and methodology of satellite based SAR for hazard mapping. Proceedings of IGARSS 2001, Sydney, Australia, pp. 3262–3264.

Wilhelm, F., 1975. Schnee- und Gletscherkunde. De Gruyter, Berlin.

Wirz, V., Schirmer, M., Gruber, S., Lehning, M., 2011. Spatio-temporal measurements and analysis of snow depth in a rock face. Cryosphere 5, 893–905.

Zischg, A., Fuchs, S., Keiler, M., Stötter, J., 2005. Temporal variability of damage potential on roads as a conceptual contribution towards a short-term avalanche risk simulation. Nat. Hazard. Earth Syst. Sci. 5, 235–242.

Index

Note: Page numbers followed by "*f*" and "*t*" refer to figures and tables, respectively.

B

D

F

H

N

Q

R

T

U

X

Z

Printed in the United States
By Bookmasters